METHODS AND PROTOCOLS IN FOOD SCIENCE

Series Editor
Anderson S. Sant'Ana
University of Campinas, Campinas, Brazil

For further volumes:
http://www.springer.com/series/16556

Methods and Protocols in Food Science series is devoted to the publication of research protocols and methodologies in all fields of food science.

Volumes and chapters will be organized by field and presented in such way that the readers will be able to reproduce the experiments in a step-by-step style. Each protocol will be characterized by a brief introductory section, followed by a short aims section, in which the precise purpose of the protocol will be clarified.

Plant-Based Proteins

Production, Physicochemical, Functional, and Sensory Properties

Edited by

Yonghui Li

Department of Grain Science and Industry, Kansas State University, Manhattan, KS, USA

Humana Press

Editor
Yonghui Li
Department of Grain Science and Industry
Kansas State University
Manhattan, KS, USA

ISSN 2662-950X ISSN 2662-9518 (electronic)
Methods and Protocols in Food Science
ISBN 978-1-0716-4271-9 ISBN 978-1-0716-4272-6 (eBook)
https://doi.org/10.1007/978-1-0716-4272-6

This Humana imprint is published by the registered company Springer Science+Business Media, LLC, part of Springer Nature.
The registered company address is: 1 New York Plaza, New York, NY 10004, U.S.A.

Preface to the Series

Methods and Protocols in Food Science series is devoted to the publication of research protocols and methodologies in all fields of food science. The series is unique as it includes protocols developed, validated, and used by food and related scientists as well as theoretical basis are provided for each protocol. Aspects related to improvements in the protocols, adaptations, and further developments in the protocols may also be approached.

Methods and Protocols in Food Science series aims to bring the most recent developments in research protocols in the field as well as very well established methods. As such the series targets undergraduates, graduates, and researchers in the field of food science and correlated areas. The protocols documented in the series will be highly useful for scientific inquiries in the field of food sciences, presented in such way that the readers will be able to reproduce the experiments in a step-by-step style.

Each protocol will be characterized by a brief introductory section, followed by a short aims section, in which the precise purpose of the protocol is clarified. Then, an in-depth list of materials and reagents required for employing the protocol is presented, followed by a comprehensive and step-by-step procedures on how to perform that experiment. The next section brings the dos and don'ts when carrying out the protocol, followed by the main pitfalls faced and how to troubleshoot them. Finally, template results will be presented and their meaning/conclusions addressed.

The Methods and Protocols in Food Science series will fill an important gap, addressing a common complain of food scientists, regarding the difficulties in repeating experiments detailed in scientific papers. With this, the series has a potential to become a reference material in food science laboratories of research centers and universities throughout the world.

Campinas, Brazil *Anderson S. Sant'Ana*

Preface

The growing interest in plant-based proteins as a sustainable and ethical alternative to animal-based proteins has spurred significant advancements in the field of food science. As consumer demand for plant-based products continues to rise, so too does the need for more consistent and standardized methodologies that can guide researchers, product developers, and industry professionals in their efforts to innovate and improve these products. This book, *Plant-Based Proteins: Production, Physicochemical, Functional, and Sensory Properties*, is designed to fill this gap by offering a comprehensive resource of protocols and methodologies specific to plant-based proteins.

The inspiration for this book emerged from the recognition that, while many active studies and publications explore various aspects of plant-based proteins—from production and processing to their physical and functional properties—the lack of a consolidated handbook of protocols has led to inconsistencies that can hinder research progress and product development. This book aims to address these challenges by providing clear, step-by-step protocols that can be reliably reproduced and adapted across different research settings.

The book is structured into five key parts as summarized below:

Part I: Introduction and Regulatory Aspects of Plant-Based Proteins (Chapters 1 and 2)

This part provides a foundational overview of plant-based proteins, including their importance, sources, chemical and functional properties, and applications. It also addresses the regulatory landscape and international testing standards, ensuring that readers are familiar with the compliance requirements necessary for global markets.

Part II: Production and Safety of Plant-Based Proteins (Chapters 3, 4, 5, 6, 7 and 8)

Focusing on the methodologies and safety protocols, this part covers the entire production spectrum of plant-based proteins. Topics include the milling of protein-rich grains, dry and wet fractionation methods, Osborne sequential extraction, and techno-economic analysis. Additionally, it emphasizes microbial safety through detailed microbial analysis protocols, ensuring the production of high-quality, safe plant-based proteins suitable for food applications.

Part III: Biochemical and Nutritional Analysis of Plant-Based Proteins (Chapters 9, 10, 11, 12, 13 and 14)

This part delves into the biochemical and nutritional aspects of plant-based proteins. It includes protocols for quantifying crude and soluble protein content, analyzing free amino and sulfhydryl groups, determining amino acid composition, assessing protein digestibility through in vitro gastrointestinal digestion, and identifying various anti-nutritional factors. These analyses provide a comprehensive understanding of the nutritional value of plant-based proteins.

Part IV: Physicochemical Characterization of Plant-Based Proteins (Chapters 15, 16, 17, 18, 19, 20, 21, 22 and 23)

Detailed physicochemical characterization methods are presented in this part, covering particle size and distribution analysis, bulk and flow properties, molecular size distribution, polyacrylamide gel electrophoresis, and secondary structural analysis. It also includes surface hydrophobicity, zeta potential, and thermal analysis, offering a thorough examination of the physical and chemical properties of plant-based proteins.

Part V: Functional and Sensory Properties of Plant-based Proteins in Food Systems (Chapters 24, 25, 26, 27, 28, 29, 30, 31, 32 and 33)

This part explores the functional and sensory attributes crucial for the application of plant-based proteins in food systems. Topics include solubility, foaming capacity and stability, emulsifying properties, water and oil holding capacity, gelation, rheological properties, and volatile compounds. It also covers twin-screw extrusion processing for textured plant protein meat analogs and sensory analysis methods for evaluating meat analogs. These protocols are essential for developing high-quality, consumer-acceptable plant-based food products.

This book would not have been possible without the contributions of many experts in the field. I extend my deepest gratitude to all the authors who have shared their expertise and insights in the form of detailed protocols. Their dedication to advancing the science of plant-based proteins is evident in every chapter. I would also like to thank the reviewers who provided invaluable feedback during the development of this book. Lastly, I am grateful to the team at Springer-Nature for their support and guidance throughout this project.

It is my hope that this book will serve as a practical guide for anyone engaged in the research, development, or commercialization of plant-based proteins. I also hope that this book will contribute to more consistent and impactful research outcomes, ultimately fostering continued growth and innovation in the plant-based protein industry.

Manhattan, KS, USA ***Yonghui Li***

Contents

Contributors

SAJID ALAVI • *Department of Grain Science and Industry, Kansas State University, Manhattan, KS, USA*

SELENE ALVARADO MARTINEZ • *School of Exercise and Nutritional Sciences, San Diego State University, San Diego, CA, USA*

JAYENDRA AMAMCHARLA • *Food Science and Nutrition, University of Minnesota, Minneapolis, MN, USA*

RANIA MARIE BUENAVISTA • *Department of Grain Science and Industry, Kansas State University, Manhattan, KS, USA*

MARTA CARRILLO • *School of Exercise and Nutritional Sciences, San Diego State University, San Diego, CA, USA*

LIUYI CHANG • *Department of Plant Sciences, North Dakota State University, Fargo, ND, USA*

BINGCAN CHEN • *Department of Plant Sciences, North Dakota State University, Fargo, ND, USA*

GENGJUN CHEN • *Department of Grain Science and Industry, Kansas State University, Manhattan, KS, USA*

QIONG CHEN • *Department of Plant Sciences, North Dakota State University, Fargo, ND, USA*

FEI CHENG • *The Scoular Company, Minneapolis, MN, USA*

COREY COE • *School of Agriculture and Food Systems, West Virginia University, Morgantown, WV, USA*

SHIVAPRASAD DODDABEMATTI PRAKASH • *Department of Grains Science and Industry, Kansas State University, Manhattan, KS, USA*

ZHENJIAO DU • *Department of Grain Science and Industry, Kansas State University, Manhattan, KS, USA*

BAOCHEN FANG • *Department of Plant Sciences, North Dakota State University, Fargo, ND, USA*

SAHIL GUPTA • *General Mills Inc., Minneapolis, MN, USA*

YANHONG HE • *Department of Food Science and Technology, Virginia Polytechnic Institute and State University, Blacksburg, Virginia, USA*

SHAN HONG • *Department of Grain Science and Industry, Kansas State University, Manhattan, KS, USA*

RUIJIA HU • *Department of Grain Science and Industry, Kansas State University, Manhattan, KS, USA*

GUO HUANG • *College of Food Science, Northeast Agricultural University, Harbin, China*

WENTAO JIANG • *School of Agriculture and Food Systems, West Virginia University, Morgantown, WV, USA*

XINGYI JIANG • *Department of Health, Nutrition, and Food Sciences, Florida State University, Tallahassee, FL, USA*

QING JIN • *School of Food and Agriculture, University of Maine, Orono, Maine, USA*

PRABHAKAR KASTURI • *Measurement Sciences, PepsiCo R&D, Plano, TX, USA*

BHUPINDER KAUR • *Food Processing Center, University of Nebraska-Lincoln, Lincoln, NE, USA; Department of Food Science and Technology, University of Nebraska-Lincoln, Lincoln, NE, USA*

NANDAN KUMAR • *Department of Grain Science and Industry, Kansas State University, Manhattan, KS, USA*

HYUKJIN KWON • *Department of Grain Science and Industry, Kansas State University, Manhattan, KS, USA*

YONGHUI LI • *Department of Grain Science and Industry, Kansas State University, Manhattan, KS, USA*

CHANGQI LIU • *School of Exercise and Nutritional Sciences, San Diego State University, San Diego, CA, USA*

KESHUN LIU • *United States Department of Agriculture, Agricultural Research Service, National Small Grains and Potato Germplasm Research Unit, Aberdeen, ID, USA*

CARLY LONG • *School of Agriculture and Food Systems, West Virginia University, Morgantown, WV, USA*

KAUSTAV MAJUMDER • *Department of Food Science and Technology, University of Nebraska-Lincoln, Lincoln, NE, USA*

ERIC NKURIKIYE • *Grain Science and Industry Department, Kansas State University, Manhattan, KS, USA*

MANOJ KUMAR PULIVARTHI • *Department of Grains Science and Industry, Kansas State University, Manhattan, KS, USA*

BIPIN RAJPUROHIT • *Department of Grain Science and Industry, Kansas State University, Manhattan, KS, USA*

JIAJIA RAO • *Department of Plant Sciences, North Dakota State University, Fargo, ND, USA*

QINCHUN RAO • *Department of Health, Nutrition, and Food Sciences, Florida State University, Tallahassee, FL, USA*

GUNVANTSINH RATHOD • *Idaho Milk Products, Jerome, ID, USA*

ALI RAZA • *School of Exercise and Nutritional Sciences, San Diego State University, San Diego, CA, USA*

JARED RIVERA • *Department of Grain Science and Industry, Kansas State University, Manhattan, KS, USA*

SUCHISMITA ROY • *Food Science and Nutrition, University of Minnesota, Minneapolis, MN, USA*

CANGLIANG SHEN • *School of Agriculture and Food Systems, West Virginia University, Morgantown, WV, USA*

KALIRAMESH SILIVERU • *Department of Grains Science and Industry, Kansas State University, Manhattan, KS, USA*

MICHAEL STUMP • *Department of Grain Science and Industry, Kansas State University, Manhattan, KS, USA*

XIAONAN SUI • *College of Food Science, Northeast Agricultural University, Harbin, China*

MARTIN J. TALAVERA • *Department of Food, Nutrition, Dietetics and Health, Kansas State University, Olathe, KS, USA*

XIAOSHUAI WANG • *College of Food Science, Northeast Agricultural University, Harbin, China*

RUOSHI XIAO • *Department of Grain Science and Industry, Kansas State University, Manhattan, KS, USA*

ZILE XIAO • *School of Food Science and Biotechnology, Zhejiang Gongshang University, Hangzhou, China*

Zhuohong Xie • *US Pharmacopeia, Rockville, MD, USA*
Minwei Xu • *Department of Plant Sciences, North Dakota State University, Fargo, ND, USA*
Yixiang Xu • *Healthy Processed Foods Research Unit, Western Regional Research Center, USDA-ARS, Albany, CA, USA*
Jun Yang • *Measurement Sciences, PepsiCo R&D, Plano, TX, USA*
Qin Zhang • *College of Food Science, Northeast Agricultural University, Harbin, China*
Yan Zhang • *College of Food Science, Northeast Agricultural University, Harbin, China; College of Horticulture and Landscape Architecture, Northeast Agricultural University, Harbin, China*
Yan Zhang • *College of Food Science, Northeast Agricultural University, Harbin, China; College of Horticulture and Landscape Architecture, Northeast Agricultural University, Harbin, China*
Yue Zhang • *School of Food Science and Biotechnology, Zhejiang Gongshang University, Hangzhou, China*
Jing Zhao • *School of Exercise and Nutritional Sciences, San Diego State University, San Diego, CA, USA*
Rongxue Zhou • *College of Food Science, Northeast Agricultural University, Harbin, China*

Part I

Introduction and Regulatory Aspects of Plant-Based Proteins

Chapter 1

Overview of Plant-Based Proteins

Zhenjiao Du, Bipin Rajpurohit, Nandan Kumar, and Yonghui Li

Abstract

Today, we are facing an unprecedented challenge to produce enough protein to feed the growing population in a way that is environmentally sustainable, easily accessible, and affordable while also meeting changing dietary preferences. Plant proteins are gaining attention because of their superiority in these regards. However, their relatively lower nutritional quality and functionality limit their use in the food industry. Progress has been made in exploring new plant protein sources, processing innovation, functionality modifications, and diversifying applications thanks to the efforts from both academia and industry. This chapter summarizes the nutritional composition and protein class distribution in popular plant protein sources. We also highlight the advantages and disadvantages of various production and modification techniques. Finally, we briefly introduce the application of plant proteins in end-products, with further discussion of meat analogues, a popular application. Overall, plant proteins have exhibited a great potential as alternative protein sources, but more research is needed to address the challenges in flavor, color, functionality, and nutrition of plant protein-based products to satisfy consumers.

Key words Plant protein sources, Protein functionality, Protein nutrition, Plant protein production, Meat analogues

1 Introduction

The world's population is projected to reach 8.5 billion in 2030 and 9.7 billion in 2050 [1]. Producing enough food to meet the needs of all people is a significant challenge. Protein is one of the six essential nutrients (carbohydrates, proteins, fats, vitamins, minerals, and water) required by the human body, which provides amino acids that are the building blocks for growth and maintenance of health. Plant-based proteins are becoming increasingly popular as a sustainable alternative to animal-based proteins. They have several advantages, including worldwide accessibility, lower environmental impacts, and higher productivity with the same resources while aligning with changes in consumers' dietary choices and animal welfare concerns [2]. However, plant-based proteins generally have relatively lower functionality (e.g., poor solubility and textural

Yonghui Li (ed.), *Plant-Based Proteins: Production, Physicochemical, Functional, and Sensory Properties*, Methods and Protocols in Food Science, https://doi.org/10.1007/978-1-0716-4272-6_1,

properties) and nutritional quality (e.g., deficiency in essential amino acids) which can limit their use in the food industry. Progress has been made in overcoming these shortcomings and commercializing novel protein sources and related food products. This chapter provides an overview of plant-based proteins from cereals, pseudocereals, pulses, oilseeds, tubers, roots, leaves, and green biomass, including their nutritional composition, production methods, functional properties, modifications, and applications in plant protein-based food development.

2 Sources and Osborne Classification of Plant Proteins

Most plant proteins are found in cereals, pseudocereals, pulses, oilseeds, roots, tubers, leaves, etc [3]. These sources are primarily composed of proteins, lipids, starch, dietary fiber, non-starch polysaccharides, etc [4]. Additionally, almost all of them contain certain bioactive components in small quantities, such as polyphenols, organic acid, flavonoids, resistant starch, and quinones. Compared to animal foods, they are free from cholesterol and saturated fat [4, 5]. The nutritional composition of common plant sources is listed in Table 1. Pulses (e.g., pea, chickpea) and oilseed meals (e.g., soybean, canola) generally have higher contents of proteins, while cereals and pseudocereals have lower protein content. The starch content of pulses is generally lower than that of cereals and pseudocereals. To evaluate the nutritional quality of proteins, essential amino acid (EAA) profiles, protein digestibility, and bioavailability must be considered. The amino acid composition of the various plant-based proteins is presented in Table 2. Parameters, such as amino acid score (AAS), protein digestibility-corrected amino acid score (PDCAAS), and digestible indispensable amino acid score (DIAAS), are proposed based on human amino acids needed to indicate the nutritional quality [6, 7]. Most plant proteins have deficiencies in specific EAAs, such as lysine in cereals and sulfur-containing amino acids in legumes. One solution is to blend plant protein such as pea-rice blends [7]. Blending not only contributes to a balanced amino acid profile but can also improve the physicochemical behavior and functionality of the blends and ultimately improve the quality of the end-product. However, the improvement in functionality is highly dependent on various factors (e.g., processing conditions, source proteins, ratios, etc.) [7]. There are also several exceptions whose amino acid compositions are also well-balanced similar to animal proteins, such as soybean, quinoa, buckwheat, chia seeds, amaranth, and potato [7–11].

Plant proteins can be classified into four primary classes based on their solubility and extractability, as classification system initially proposed by Osborne: albumins, globulins, prolamins, and glutelins [12, 13]. The composition of these four classes in the common

Table 1
Nutritional composition of common plants (%)

Plant types	Protein sources	Protein	Lipid	Ash	Carbohydrate	Fiber	Starch	Description	NDB number
Cereal	Wheat	15.1	2.73	1.56	71.2	10.6	NA	Flour, whole wheat, unenriched	20,080
	Rice	7.04	1.03	0.42	80.3	2.77	74.4	Rice, white, long grain, unenriched, raw	20,444
	Sorghum	8.27	3.59	1.34	77.4	8.16	66.3	Flour, sorghum	100,289
	Barley	8.72	2.45	1.53	77.4	16.2	56.9	Flour, barley	20,130
	Corn	6.2	1.74	0.46	80.8	4.3	NA	Flour, corn, yellow, fine meal, enriched	100,251
	Oat	13.2	6.31	1.75	69.9	12.9	53.4	Flour, oat, whole grain	100,274
	Triticale	13.2	1.81	1.85	73.1	14.6	NA	Triticale flour, whole grain	20,070
	Rye	8.4	1.91	1.4	77.2	17.9	54.1	Flour, rye	100,290
Pseudocereal	Quinoa	11.0	6.6	2.0	69.5	6.95	58.1	Flour, quinoa	100,288
	Millet	10.0	4.19	1.13	74.4	3.33	67.1	Millet, whole grain	20,031
	Amaranth	13.2	6.24	2.29	68.8	7.2	55.7	Flour, amaranth	100,287
	Buckwheat	8.88	2.48	1.57	75.0	11.4	59.0	Flour, buckwheat	20,011
	Chia	16.5	30.7	4.8	42.1	34.4	NA	Seeds, chia seeds, dried	12,006
Pulse	Chickpea	22.4	6.69	2.82	57.8	10.8	NA	Chickpea flour (besan)	16,157
	Lentil	24.6	1.06	2.71	63.4	10.7	49.9	Lentils, raw	16,069
	Lupin	36.2	9.74	3.28	40.4	18.9	NA	Lupins, mature seeds, raw	16,076
	Pea	23.1	3.89	2.67	61.6	22.2	NA	Peas, green, split, mature seeds, raw	16,085
	Cowpea	23.8	2.07	3.39	59.6	10.7	NA	Cowpeas, catjang, mature seeds, raw	16,060
	Navy bean	22.3	1.5	3.32	60.8	15.3	32.9	Beans, navy, mature seeds, raw	16,037
	Mung bean	23.9	1.15	3.32	62.6	16.3	NA	Mung beans, mature seeds, raw	16,080
	Pigeon pea	21.7	1.49	3.45	62.8	15.0	NA	Pigeon peas (red gram), mature seeds, raw	16,101
	Fava bean	26.1	1.53	3.08	58.3	25.0	NA	Broad beans (fava beans), mature seeds, raw	16,052

(continued)

Table 1 (continued)

Plant types	Protein sources	Protein	Lipid	Ash	Carbohydrate	Fiber	Starch	Description	NDB number
Oilseed	Soybean	38.6	20.7	4.92	27.9	NA	NA	Flour, soy, full fat	16,115
	Sunflower	18.9	48.4	3.31	24.5	7.2	NA	Seeds, sunflower seed, kernel, raw	12,036
	Peanut	23.2	43.3	2.2	26.5	8.0	NA	Peanuts, raw	16,087
	Safflower	16.2	38.4	5.47	34.3	NA	NA	Seeds, safflower seed kernels, dried	12,021
	Sesame	17.7	49.7	4.45	23.4	11.8	NA	Seeds, sesame seeds, whole, dried	12,023
	Cottonseed	32.6	36.3	4.58	21.9	5.5	NA	Seeds, cottonseed kernels, roasted (glandless)	12,160
	Canola	38.6	0.99	6.7	NA	32.4	NA	Canadian canola meal (12% moisture)	[a]
	Flax	18.3	NA	42.2	28.9	27.3	NA	Flax seeds	NA
	Hemp	31.6	48.8	6.06	8.6	4.0	NA	Seeds, hemp seed, hulled	12,012
Roots, tubers, leaves, and others	Potato	8.11	0.95	2.71	79.9	16.6	62.7	Flour, potato	11,413
	Sweet potato	1.58	0.38	1.18	17.3	4.44	NA	Sweet potatoes, orange flesh, without skin, raw	100,286
	Cassava	0.92	0.49	0.72	87.3	4.8	79.7	Flour, cassava	100,291
	Pumpkin leaves	3.15	0.4	1.24	2.33	NA	NA	Pumpkin leaves, raw	11,418
	Spinach leaves	2.91	0.6	1.42	2.64	1.6	NA	Spinach, mature	11,457
	Broccoli	2.57	0.34	0.83	6.27	2.4	0	Broccoli, raw	11,090
	Amaranth leaves	2.46	0.33	1.5	4.02	NA	NA	Amaranth leaves, raw	11,003
	Seaweed	6.21	0.3	3.93	80.9	7.7	NA	Seaweed, agar, dried	11,663

Data source: US Department of Agriculture, Agricultural Research Service-FoodData Central, 2023. https://fdc.nal.usda.gov/fdc-app.html#/

Note: *NA* not available

[a]This entry is collected from the Canola Council of Canada at https://www.canolacouncil.org/about-canola/meal/

Table 2
Amino acid composition of common plant proteins (mg/g protein), grouped by food category: cereal, pseudocereal, pulse, oilseed, and root, tuber, leaf, and others

Amino acids		Cereal							
		Barley (flour or meal)	Corn (grain/whole meal)	Oat (meal)	Rice (white, long grain, regular, raw, unenriched)	Rye (whole meal)	Sorghum	Triticale (flour, whole grain)	Wheat (flour, whole grain)
Essential	Ile	36.5	36.8	40.5	43.2	37.6	39.2	36.7	33.6
	Leu	67.9	125.3	77.8	82.6	66.2	133.1	69.7	68.0
	Lys	37.2	26.7	39.8	36.2	36.5	20.2	28.0	27.2
	Met	19.2	19.2	18.0	23.6	15.6	13.9	15.6	17.3
	Phe	56.1	48.8	53.7	53.4	47.5	49.0	48.8	51.7
	Thr	33.9	36.0	35.5	35.8	35.9	30.2	31.0	27.8
	Trp	16.7	7.0	–	11.6	–	12.2	12.0	13.2
	Val	49.0	48.5	54.7	61.0	51.0	50.1	46.6	42.7
	His	22.5	27.2	22.5	23.6	23.7	21.4	23.8	27.0
	Total	**339**	**375.5**	**342.5**	**371**	**314.0**	**369.3**	**312.0**	**308.5**
Non-essential	Cys	22.1	15.5	28.6	20.5	20.5	15.0	21.1	20.8
	Tyr	28.7	38.2	35.3	33.4	20.6	26.7	29.3	20.8
	Arg	50.1	41.9	67.4	83.3	49.2	30.7	51.4	49.1
	Ala	39.0	75.4	48.7	57.9	45.7	93.4	37.2	37.0
	Asp+Asn	62.4	62.7	82.7	94.0	76.8	65.0	60.2	54.7
	Glu + Gln	261.0	189.4	224.5	195.0	259.6	211.5	306.8	328.0
	Gly	36.2	37.0	50.5	45.6	46.5	29.8	42.8	43.1
	Pro	119.0	89.4	55.6	47.0	100.7	81.1	90.9	157.6
	Ser	42.2	49.8	50.5	52.6	46.4	41.1	45.4	47.0
	Total	**660.6**	**599.4**	**643.8**	**629.2**	**666.1**	**594.4**	**685.0**	**758.2**
Total		999.6	974.9	986.2	1000.1	980.1	963.7	997.0	1066.7
Source		USDA-20130	FAO	FAO	USDA-20444	FAO	FAO	USDA-20070	FAO

(continued)

Table 2 (continued)

Amino acids		Pseudocereal				
		Amaranthus (*Amaranthus hybridus*)	Buck wheat (*Fagopyrum sagittatum*, hulled, groats, dark flour)	Chia (seeds, dried)	Millet (grain, *Pennisetum* spp.)	Quinoa (*Chenopodium quinoa*)
Essential	Ile	41.9	34.0	48.5	40.9	36.0
	Leu	62.0	59.0	83.0	95.6	60.0
	Lys	58.4	38.0	58.8	34.2	56.0
	Met	20.7	15.0	35.6	24.6	20.0
	Phe	42.6	38.0	61.8	48.1	41.0
	Thr	40.7	36.0	43.0	38.6	35.0
	Trp	–	–	26.4	19.5	–
	Val	50.7	67.0	57.6	55.2	45.0
	His	27.2	20.9	32.2	24.4	24.0
	Total	**344.3**	**308.0**	**447.0**	**381.1**	**317**
Non-essential	Cys	0	24.0	24.7	23.6	0
	Tyr	39.6	24.0	34.1	32.5	28.0
	Arg	87.3	98.0	129.7	52.8	70.1
	Ala	40.2	47.0	63.0	79.3	47.0
	Asp+Asn	97.8	88.9	102.4	80.1	73.0
	Glu + Gln	181.0	173.3	212.1	185.7	119.0
	Gly	102.6	63.9	57.2	37.5	52.0
	Pro	42.6	43.0	47.0	61.3	31.0
	Ser	83.1	50.0	63.6	48.6	37.0
	Total	**674.2**	**612.0**	**733.9**	**601.3**	**457.1**
Total		**1018.5**	920.0	**1180.8**	**982.5**	**774.1**
Source		FAO	FAO	USDA-12006	FAO	FAO

Amino acids		Pulse										
		Chickpea (*Cicer arietinum*)	Common bean	Cowpea (mature seeds, raw)	Fava bean (mature seeds, raw)	Lentil (raw)	Lima bean	Lupin (mature seeds, raw)	Mung bean (mature seeds, raw)	Navy bean (mature seeds, raw)	Pea (green, split, mature seeds, raw)	Pigeon pea (mature seeds raw)
Essential	Ile	44.3	41.9	40.7	40.2	43.1	49.6	44.8	42.3	42.7	42.6	36.2
	Leu	74.9	76.2	76.9	75.1	72.8	81.4	75.7	77.4	77.1	72.7	71.4
	Lys	68.5	72.1	67.6	64.0	69.9	74.4	53.3	69.5	57.4	76.6	70.0
	Met	10.4	10.6	14.3	8.2	8.5	12.5	7.0	12.0	12.2	8.4	11.2
	Phe	57.3	52.2	58.4	42.1	49.6	60.7	39.8	60.3	52.0	49.8	85.7
	Thr	37.6	39.7	38.2	35.6	35.9	41.8	36.7	32.7	31.9	35.2	35.3
	Trp	–	0	12.4	9.5	9.0	10.1	8.0	10.9	11.1	6.9	9.8
	Val	45.4	46.0	47.9	44.4	49.6	51.5	41.7	51.9	55.6	45.0	43.2
	His	26.4	28.4	31.1	25.4	28.2	31.5	28.5	29.1	22.7	25.4	35.7
	Total	**364.8**	**367.1**	**387.4**	**344.5**	**366.5**	**413.5**	**335.5**	**385.9**	**362.8**	**362.6**	**398.5**
Non-essential	Cys	11.8	8.5	11.1	12.8	13.1	10.1	12.3	8.8	8.4	11.8	11.5
	Tyr	29.3	25.3	32.4	31.7	26.7	32.3	37.6	29.9	21.7	22.4	24.8
	Arg	94.1	56.9	69.3	92.3	77.2	59.3	107.2	69.9	45.7	82.3	59.9
	Ala	43.4	41.9	45.8	41.0	41.9	46.5	35.9	43.9	40.7	45.5	44.8
	Asp+Asn	116.0	119.8	121.0	111.9	110.6	122.9	107.2	115.5	116.6	110.4	99.1
	Glu + Gln	158.6	148.0	189.9	170.1	155.3	130.9	240.1	178.2	139.0	167.5	231.8
	Gly	40.2	38.0	41.4	42.1	40.7	41.9	42.5	39.9	35.9	43.7	37.0
	Pro	42.1	35.7	45.0	42.1	41.9	46.9	40.9	46.0	50.2	45.0	44.0
	Ser	50.9	55.6	50.0	46.0	46.3	65.4	51.7	49.4	52.9	46.3	47.5
	Total	**586.2**	**529.7**	**605.8**	**590.1**	**553.7**	**556.3**	**675.3**	**581.5**	**511.2**	**574.9**	**600.3**
Total		951.0	896.8	993.3	934.6	920.2	969.8	1010.8	967.4	874.0	937.5	998.8
Source		FAO	FAO	USDA-16060	USDA-16052	USDA-16069	FAO	USDA-16076	USDA-16080	USDA-16037	USDA-16085	USDA-16101

(continued)

Table 2
(continued)

Amino acids		Oilseed							
		Soybean (flour, full fat)	Cottonseed (kernels, roasted, glandless)	Flax (seeds)	Hemp (seed, hulled)	Peanut (raw)	Safflower (seed kernel, dried)	Sesame (seeds, whole, dried)	Sunflower (seed, kernel, dried)
Essential	Ile	45.1	35.9	49.0	40.8	35.2	44.3	43.1	54.8
	Leu	79.3	68.4	67.8	68.4	64.7	71.0	76.8	79.8
	Lys	55.4	50.6	47.1	40.5	35.9	33.0	32.1	45.0
	Met	12.6	16.2	20.2	29.5	12.3	17.5	33.1	23.8
	Phe	55.7	62.3	52.3	45.9	53.5	49.8	53.1	56.3
	Thr	38.3	36.8	41.9	40.2	34.2	36.2	41.6	44.6
	Trp	11.7	15.1	16.2	11.7	9.7	11.3	21.9	16.7
	Val	45.1	51.2	58.5	56.3	41.9	63.0	55.9	63.5
	His	24.6	31.6	25.8	30.7	25.3	27.9	29.5	30.4
	Total	**367.7**	**368.1**	**378.7**	**364.0**	**312.6**	**353.8**	**387.2**	**414.9**
Non-essential	Cys	19.8	26.2	18.6	21.3	12.8	19.2	20.2	21.7
	Tyr	36.3	35.9	26.9	39.9	40.7	32.8	42.0	32.0
	Arg	76.2	135.0	104.9	144.0	119.4	108.0	148.6	115.4
	Ala	55.4	46.3	50.5	48.4	39.5	47.7	52.4	53.8
	Asp+Asn	97.4	108.9	112.0	115.8	122.1	111.7	93.2	117.8
	Glu + Gln	176.2	250.3	220.8	198.4	208.9	228.4	223.7	268.3
	Gly	35.0	48.5	68.3	50.9	60.1	62.3	68.9	70.2
	Pro	55.2	42.6	44.0	50.6	44.2	44.8	45.8	56.7
	Ser	53.6	50.0	53.0	54.1	49.2	50.1	54.6	51.9
	Total	**605.1**	**743.7**	**699.1**	**723.5**	**696.9**	**705.1**	**749.4**	**787.8**
Total		**972.8**	**1111.8**	**1077.8**	**1087.4**	**1009.5**	**1058.9**	**1136.7**	**1202.7**
Source		USDA-16115	USDA-12160	USDA-12220	USDA-12012	USDA-16087	USDA-12021	USDA-12023	USDA-12036

Amino acids		Roots, tubers, leaves, and others						
		Amaranth leaves (raw)	Broccoli (raw)	Cassava (meal)	Potato (flour)	Pumpkin leaves	Spinach (raw)	Sweet potato
Essential	Ile	48.4	23.0	28.8	43.3	54.5	51.4	36.9
	Leu	79.3	45.7	40.0	61.6	100.0	78.0	54.6
	Lys	51.6	47.9	41.9	59.9	63.5	60.8	34.6
	Met	14.6	13.5	13.8	15.5	20.0	18.5	16.9
	Phe	54.1	41.5	25.6	45.8	62.5	45.1	39.2
	Thr	40.2	31.2	26.9	40.6	51.0	42.7	38.5
	Trp	12.6	11.7	11.9	16.7	0	13.6	–
	Val	55.7	44.3	33.8	51.6	62.5	56.3	45.4
	His	21.1	20.9	21.3	24.1	22.5	22.4	13.8
	Total	**377.6**	**284.8**	**243.8**	**359.0**	**436.5**	**388.8**	**280.0**
Non-essential	Cys	11.8	9.9	14.4	10.1	9.5	12.2	10.8
	Tyr	32.5	17.7	16.3	32.5	49.0	37.8	23.8
	Arg	49.2	67.7	111.3	54.2	73.0	56.6	49.2
	Ala	56.5	36.9	38.1	47.5	63.5	49.7	47.7
	Asp+Asn	93.1	115.2	66.3	169.6	101.5	83.9	133.1
	Glu + Gln	118.7	192.2	163.8	135.1	122.5	119.9	87.7
	Gly	53.7	31.6	26.3	37.1	59.0	46.9	37.7
	Pro	49.2	39.0	28.1	38.4	45.5	39.2	35.4
	Ser	45.1	42.9	33.1	33.9	45.5	36.4	41.5
	Total	**509.8**	**553.2**	**497.5**	**558.4**	**569.0**	**482.5**	**466.9**
Total		**887.4**	**837.9**	**741.3**	**917.4**	**1005.5**	**871.3**	**746.9**
Source		USDA-11003	USDA-11090	FAO	USDA-11413	FAO	USDA-11457	FAO

Note: The data source for the entries is either FAO (https://www.fao.org/3/ac854t/AC854T00.htm#TOC) or USDA (https://fdc.nal.usda.gov/). For data sourced from USDA, the NDB number is provided

plant proteins is summarized in Table 3. Albumins and globulins are predominately present in most pulses, oilseeds, and some pseudocereals, whereas prolamins and glutelins make up the largest portions in cereals and some pseudocereals. Only a few data are reported regarding the protein classes in roots, tubers, and leaves.

Albumins are compact globular proteins that are soluble in water and have a relatively smaller molecular weight compared to the other three protein classes and have a diverse range of amino acid profiles[4]. Globulins, on the other hand, are insoluble in water but soluble in dilute salt solutions. Unlike albumin, globulins have a relatively low content of sulfur-containing amino acids, which is why most legumes, where globulins are the dominant protein class, are deficient in cystine and methionine [4]. Prolamins are insoluble in both water and salt solutions but extractable in concentrated aqueous alcohol solutions (e.g., 60–70% v/v). Glutelins are not soluble in neutral aqueous solutions, salt, or alcohol, but may be extractable in dilute aqueous acid or alkali solutions.

3 Plant Protein Production

The process of producing plant protein can be divided into two broad categories: wet fractionation and dry fractionation. Wet fractionation is commonly used to produce protein isolates and concentrates, as it yields high-purity products [31]. Conventional extraction techniques in wet fractionation are still widely used in industries owing to their cost-effectiveness. Alkali extraction is the most popular conventional method, in which NaOH or KOH is used for adjusting pH, disrupting the cell wall, breaking disulfide bonds, ionizing acidic and neutral amino acids, and finally achieving high extraction yield [32–34]. Other conventional methods including organic solvent, salt, and acidic solvent-based extraction also play an important role in protein extraction, although their extraction yields are generally lower than alkali extraction [33]. However, conventional extraction techniques are energy-intensive and not eco-friendly due to the high consumption of organic solvents, alkalis, and acids [33]. In addition, the extreme extraction conditions (e.g., pH) can cause excessive protein denaturation and aggregation, loss of functionality, lower digestibility, and even the production of toxic compounds, e.g., lysinoalanine [13, 33]. Driven by the need for high-quality proteins, advanced extraction technologies, such as enzymatic-assisted extraction, subcritical water extraction, reverse micelle extraction, aqueous two-phase system extraction, and physical treatment-assisted extraction (e.g., ultrasound, pulsed electric field, microwave, and high-pressure homogenization), are gaining increasing attention and being widely studied [32, 33]. Among these methods, enzyme-based extraction and physical treatments are mainly employed to assist conventional

Table 3
Distribution of four major protein classes in common plant proteins (%)

Plant types	Protein sources	Albumin	Globulin	Prolamin	Glutenin	References
Cereal	Wheat	6.0–10.0	5.0–8.0	35.0–40.0	40.0	[4]
	Rice	4.0–22.0	5.0–13.0	1.0–5.0	60.0–80.0	[14]
	Sorghum	2.0–7.0	2.0–10.0	35.0–60.0	20.0–35.0	[4]
	Barley	3.0–5.0	10.0–20.0	35.0–45.0	35.0–45.0	[4]
	Corn	4.0	4.0	60.0	26.0	[4]
	Oat	1.6	15.7	23.8	54.7	[15]
	Triticale	6.3	19.3	36.2	22.8	[15]
	Rye	6.7	27.0	35.6	21.5	[15]
Pseudocereal	Quinoa	25.0–35.0	37.0–38.0	0.5–7.0	NA	[16]
	Foxtail millet	2.1	1.9	31.8	31.7	[17]
	Perl millet	6.4–9.1	11.6–16.8	33.1–49.5	31.3–45.3	[18]
	Amaranth	40.0–52.0	15.9–20.0	2.0–3.0	25.0–31.1	[9]
	Buckwheat	43.8	7.82	10.5	14.6	[19]
	Chia	17.3	52.0	12.7	14.5	[20]
Pulse	Chickpea	8.0–12.0	53.0–60.0	3.0–7.0	19.0–25.0	[4]
	Lentil	16.0–26.0	47.0–70.0	3.0–4.0	11.0–15.0	[21]
	Lupin	25.0	75.0	NA	NA	[4]
	Pea	18.0–25.0	55.0–65.0	4–5	3.0–4.0	[22]
	Cowpea	8.2–11.9	50.0–70.0	2.3–5.0	14.4–15.6	[23]
	Mung bean	16.3	62.0	0.9	13.3	[22]
	Pigeon pea	10.2	59.9	3.0	17.4	[24]
	Fava bean	18.0–22.0	69.0–78.0	3.0–4.0	10.0–18.0	[21]
	Dry bean	18.7	39.9	0	7.2	[15]
Oilseed	Soybean	32.8	35.0	4.0	12.35	[22]
	Sunflower seed	25.0–30.0	60.0–80.0	NA	NA	[25]
	Peanut	10.0	90.0	NA	NA	[26]
	Sesame	8.6	67.3	1.4	6.9	[27]
	Cottonseed	6.3	38.1	0	2.4	[15]
	Canola	20.0	60.0	2.0–5.0	15.0–20.0	[4]
	Flax	40.0–42.0	64.0–85.0	NA	NA	[28]
	Hemp	25.0–37.0	67.0–75.0	NA	NA	[29]
Tuber	Potato	30.0–40.0	35.0–40.0	NA	NA	[30]

Note: *NA* not available

approaches in order to increase the extraction yield, improve purity, and maintain protein functionality by reducing or avoiding extreme extraction conditions [34].

After extraction, the protein must be separated from the wet fractions. Isoelectric precipitation is the most common method and is generally combined with the alkali/acid extraction methods, which take advantage of the fact that protein will aggregate and precipitate at their isoelectric point [13]. However, this involves the addition of harsh acids or bases, which can further damage the protein structure and functionality. An alternative is ultrafiltration and/or diafiltration which can separate the proteins by molecular weight without sacrificing protein concentration and functionality, though the membrane is vulnerable to blocking and contamination. It is important to note that different production methods can modulate the extracted protein's molecular weight, structure, and functionality. Therefore, when employing novel technologies for plant protein production to substitute the previous approaches, it is important to consider the downstream applications [13].

In cases where protein-rich ingredients with high functionality are needed instead of high purity, dry fractionation processes can be used. Two widely studied dry fractionation methods are air classification and electrostatic separation. Air classification separates particles based on their size and density to produce protein and starch-enriched flours, while electrostatic separation separates particles based on differences in their dielectric properties, where the protein can be charged to a much greater extent than carbohydrates allowing for their separation [31, 35]. There have also been attempts in hybrid approaches combining dry and wet fractionation together to produce high-quality and high-purity protein concentrates and isolates [35].

4 Protein Functionality, Chemistry, and Modification

Proteins are macromolecules composed of amino acids linked by peptide bonds and folded into three-dimensional structures through covalent and noncovalent bonds. The type, number, sequence, and the spatial arrangement of these amino acid residues, along with their interactions, collectively determine numerous protein properties. These properties include hydrophobicity, net charge, molecular weight, size, shape, etc., ultimately determining their functionality. Functionality properties include solubility, water/oil holding, emulsification properties, foaming, thermal stability, gelation, etc [13, 36]. Generally, plant proteins have inferior functionality compared to animal-based proteins [2, 13]. In addition, the extreme protein extraction process condition also imposes an impact on the functionality. Thus, modification methods of plant proteins for improved functionality have become increasingly popular in recent years [37].

The modification strategies involve the application of physical, chemical, and enzymatic processing or a combination of these methods. Physical techniques for modifications could include heating, sonication, microwave treatment, high-pressure treatment, extrusion, and cold plasma technology, among others. These techniques lead to changes in protein size and distribution, unfolding, disaggregation, or permanent denaturation of the protein structure [2]. Chemical modifications include reactions between proteins and chemical reagents, as well as structural changes through pH alteration. These reactions involve the derivatization of protein side chains (e.g., amino, carboxyl, disulfide, imidazole, indole, phenolic, sulfhydryl, thioether, and guanidine) to modulate biophysical properties and functionality. These reactions include acylation, succinylation, glycosylation, deamidation, and phosphorylation [2]. In pH-shifting, the extreme acid-base environment results in protein conformational changes, denaturing and unfolding proteins and allowing novel protein interaction to emerge [13]. Various resources and platforms are available to study these protein interactions. Understanding these interactions is crucial for developing new and improved plant-based protein products with superior functionality and nutritional value [38, 39].

Compared to physical and chemical modification, enzymatic processing is gentler and more controllable under environmental conditions. Enzymatic processing includes enzymatic hydrolysis and enzymatic cross-linking, which involve breaking or forming covalent bonds, respectively. Protein hydrolysis is generally achieved by proteolytic enzymes to reduce the molecular weight, expose hydrophobic groups, and increase the protein flexibility, which can potentially improve the protein functionality. It is important to note that only limited hydrolysis contributes to improved functionality, while excessive hydrolysis will lead to complete digestion and loss of functionality. Transglutaminase and some oxidative enzymes are popular choice for inducing cross-linking to build polypeptides into stronger structure and improve the textural properties of the proteins. So far, transglutaminase (EC 2.3.2.13) is the only commercially available food-grade cross-linking enzyme and has been reported to improve texture, viscosity, emulsion, gelation, and foaming [2].

5 Protein Applications

The studies on functionality and modifications of plant proteins are driven by the needs of the food industry. Plant proteins have been extensively studied and used in meat analogues, functional extenders, yogurts, baked goods, pasta, beverages, edible films, dairy alternatives, and more. Among these, meat analogues have gained most attention and are well commercialized recently due to their great potential to substitute traditional meat protein sources and their health claims. Creating a real meat-like texture is one of the major challenges in producing meat analogues, as most plant proteins exist in globular structures, rather than the fibrous structures found in meat [13]. Extrusion is the most common technique for creating a fibrillar structure, using hydration, high temperature, high pressure, and shearing forces [30].

Depending on the moisture level during extrusion, there are two types of extrusion: low-moisture and high-moisture extrusion. Low-moisture extrusion (water content around 30%) is primarily used to produce low-moisture textured vegetable protein (LM-TVP), which has a porous, spongy structure, high water-binding capacity, and low water and fat content [30, 40]. LM-TVP needs further hydration and processing, and its mouthfeel is not very satisfactory. It is commonly used to form patties or as meat extenders to improve viscoelasticity, moisture retention, firmness, and juiciness [21, 41]. High-moisture textured vegetable protein (HM-TVP) can be produced by high-moisture extrusion (water content above 40%). Its fibrous structure makes its mouthfeel highly similar to cooked meat, but the shelf life is inferior to LM-TVP due to its high moisture [41]. In addition to extrusion, shear cell technology and electrospinning can also be used to produce HM-TVP, but only extrusion is currently used industrially [41]. Wheat gluten, soybean proteins, and pea proteins are the primary sources for TVP production, but the allergenicity and genetic modifications in soybean proteins and the imbalanced EEA profile in wheat gluten may narrow down the application in the future [41, 42]. Pea proteins are superior in regard to being hypoallergenic and not genetically modified [21]. Other plant protein sources, such as mung bean, are gaining increasing attention to enrich the protein sources for meat analogue production [42, 43].

Beyond the aforementioned applications, there is potential to incorporate plant proteins in infant and children's formula, breakfast cereals, and extruded snack products and to explore new applications such as flavor development, personalized and precision nutrition, bioactive peptide production, and innovative technologies such as 3D printing and artificial intelligence (AI) for novel application development [44].

6 Conclusion

This chapter provides an overview of plant proteins, including their sources, classes, production approaches, functionality, modifications, and applications. Additionally, it summarizes the nutritional composition of common plant protein sources and discusses the basic concepts of popular techniques for protein production and modifications, along with their advantages and disadvantages. This chapter also introduces the application scenarios of plant protein, with a focus on meat analogues. Plant proteins are an underestimated resource that is expected to play a larger role in feeding the growing population in the future. Further research is needed on unexplored plant protein sources, balanced nutritional quality formula design, gentle and economical production approaches, and novel modification methods in order to continue growing the plant-based protein sector and continue to produce plant protein-based products with satisfactory flavor, texture, color, and high consumer acceptance.

Acknowledgments

This is contribution no. 24-259-B from the Kansas Agricultural Experimental Station. This work was supported in part by the USDA National Institute of Food and Agriculture Hatch project (Grant Accession No. 7003330).

References

1. United Nations Department of Economic and Social Affairs, Population Division (2022) World population prospects 2022: summary of results. United Nations, New York; p UN DESA/POP/2022/TR/NO. 3
2. Akharume FU, Aluko RE, Adedeji AA (2021) Modification of plant proteins for improved functionality: a review. Compr Rev Food Sci Food Saf 20(1):198–224. https://doi.org/10.1111/1541-4337.12688
3. Schweiggert-Weisz U, Eisner P, Bader-Mittermaier S, Osen R (2020) Food proteins from plants and fungi. Curr Opin Food Sci 32:156–162. https://doi.org/10.1016/j.cofs.2020.08.003
4. Day L (2013) Proteins from land plants—potential resources for human nutrition and food security. Trends Food Sci Technol 32(1):25–42. https://doi.org/10.1016/j.tifs.2013.05.005
5. Iqbal S (2022) High protein foods: a comparison of animal origin vs plant origin. In: Manickavasagan A, Lim L-T, Ali A (eds) Plant protein foods. Springer International Publishing, Cham, pp 1–25. https://doi.org/10.1007/978-3-030-91206-2_1
6. FAO/WHO/UNU (2007) Protein and amino acid requirements in human nutrition: report of a joint FAO/WHO/UNU expert consultation; World Health Organization technical report series; no. 935. World Health Organization, Geneva, Switzerland
7. Jiménez-Munoz LM, Tavares GM, Corredig M (2021) Design future foods using plant Protein blends for best nutritional and technological functionality. Trends Food Sci Technol 113:139–150. https://doi.org/10.1016/j.tifs.2021.04.049
8. Dakhili S, Abdolalizadeh L, Hosseini SM, Shojaee-Aliabadi S, Mirmoghtadaie L (2019) Quinoa protein: composition, structure and functional properties. Food Chem 299:125161. https://doi.org/10.1016/j.foodchem.2019.125161
9. Rivero Meza SL, Hirsch Ramos A, Cañizares L, de O. Raphaelli C, Bueno Peres B, Gaioso CA, Egea I, Estrada Y, Flores FB, de Oliveira M (2023) A review on amaranth protein: composition, digestibility, health benefits and food industry utilisation. Int J Food Sci Technol 58(3):1564–1574. https://doi.org/10.1111/ijfs.16056
10. López DN, Galante M, Robson M, Boeris V, Spelzini D (2018) Amaranth, quinoa and chia protein isolates: physicochemical and structural properties. Int J Biol Macromol 109:152–159. https://doi.org/10.1016/j.ijbiomac.2017.12.080
11. Jin J, Ohanenye IC, Udenigwe CC (2022) Buckwheat proteins: functionality, safety, bioactivity, and prospects as alternative plant-based proteins in the food industry. Crit Rev Food Sci Nutr 62(7):1752–1764. https://doi.org/10.1080/10408398.2020.1847027
12. Osborne TB (1924) The vegetable proteins. Longmans, Green and Company
13. Sim SYJ, Srv A, Chiang JH, Henry CJ (2021) Plant proteins for future foods: a roadmap. Food Secur 10(8):1967. https://doi.org/10.3390/foods10081967
14. Hoogenkamp, H.; Kumagai, H.; Wanasundara, J. P. D. Chapter 3 - Rice protein and rice protein products. In Sustainable protein sources; Nadathur, S. R., Wanasundara, J. P. D., Scanlin, L., Eds.; Academic Press: San Diego, 2017; pp. 47–65. doi: https://doi.org/10.1016/B978-0-12-802778-3.00003-2
15. Nikokyris PN, Kandylis K (1997) Feed protein fractions in various solvents of ruminant feedstuffs. J Sci Food Agric 75(2):198–204. https://doi.org/10.1002/(SICI)1097-0010(199710)75:2<198::AID-JSFA863>3.0.CO;2-C
16. Alrosan M, Tan T-C, Easa AM, Gammoh S, Alu'datt, M. H. (2022) Recent updates on lentil and quinoa protein-based dairy protein alternatives: nutrition, technologies, and challenges. Food Chem 383:132386. https://doi.org/10.1016/j.foodchem.2022.132386
17. Sachdev N, Goomer S, Singh LR (2021) Foxtail millet: a potential crop to meet future demand scenario for alternative sustainable

protein. J Sci Food Agric 101(3):831–842. https://doi.org/10.1002/jsfa.10716
18. Chandna M, Matta NK (1990) Characterization of pearl millet protein fractions. Phytochemistry 29(11):3395–3399. https://doi.org/10.1016/0031-9422(90)85245-B
19. Guo X, Yao H (2006) Fractionation and characterization of Tartary buckwheat flour proteins. Food Chem 98(1):90–94. https://doi.org/10.1016/j.foodchem.2005.05.055
20. Sandoval-Oliveros MR, Paredes-López O (2013) Isolation and characterization of proteins from chia seeds (Salvia Hispanica L.). J Agric Food Chem 61(1):193–201. https://doi.org/10.1021/jf3034978
21. Vatansever S, Tulbek MC, Riaz MN (2020) Low- and high-moisture extrusion of pulse proteins as plant-based meat ingredients: a review. Cereal Foods World 65(4). https://doi.org/10.1094/CFW-65-4-0038
22. Shrestha S, van't Hag L, Haritos VS, Dhital S (2022) Lentil and Mungbean protein isolates: processing, functional properties, and potential food applications. Food Hydrocoll 135:108142. https://doi.org/10.1016/j.foodhyd.2022.108142
23. Jayathilake C, Visvanathan R, Deen A, Bangamuwage R, Jayawardana BC, Nammi S, Liyanage R (2018) Cowpea: an overview on its nutritional facts and health benefits. J Sci Food Agric 98(13):4793–4806. https://doi.org/10.1002/jsfa.9074
24. Saxena KB, Kumar RV, Rao PV (2002) Pigeonpea nutrition and its improvement. J Crop Prod 5(1–2):227–260. https://doi.org/10.1300/J144v05n01_10
25. Kaur R, Ghoshal G (2022) Sunflower protein isolates-composition, extraction and functional properties. Adv Colloid Interf Sci 306:102725. https://doi.org/10.1016/j.cis.2022.102725
26. Cui S, McClements DJ, Xu X, Jiao B, Zhou L, Zhou H, Xiong L, Wang Q, Sun Q, Dai L (2023) Peanut proteins: extraction, modifications, and applications: a comprehensive review. Grain Oil Sc Technol 6:135. https://doi.org/10.1016/j.gaost.2023.07.001
27. Rivas RN, Dench JE, Caygill JC (1981) Nitrogen extractability of sesame (Sesamum Indicum L.) seed and the preparation of two protein isolates. J Sci Food Agric 32(6):565–571. https://doi.org/10.1002/jsfa.2740320607
28. Ye X-P, Xu M-F, Tang Z-X, Chen H-J, Wu D-T, Wang Z-Y, Songzhen Y-X, Hao J, Wu L-M, Shi L-E (2022) Flaxseed protein: extraction, functionalities and applications. Food Sci Technol 42:e22021. https://doi.org/10.1590/fst.22021
29. Montero L, Ballesteros-Vivas D, Gonzalez-Barrios AF, del P. Sánchez-Camargo A (2023) Hemp seeds: nutritional value, associated bioactivities and the potential food applications in the Colombian context. Front Nutr 9:1039180. https://doi.org/10.3389/fnut.2022.1039180
30. Nowacka M, Trusinska M, Chraniuk P, Drudi F, Lukasiewicz J, Nguyen NP, Przybyszewska A, Pobiega K, Tappi S, Tylewicz U, Rybak K, Wiktor A (2023) Developments in plant proteins production for meat and fish analogues. Molecules 28(7):2966. https://doi.org/10.3390/molecules28072966
31. Schutyser MAI, van der Goot AJ (2011) The potential of dry fractionation processes for sustainable plant Protein production. Trends Food Sci Technol 22(4):154–164. https://doi.org/10.1016/j.tifs.2010.11.006
32. Pojić M, Mišan A, Tiwari B (2018) Eco-innovative technologies for extraction of proteins for human consumption from renewable protein sources of plant origin. Trends Food Sci Technol 75:93–104. https://doi.org/10.1016/j.tifs.2018.03.010
33. Kumar M, Tomar M, Potkule J, Verma R, Punia S, Mahapatra A, Belwal T, Dahuja A, Joshi S, Berwal MK, Satankar V, Bhoite AG, Amarowicz R, Kaur C, Kennedy JF (2021) Advances in the plant protein extraction: mechanism and recommendations. Food Hydrocoll 115:106595. https://doi.org/10.1016/j.foodhyd.2021.106595
34. Hadidi M, Aghababaei F, McClements DJ (2023) Enhanced alkaline extraction techniques for isolating and modifying plant-based proteins. Food Hydrocoll 145:109132. https://doi.org/10.1016/j.foodhyd.2023.109132
35. Assatory A, Vitelli M, Rajabzadeh AR, Legge RL (2019) Dry fractionation methods for plant protein, starch and fiber enrichment: a review. Trends Food Sci Technol 86:340–351. https://doi.org/10.1016/j.tifs.2019.02.006
36. Berrazaga I, Micard V, Gueugneau M, Walrand S (2019) The role of the anabolic properties of plant- versus animal-based protein sources in supporting muscle mass maintenance: a critical review. Nutrients 11(8):1825. https://doi.org/10.3390/nu11081825
37. Shen Y, Hong S, Du Z, Chao M, O'Quinn T, Li Y (2022) Effect of adding modified pea protein as functional extender on the physical and sensory properties of beef patties. LWT

154:112774. https://doi.org/10.1016/j.lwt.2021.112774

38. Kumar YB, Kumar N, Vaikundamani S, Nagamani S, Mahanta HJ, Sastry GM, Sastry GN (2023) Analyzing the aromatic-aromatic interactions in proteins: A2ID 2.0. Int J Biol Macromol 253(Pt 5):127207. https://doi.org/10.1016/j.ijbiomac.2023.127207
39. Kumar YB, Kumar N, John L, Mahanta HJ, Vaikundamani S, Nagamani S, Sastry GM, Sastry GN (2023) Analyzing the cation-aromatic interactions in proteins: cation-aromatic database V2.0. Proteins 92:179. https://doi.org/10.1002/prot.26600
40. Hong S, Shen Y, Li Y (2022) Physicochemical and functional properties of texturized vegetable proteins and cooked patty textures: comprehensive characterization and correlation analysis. Food Secur 11(17):2619. https://doi.org/10.3390/foods11172619
41. Baune M-C, Terjung N, Tülbek MÇ, Boukid F (2022) Textured vegetable proteins (TVP): future foods standing on their merits as meat alternatives. Future Foods 6:100181. https://doi.org/10.1016/j.fufo.2022.100181
42. Samard S, Ryu G-H (2019) Physicochemical and functional characteristics of plant protein-based meat analogs. J Food Process Preserv 43(10):e14123. https://doi.org/10.1111/jfpp.14123
43. Brishti FH, Chay SY, Muhammad K, Ismail-Fitry MR, Zarei M, Saari N (2021) Texturized mung bean protein as a sustainable food source: effects of extrusion on its physical, textural and protein quality. Innovative Food Sci Emerg Technol 67:102591. https://doi.org/10.1016/j.ifset.2020.102591
44. Rajpurohit B, Li Y (2023) Overview on pulse proteins for future foods: ingredient development and novel applications. J Future Foods 3(4):340–356. https://doi.org/10.1016/j.jfutfo.2023.03.005

Chapter 2

Regulations and International Testing Standards

Zhuohong Xie and Fei Cheng

Abstract

The global demand for plant-based proteins is increasing, driven by consumer interest in healthier and more sustainable diets. This chapter provides an overview of the US regulations and international testing standards related to plant-based proteins, focusing on market approval/recognition, labeling requirements, testing standards resources, and the relationship between regulations and testing standards. The information can help industry, academia, and consumers understand the requirements for introducing new plant-based protein products to the market or developing new applications for the existing protein ingredients.

Key words Plant-based protein, Protein regulations, Protein standards, Approval, Specifications

1 Introduction

The pursuit of nutrition, sustainability, and ethical products (fairness and animal welfare) by consumers is driving an increased demand for protein-rich ingredients. While animal proteins such as meat and dairy products remain the dominant players in this market, consumption of plant-based proteins has been increasing as an alternative to animal-based proteins.

Some plant-based protein ingredients have been used in the human diet for centuries. However, proteins extracted from new sources, proteins derived from innovative processing techniques, or proteins with different characteristics and nutrition profiles are emerging as new food ingredients. This poses challenges for regulations and standards in terms of safety evaluation, nutritional profile, market approval or notification, validated test methods, and specifications. This chapter will provide an overview of the regulations and standards related to approval, labeling, and testing requirements for plant-based proteins. The focus will be on the United States, but the same guiding principle could be applied internationally.

Yonghui Li (ed.), *Plant-Based Proteins: Production, Physicochemical, Functional, and Sensory Properties*, Methods and Protocols in Food Science, https://doi.org/10.1007/978-1-0716-4272-6_2,

2 Market Approval/Notification Regulations

In the United States, protein ingredients for human foods have two major legal pathways to the market: food additive petitions and the general recognition of safety (generally recognized as safe, GRAS) mechanisms (including GRAS notifications and GRAS self-determinations).[1]

2.1 Food Additive Petition

A food additive petition is a request submitted to the US Food and Drug Administration (FDA) by a person or company seeking approval for a new food additive or a new use for an already approved food additive. The petition should include the identity and composition of the additive, proposed use, use level, data establishing the intended effect, quantitative detection methods, estimated exposure from the proposed use, full reports of all safety studies, proposed tolerances (if needed), environmental information, consistent information related to chemistry, toxicology, environmental science, and any other pertinent studies (e.g., microbiology). FDA reviews the petition to determine if the data establishes the identity of the substance, justifies its intended effect on the food, and establishes that its intended use in food is safe. The regulations for food additive petitions are found in Title 21 of the Code of Federal Regulations (CFR), Part 171 for human food and Part 571 for animal food [1, 2].

2.2 GRAS Notification

GRAS stands for "generally recognized as safe" and refers to a substance that is generally recognized, among qualified experts, to be safe under the conditions of its intended use. The GRAS notification package must have specific information about the substance including its intended use, identity, method of manufacture, specifications, dietary exposure, self-limiting levels of use, common use in food before 1958 (if applicable), and a comprehensive discussion of the basis for the GRAS determination. The GRAS notification program is described in CFR Title 21, Part 170, Subpart B and Subpart E for human food, and Part 570, Subpart E for animal food. Notification of an ingredient's GRAS status to the FDA is voluntary. Many substances are marketed on the basis of a GRAS self-determination in which no notification of the substance's status is made to FDA. GRAS notifications submitted to FDA are not "approved" by the Agency but have three possible outcomes: "FDA has no questions," "Notice does not provide a basis for a

[1] The color additive petition is not currently applicable, as there is no use of protein for coloring foods. The list of prior-sanctioned substances before 1958 does not include any protein ingredients. The list of specific prior-sanctioned substances can be found under the CFR Title 21, Part 181. Therefore, these two pathways are not discussed.

GRAS determination" which will require additional follow-up with FDA, or "At notifier's request, FDA ceased to evaluate the notice" which is a neutral outcome for the submitter.

2.3 Which Pathway Should Manufacturers Choose?

Manufacturers bringing a new protein ingredient into the market could opt for either GRAS or food additive petition approach. There are a few key differences between these two processes [3]:

(a) Premarket approval: a food additive petition requires premarket approval by FDA. GRAS does not require premarket approval, and manufacturers can voluntarily notify FDA. Alternatively, manufacturers can market their substances on the basis of a GRAS self-determination.

(b) Data and information: Data and information supporting a food additive petition may be non-published. The GRAS package consists of mostly published data, information, and methods as primary evidence, as well as unpublished scientific data as corroborative evidence. The data and information submitted to FDA in support of a GRAS notification are available for public disclosure [4, 5].

(c) Publication: The lists of approved food additives can be found in 21 CFR 170–199 [6]. The list of GRAS substances can be found in 21 CFR 184 or 584 (affirmation list [7, 8]) or the inventory of GRAS notices [9]. There is no comprehensive list of substances with GRAS self-determinations.

Most manufacturers choose a GRAS process because it does not require FDA's premarket approval and has a faster processing time. Animal feeds follow a similar food additives and GRAS process.

3 Labeling Requirements

Similar to other products, protein ingredients, when presented in finished goods, are required to be disclosed and properly labeled. Details can be found in 21 CFR 101 [10].

Structure/function claims: "Structure/function claims may describe the role of a nutrient or dietary ingredient intended to affect the normal structure or function of the human body" [11]. Food manufacturers are not required to notify FDA about their structure/function claims, and disclaimers are not required for claims on foods. For dietary supplements, manufacturers should substantiate that the claim is truthful and not misleading and notify the FDA no later than 30 days after marketing the dietary supplements. Potential structure/function claims for proteins might include that diets high in protein can help curb hunger [12].

Health claims: "Health claims in food labeling are claims that have been reviewed by FDA and are allowed on food products to show that a food or food component may reduce the risk of a disease or a health-related condition" [13]. There is one health claim authorized by the FDA related to protein which can be found in 21 CFR 101.82: Soy Protein and Risk of Coronary Heart Disease (CHD). Based on some scientific studies, soy protein consumption is associated with a reduced risk of heart disease [14]. However, FDA has recently reevaluated the evidence for this claim and concluded that it is not strong enough to support the claim [15]. FDA has proposed to revoke the soy protein health claim, but this proposal has not yet been finalized. More research is warranted to confirm the health benefits and corroborate this health claim.

Nutrient content claims: Nutrient content claims are direct statements about the amount of nutrients, such as proteins, contained in a product. In the case of children (above 4 years of age) through to adults, a protein-containing product with 5 g or 10% of the daily value (DV) for protein can display the "good source of protein" nutrient content claim.

As described in 21CFR101.9 [16], protein content claims must be substantiated by both protein quantity and quality. FDA requires food manufacturers to use the protein digestibility-corrected amino acid score (PDCAAS) to calculate the DV for the protein in their products. PDCAAS is a measure of the quality of a protein related to human nutritional needs. A higher score of PDCAAS indicates a higher-quality protein, with 1.0 being the highest value achievable. While PDCAAS scores for animal-derived proteins are usually 1.0, the scores for many plant proteins typically range between 0.4 and 0.9 [17]. The scores vary greatly due to plant genetics, environmental conditions, processing, and post-harvesting steps. The presence of fibers and anti-nutrient components can negatively affect the score. The corrected amount of protein can be calculated by crude protein times PDCAAS score. If a food contains a certain corrected level of protein in %DV, it can be labeled "high" in protein, "good" or "excellent" source of protein. For products intended for infants and children under 4 years of age, the %DV for protein must be listed on the label of the finished good. For products intended for children above 4 years of age through adults where no claim is made about protein on the label, the %DV for protein is not required [18].

In current US food regulations [19], there are nine major food allergens. The current food allergens are milk, eggs, fish, shellfish, tree nuts, peanuts, wheat, soybeans, and sesame. Protein products derived from one of the nine major food allergens or containing any protein derived from a major food allergen are required to clearly identify the source of the allergen.

4 Testing Standards

Testing standards are important to ensure that protein products meet quality requirements expected by stakeholders and that they are free of harmful microorganisms and chemicals. There are two types of testing standards: food safety/microbiological standards and food quality/chemical standards. Microbiological standards establish the maximum levels of certain microorganisms, such as bacteria and fungi, that are allowed in food. Chemical standards establish identity and the maximum levels of certain chemicals that are allowed in food. Microbiological standards can be found in the following publications: Bacteriological Analytical Manual (BAM) published by FDA, the Official Methods of Analysis of AOAC International (OMA AOACI), AACC Approved Methods of Analysis published by the Cereals and Grains Association, International Organization for Standardization (ISO), and the Food Chemicals Codex (FCC). USP (United States Pharmacopeia) general test chapters are sometimes adopted by manufacturers to test food products. For chemical standards, the OMA AOACI, American Oil Chemists' Society (AOCS), Cereals and Grains Association, ISO, Codex Alimentarius, Joint FAO/WHO Expert Committee on Food Additives (JECFA), and FCC provide testing methods. Codex Alimentarius, JECFA, and FCC also provide specifications, which are acceptable levels of quality and purity of food-grade ingredients. As an example, the FCC Pea Protein monograph standard has the following sections: title, description, function, packaging and storage, tests, and acceptance criteria. The tests, procedures, and acceptance criteria can include identification (amino acid composition and peptide identification), assay (by nitrogen), (elemental) impurities, ashes, lipid (fat), and loss on drying [20]. Guideline documents can also be found in these compendia, such as the FCC Guidance on Developing and Validating Non-Targeted Methods for Adulteration Detection or the AOAC Guidelines for Single Laboratory Validation of Chemical Methods for Dietary Supplements and Botanicals. These resources for method development and validation provide analysts with valuable practical guidance to address common concerns.

The protein content of food products is typically measured by the Kjeldahl method or the Dumas method. These methods measure nitrogen content, which is directly correlated to protein content using a nitrogen conversion factor. A general nitrogen conversion factor of 6.25 is provided in 21CFR101.9 [16]. For the conversion factors of specific proteins, the regulation refers to OMA AOACI [21]. OMA currently indicates the conversion factors for a few plant-based commodities, including 5.7 for flour, wheat, and its products, 5.18 for almonds, 5.46 for peanuts and

Brazil nuts, and 5.30 for tree nuts and coconut. As more plant protein sources emerge, the nitrogen conversion factors for those protein ingredients will need to be established.

5 Relationship Between Regulations and Standards

Regulations establish minimum requirements for a protein ingredient, usually related to the intended uses and usage levels. Standards provide specifications that include test methods, procedures, and acceptance criteria to ensure that foods are safe and of high quality. Standards are typically developed through collaborative and consensus-based processes, often through global standard development organizations. In many cases, they can be created and revised more often than regulations. Regulations are typically much more rigid and are changed less often. Regulations are legally binding, whereas some public standards may not be enforceable. To ensure they remain up-to-date and protect consumers, regulations commonly cite various public standards without directly listing the specifications or test methods.

6 Summary

Newer protein ingredients are being developed to meet changing consumer preferences. The regulatory and standard landscape for plant-based proteins is complex and evolving. To ensure that plant-based protein ingredients and products are safe and of high quality, regulators and standard-setting organizations should work together in developing regulations and standards that meet emerging needs. Ingredient producers and final product manufacturers should navigate the regulations and standards comprehensively.

Acknowledgments

We would like to thank Kristie Laurvick for her feedback.

References

1. Code of Federal Regulations (2023) Title 21, Part 171. Food additive petitions. U.-S. Government Printing Office, Washington, DC. https://www.ecfr.gov/current/title-21/chapter-I/subchapter-B/part-171. Accessed 29 Oct 2023
2. Code of Federal Regulations (2023) Title 21, Part 571. Food additive petitions. U.-S. Government Printing Office, Washington, DC. https://www.ecfr.gov/current/title-21/chapter-I/subchapter-E/part-571. Accessed 29 Oct 2023
3. Gaynor P (2005) How US FDA's GRAS notification program works. Foodsafety magazine
4. Code of Federal Regulations (2023) Title 21, Part 20. Public information. U.-S. Government Printing Office, Washington, DC. https://www.ecfr.gov/current/title-21/

chapter-I/subchapter-A/part-20. Accessed 29 Oct 2023

5. Code of Federal Regulations (2023) Title 21, Part 170.275. Public disclosure of a GRAS notice. U.S. Government Printing Office, Washington, DC. https://www.ecfr.gov/current/title-21/chapter-I/subchapter-B/part-170/subpart-E/section-170.275. Accessed 29 Oct 2023
6. Code of Federal Regulations (2023) Title 21, Part 170–199. Food for human consumption. U.S. Government Printing Office, Washington, DC. https://www.ecfr.gov/current/title-21/chapter-I/subchapter-B. Accessed 29 Oct 2023
7. Code of Federal Regulations (2023) Title 21, Part 184. Direct food substances affirmed as generally recognized as safe. U.S. Government Printing Office, Washington, DC. https://www.ecfr.gov/current/title-21/chapter-I/subchapter-B/part-184. Accessed 29 Oct 2023
8. Code of Federal Regulations (2023) Title 21, Part 584. Food substances affirmed as generally recognized as safe in feed and drinking water of animals. U.S. Government Printing Office, Washington, DC. https://www.ecfr.gov/current/title-21/chapter-I/subchapter-E/part-584. Accessed 29 Oct 2023
9. GRAS Notice Inventory (2018) https://www.fda.gov/food/generally-recognized-safe-gras/gras-notice-inventory. Accessed 29 Oct 2023
10. Code of Federal Regulations (2023) Title 21, Part 101. Food labeling. U.S. Government Printing Office, Washington, DC. https://www.ecfr.gov/current/title-21/chapter-I/subchapter-B/part-101. Accessed 29 Oct 2023
11. Structure/Function Claims (2022) https://www.fda.gov/food/food-labeling-nutrition/structurefunction-claims. Accessed 29 Oct 2023
12. US Dairy Export Council (2010) Navigating protein and satiety messages. https://www.thinkusadairy.org/assets/Documents/Customer%20Site/C3-Using%20Dairy/C3.7-Resources%20and%20Insights/05-Marketing%20Trends%20and%20Information/NavigatingProteinSatiety_FINAL_09-30-10.pdf. Accessed 29 Oct 2023
13. Questions and Answers on Health Claims in Food Labeling (2017) https://www.fda.gov/food/food-labeling-nutrition/questions-and-answers-health-claims-food-labeling. Accessed 29 Oct 2023
14. Authorized Health Claims That Meet the Significant Scientific Agreement (SSA) Standard | FDA (2022) https://www.fda.gov/food/food-labeling-nutrition/authorized-health-claims-meet-significant-scientific-agreement-ssa-standard. Accessed 29 Oct 2023
15. Petersen KS (2019) The dilemma with the soy protein health claim. J Am Heart Assoc 8(13): e013202. https://www.ahajournals.org/doi/full/10.1161/JAHA.119.013202
16. Code of Federal Regulations (2023) Title 21, Part 101.9 Nutrition labeling of food. U.S. Government Printing Office, Washington, DC. 21CFR101.9. https://www.ecfr.gov/current/title-21/chapter-I/subchapter-B/part-101. Accessed 29 Oct 2023
17. World Health Organization (1991) Protein quality evaluation: report of the Joint FAO/-WHO Expert Consultation, Bethesda, MD, USA, 4–8 December 1989, vol 51. Food & Agriculture Org
18. How to Understand and Use the Nutrition Facts Label (2023) https://www.fda.gov/food/nutrition-facts-label/how-understand-and-use-nutrition-facts-label. Accessed 29 Oct 2023
19. Food Allergies (2023) https://www.fda.gov/food/food-labeling-nutrition/food-allergies. Accessed 29 Oct 2023
20. Food Chemicals Codex (2023) https://www.foodchemicalscodex.org/. Accessed 29 Oct 2023
21. AOAC Official Methods of Analysis, 22nd Edition (2023) Association of Official Analytical Chemist, Washington, DC. https://www.aoac.org/official-methods-of-analysis/

Part II

Production and Safety of Plant-Based Proteins

Chapter 3

Roller and Hammer Milling of Pulses

Manoj Kumar Pulivarthi, Shivaprasad Doddabematti Prakash, and Kaliramesh Siliveru

Abstract

The milling process plays a crucial role in producing value-added ingredients from pulses such as lentils and will tremendously alter the future of pulses and cereal industries. This chapter presents a detailed guide on reducing the particle-size of red lentils through two distinct milling methods: roller milling and hammer milling. Additionally, an analysis of the particle-size distribution (PSD) of the milled fractions reveals a bimodal distribution for roller-milled red lentils and a unimodal distribution for hammer-milled red lentils.

Key words Roller mill, Hammer mill, Milling, Particle-size distribution, Red lentils

1 Introduction

The milling of pulses can be categorized into three primary methods. The first involves "dehulling" or "decortication," which focuses on removing the tough seed coat to create polished seeds. The second method is "splitting," where the pulse's two cotyledons are separated, resulting in split pulses. It's important to note that dehulling and splitting are often carried out simultaneously within the same milling process for many pulse varieties. Lastly, "flour milling" or "grinding" is the third method, which aims to reduce either whole seeds or cotyledons to a fine powder known as flour. These distinct categories of milling methods play essential roles in the preparation of pulses for a wide range of applications, enhancing their versatility and utilization [1].

Various milling technologies, such as impact, hammer, roller, stone, and attrition milling, have proven to be effective in the production of pulse flours [2]. The choice of milling techniques and their associated parameters has a direct impact on the physicochemical attributes of the flour, including particle-size distribution, protein quality, and starch damage. These attributes play a pivotal role in determining flour hydration, dough structure, and,

Yonghui Li (ed.), *Plant-Based Proteins: Production, Physicochemical, Functional, and Sensory Properties*, Methods and Protocols in Food Science, https://doi.org/10.1007/978-1-0716-4272-6_3,

ultimately, consumer satisfaction through sensory characteristics like texture, crunchiness, and crispiness. It's important to note that variations in these flour properties can significantly influence their suitability for specific end-use applications [3].

Hammer mills are commonly employed in the processing industry due to their superior capacity to finely grind a wide range of materials when compared to other milling equipment [4, 5]. During size reduction, hammer mills employ a combination of impact, shear, and compression forces, with impact accounting for the largest share [6]. However, the hammer mill requires more time and a higher speed to produce fine particles (flour), consuming a lot of energy every kilogram of flour produced [7].

Roller milling, in contrast to other milling methods, offers unique advantages by enabling the production of a diverse range of flours [8, 9]. The mode of grinding action in roller mill is compression and shear. Moreover, roller mills are known for their relatively lower specific energy consumption compared to alternative technologies such as pin and attrition mills. Numerous research studies have explored the milling of pulse products using roller mills [9–11].

What sets roller milling apart is the ability to create specific flours through a gradual process involving multiple roll and sifter passes. This flexibility allows for numerous adjustments in the milling process, granting greater control over particle-size reduction and the development of flours with a wide array of functional characteristics. Consequently, roller milling technology has garnered a significant interest as a means to regulate variability and effectively manage diversity in pulse flour quality [12].

In this chapter, the milling process was conducted on whole red lentils using both a bench-top stainless steel roller mill (Model 915, Ross Machine and Mill Supply, Oklahoma City, OK, USA) and a hammer mill (Cross Beater Mill SK 300, Retsch GmbH, Haan, Germany). The key processing parameters for the roller mill included the adjustment of roller gaps and sieve sizes to achieve the desired flour characteristics. Conversely, for the hammer mill, the mill speed was carefully adjusted to optimize the milling process. These precise adjustments in the milling parameters are essential in tailoring the final flour products to meet specific requirements and applications, highlighting the importance of process control in achieving desired flour quality.

2 Materials and Equipment

1. Red lentils.
2. Roller mill (Model 915).
3. Hammer mill with cyclone system (SK 300).

4. Box siever with timer.
5. Sieves of different size openings.
6. Feeler gauge.
7. Pans.
8. Power-driven rotary tempering bins.
9. Distilled water.
10. Airtight re-sealable bags.
11. Brush.
12. Particle-size analyzer (Alpine Jet Sieve).
13. Weighing scale.

3 Methods

3.1 Roller Mill

3.1.1 Tempering

1. The tempering process is grain-specific and plays a crucial role as a preparatory step before roller milling.
2. Within this chapter, red lentils with an initial moisture content of 10.4% (wet basis %) were selected for roller milling.
3. A sequence of preliminary milling trials was carried out to investigate a range of combinations involving tempering moisture and time.
4. These trials unveiled that maintaining a moisture content of 12% for 18 h stands out as the optimal condition for achieving an efficient separation of flour and hull fractions within red lentils.
5. To attain this desired moisture level, the amount of water to be added (Q, mL) was calculated using Eq. 1.

$$Q = w\frac{(m_{\mathrm{f}} - m_{\mathrm{i}})}{100 - m_{\mathrm{f}}} \tag{1}$$

Where $w = 1000$ g—the weight of grains to be tempered in grams.

$m_{\mathrm{f}} = 12\%$—the target grain moisture in percentage.

$m_{\mathrm{i}} = 10.4\%$—the initial grain moisture in percentage.

Therefore required amount of water (Q) = 9.09 mL.

6. Initially, 1000 g red lentils were placed into the power-driven rotary bins as shown in Fig. 1.
7. A total of 9.09 mL of water was added to the bins and mixed for 15 min with the lids securely closed to ensure the uniform distribution of water (*see* **Note 1**).
8. After the mixing process, the lentils were transferred into airtight re-sealable bags and rested for 18 h at room temperature, allowing for the even absorption of water within the lentils.

Fig. 1 Motor-driven bins for tempering the grains

9. Following the exact 18-h resting period, the lentils were milled as per the process outlined in the section below.

3.1.2 Milling Process

1. The mill, collection trays, sieves, and gyratory sieving box were thoroughly cleaned using a brush to eliminate any residual mill stock from previous milling runs (*see* **Notes 2** and **3**).
2. The milling flowsheet for grinding red lentils was fine-tuned after multiple trials, and the optimized settings are detailed in Fig. 2.
3. The roller gap for the initial set of break rolls (yellow) was adjusted to a 0.025-inch gap, tailored to the size of the red lentils.
4. To adjust the roller gap, the primary roller gap adjuster (lever) located next to the hopper, as shown in Fig. 3, was utilized. The adjustment was made with the assistance of a feeler gauge (Fig. 4) (*see* **Notes 4– 6**).
5. The feed rate was set to enable the lentils to pass through the rolls in a single stream, ensuring uniform grinding.
6. After the first pass (1 BK), the mill fractions underwent a 2-min sieving process using the sifter box, which was equipped with sieves arranged in the following order, 1041 μm, 630 μm, 240 μm, 150 μm, and a pan, as shown in Fig. 2.
7. The overs from 1041 μm sieve were further passed through the second set of break rolls (blue) with a roller gap of 0.012 inches.
8. After the second pass (2 BK), the mill fractions were sieved for 2 min using the same set of sieves mentioned in **step 6**.

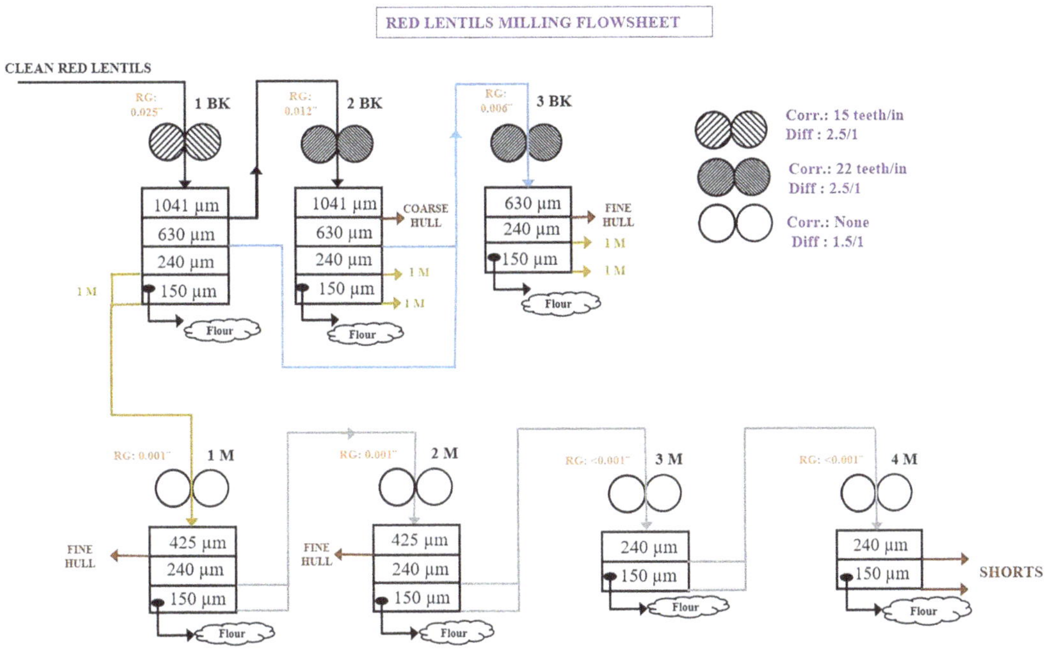

Fig. 2 Roller milling flowsheet for red lentils

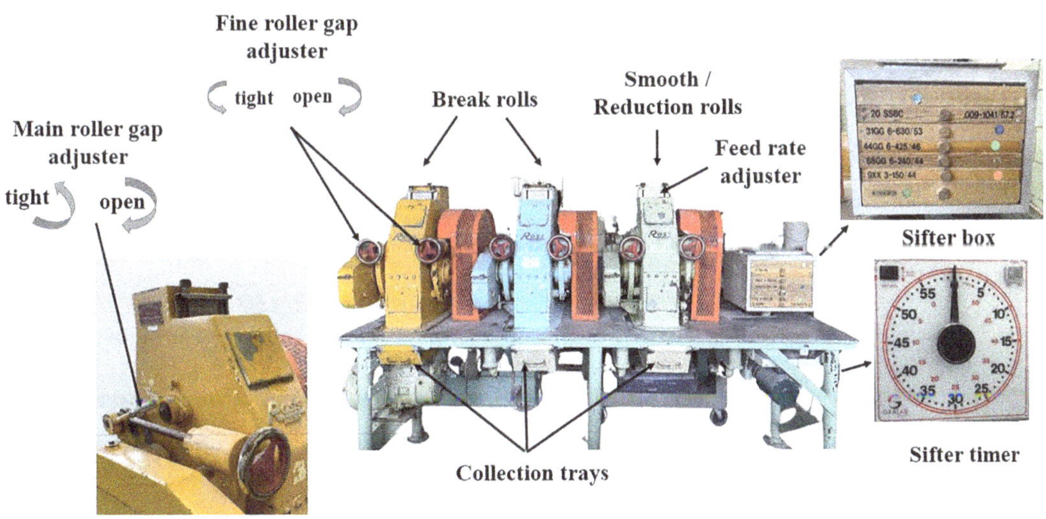

Fig. 3 A tabletop stainless steel roller mill setup along with sifter box and timer

9. The overs from 1041 µm were separated as coarse hulls, and their weight was recorded.
10. The overs obtained from 630 µm sieve during 1 BK and 2 BK passes underwent a third pass (3 BK) through the second set of break rolls (blue), with a roller gap of 0.006 inches.

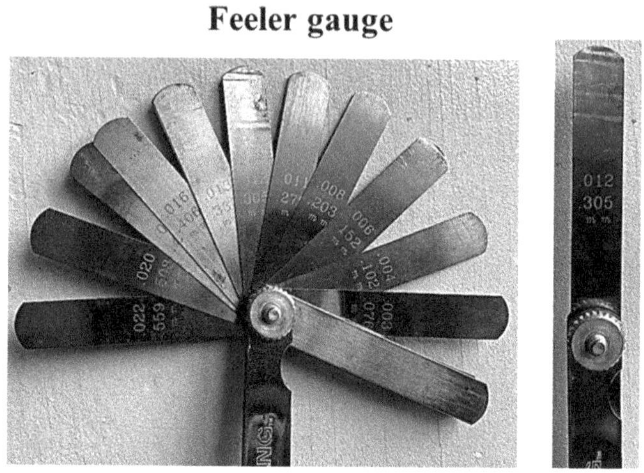

Fig. 4 Feeler gauge for adjusting the roller gap

11. After the third pass (3 BK), the mill fractions were sieved using the following set of sieves: 630 μm, 240 μm, 150 μm, and a pan.
12. The overs from 630 μm were collected as fine hulls, and their weight was recorded.
13. The overs from 240 μm and 150 μm sieves from all three passes (1 BK, 2 BK, and 3 BK) were passed through reduction/ smooth rolls (1 M) adjusted to a 0.003-inch roller gap.
14. The resultant mill fractions from 1 M were sieved for 3 min using sieves with openings of 425 μm, 240 μm, 150 μm, and a pan as shown in Fig. 2.
15. After sieving, additional fine hulls were collected from 425 μm sieve and weighed.
16. The mill fractions from 240 μm and 150 μm from 1 M were passed through reduction rolls (2 M) with a roller gap of 0.001 inches.
17. The sieving and milling steps were repeated using the same set of sieves as in **step 13** and the roller gap settings detailed in **step 15** to achieve 3 M and 4 M passes.
18. The flour fraction (<150 μm) was collected from the pan, and the shorts were collected from the 150 μm and 240 μm sieve and weighed.
19. The yield of different mill fractions, such as flour, hull (coarse hull and fine hull), and shorts, was calculated using Eq. 2 and is presented in Table 1:

$$\text{Milling Yield (\%)} = \frac{\text{Weight of flour fraction}}{\text{Weight of pulses}} \times 100 \quad (2)$$

Table 1
Milling yields from roller milling of red lentils

Sample	Hull (%)	Shorts (%)	Flour (%)	Loss (%)
Roller milling yields	10.2	4.2	82.6	3

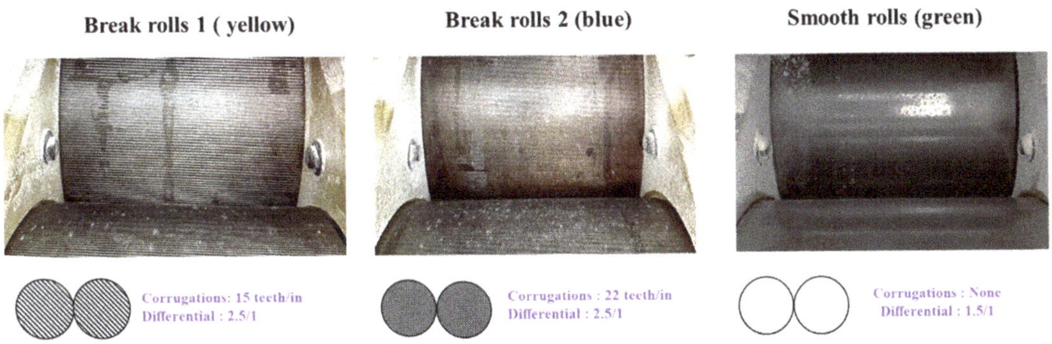

Fig. 5 A closer look at roller surfaces from break rolls to reduction rolls (left to right)

Fig. 6 Mill fractions from different passes and red lentil flour

Results

1. The variations in surface corrugations and differential speed among the three roller stands (yellow, blue, and green) are visually represented in Fig. 5.
2. The mill fractions from different passes are shown in Fig. 6.
3. A particle-size analysis, as illustrated in Fig. 7, reveals a bimodal distribution in the red lentil flour, with a total flour yield of 82.6%, as detailed in Table 1.

3.1.3 Notes

1. When conducting the mixing stage of tempering, ensure that the lids of the rotary bins are securely closed to prevent the evaporation of water.

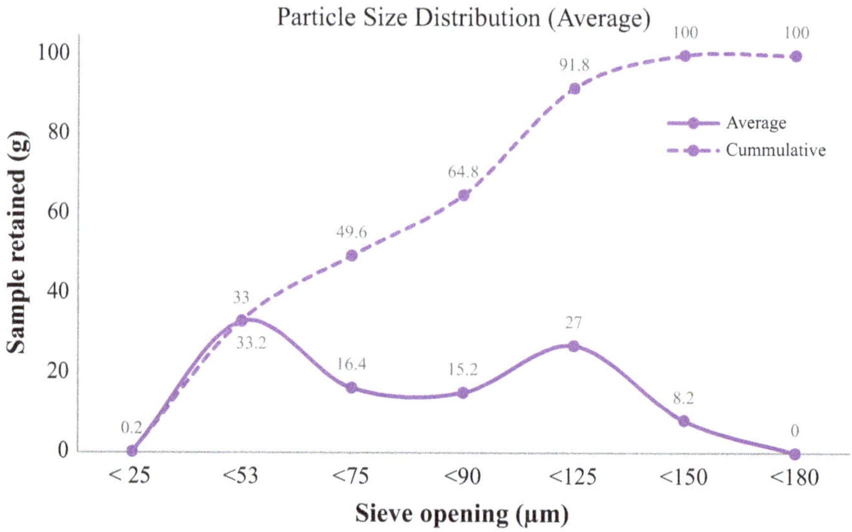

Fig. 7 Particle-size distribution of roller milled red lentil flour

2. Following the mill cleaning process with a brush, it is recommended to pass the sample through all three roller stands once. This step effectively flushes out any remnants of previous mill fractions from inaccessible areas on the mill surfaces.
3. Prior to milling, it is essential that the feed material is entirely free from foreign substances, such as metals or stones. This precautionary measure prevents any potential damage to the mill and ensures food safety.
4. While all three roller stands may be operational simultaneously, it is advisable to feed one sample at a time into one roller set. Feeding multiple samples concurrently into two or more set of rollers may lead to unintended alterations in the roller gaps, resulting in undesired granulation.
5. The initial roller gap setting is of paramount importance and is specific to the type of kernels being processed. Fine-tuning this setting is necessary to enable optimal grinding. A gap that is excessively large would allow grains to pass through without proper grinding, whereas a smaller gap would cause kernels to rebound after coming into contact with the rolls.
6. The roller gap settings and sieve selections described in this chapter are tailored specifically for red lentils. It is crucial to conduct trial runs at varying roller gaps and sieve arrangements when dealing with different grains to achieve optimization.
7. The mill settings provided for red lentils can be further adjusted to modify the granulation.

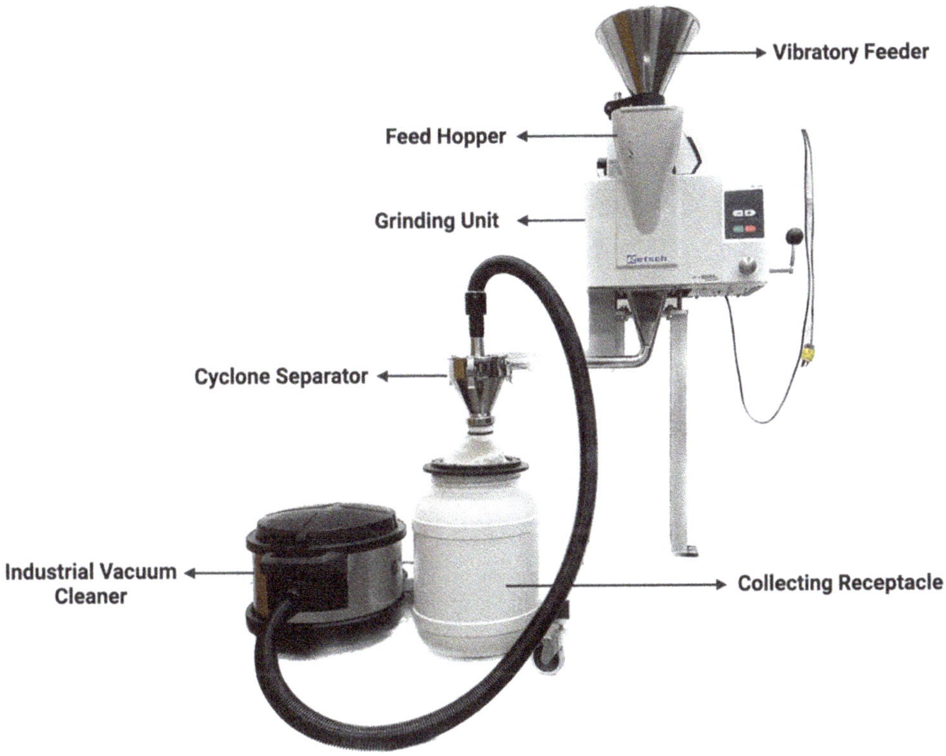

Fig. 8 Laboratory scale hammer mill

3.2 Hammer Mill

1. The vibratory feeder, rotor, sieves, and cyclone separator with collecting receptacle as shown in Fig. 8 were thoroughly cleaned using a brush to eliminate any residual mill stock from previous milling runs (*see* **Note 1**).
2. The feed material passes from the vibratory feeder into the grinding chamber through feed hopper. The sample will then be comminuted between the baffle plates of the rotor and the serrated grinding insert (*see* **Notes 2** and **3**).
3. The grinding material smaller than the mesh size of the selected bottom sieve will enter the cyclone separator with collecting receptacle where the grinded sample will be collected (*see* **Notes 4–7**).

Result

A particle-size analysis, as illustrated in Fig. 9, reveals a unimodal distribution in the red lentil flour.

3.2.1 Notes

1. Prior to milling, it is essential that the feed material is entirely free from foreign substances, such as metals or stones. This precautionary measure prevents any potential damage to the mill.

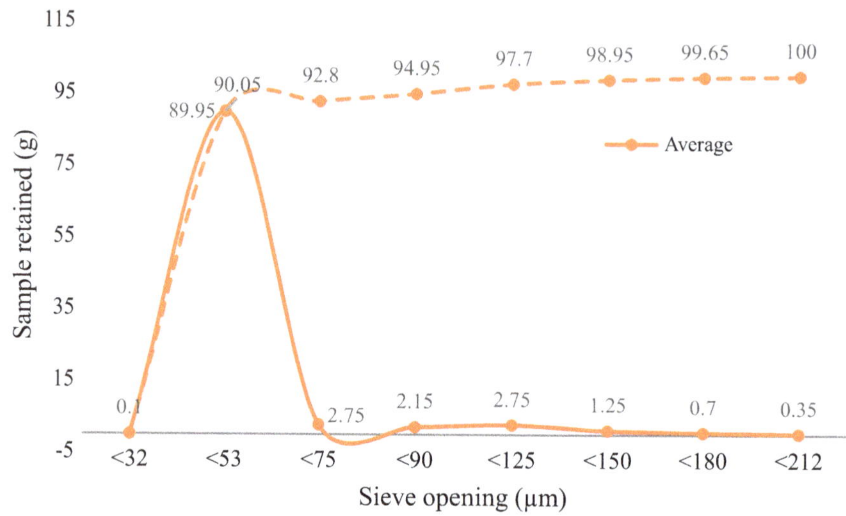

Fig. 9 Particle-size distribution of hammer-milled red lentil flour

2. The vibratory feeder is designed for uniform feed and conveying of free-flowing bulk material and powders in dry conditions. It shall not be used for any wet materials.
3. It is recommended to commence the operation by activating the vacuum cleaner first and subsequently initiating the grinding device before introducing the sample material.
4. Selection of the mesh size of the sieve being used depends on the desired particle size. In general, the achievable final fineness can be <100 μm.
5. The hammer mill used for this experiment is a laboratory scale mill and is not suitable for long-term operation. This device is designed for an 8-h single-shift operation with a duty cycle of 30%. The equipment cannot be used beyond its capacity.
6. Caution is essential when grinding larger grains and harder materials, as the high pulling power of the rotor may lead to blockages. To prevent this issue, it is advisable to closely monitor the speed at which the material is fed into the hopper.
7. Cleaning the bottom sieve, which has a smaller mesh size, may pose challenges with a brush. Instead, consider utilizing ultrasonication in a water bath for 3–4 min as an effective cleaning method. Avoid using strong acids and bases for cleaning, as they are not recommended.

References

1. Wood JA, Malcolmson LJ (2021) Pulse milling technologies. In: Pulse foods. Academic Press, pp 213–263
2. Maskus H, Bourre L, Fraser S, Ashok S, Malcolmson L (2016) Effects of grinding method on the compositional, physical, and functional properties of whole and split yellow pea flours. Cereal Foods World 61(2):59–64
3. Scanlon MG, Thakur S, Tyler RT, Milani A, Der T, Paliwal J (2018) The critical role of milling in pulse ingredient functionality. Cereal Foods World 63(5):201–206
4. Bitra VSP, Womac AR, Chevanan N, Mui PI, Igathinathane C, Sokhansanj S, Smith DR (2009) Direct mechanical energy measures of hammer mill comminution of switchgrass, wheat straw, and corn stover and analysis of their particle size distributions. Powder Technol 193:32–45. https://doi.org/10.1016/j.powtec.2009.02.010
5. Basiouny MA, El-Yamani AE (2016) Performance evaluation of two different hammer mills for grinding corn cobs. J Soil Sci Agric Eng 7:77–87. https://doi.org/10.21608/jssae.2016.39322
6. Probst KV, Ambrose RPK, Pinto RL, Bali R, Krishnakumar P, Ileleji KE (2013) The effect of moisture content on the grinding performance of corn and corncobs by hammer milling. Trans ASABE 56:1025–1033. https://doi.org/10.13031/trans.56.9996
7. Bassey J, Odesola I, Olawuyi J (2022) A comparative technique for performance evaluation of hammer mill and disk mill in yam flour processing. World J Eng Technol 10:613–625. https://doi.org/10.4236/wjet.2022.103040
8. Campbell GM (2007) Roller milling of wheat. Handb Powder Technol 12:383–419
9. Sakhare SD, Inamdar AA, Gaikwad SB (2014) Roller milling fractionation of green gram (Vigna radiata): optimization of milling conditions and chemical characterization of millstreams. J Food Sci Technol 51:3854–3861
10. Mittal R, Nagi HPS, Sharma P, Sharma S (2012) Effect of processing on chemical composition and antinutritional factors in chickpea flour. J Food Sci Eng 2(3):180
11. Pulivarthi MK, Nkurikiye E, Watt J, Li Y, Siliveru K (2021) Comprehensive understanding of roller milling on the physicochemical properties of red lentil and yellow pea flours. PRO 9(10):1836
12. Motte JC, Tyler R, Milani A, Courcelles J, Der T (2021) Pea and lentil flour quality as affected by roller milling configuration. Legum Sci 3(4):e97

Chapter 4

Dry Fractionation of Pulses to Produce Protein-Rich Fractions

Manoj Kumar Pulivarthi and Kaliramesh Siliveru

Abstract

The growing demand for plant-based protein concentrates and isolates has led to the search for sustainable alternatives to energy- and water-intensive wet fractionation processes. Dry fractionation, which involves dry milling followed by air classification, is a sustainable and energy-efficient way to enrich plant proteins from various agro-materials, such as legumes, oilseed meals, cereals, and brans. Dry milling is a crucial step because it must optimally break down the protein-starch cell matrix to separate the protein bodies from the larger starch granules. Air classification separates the fine protein-rich particles from the coarse carbohydrate-rich particles based on their size and density differences. Centrifugal air classifiers are the most common type of classifier used for this purpose in agro-materials.

This chapter outlines a clear procedure for milling and air classification of red lentils to produce protein-rich fine fractions. The classifying mill was optimized to obtain red lentil flour with a particle size (D_{90}) of 40–42 μm to achieve maximum separation of protein bodies from the cell matrix. The flour is passed through an air classifier with pre-optimized settings to obtain a protein-rich fine fraction with a particle size (D_{90}) of 20–22 μm. Detailed explanations of the milling and air classification steps are provided in this chapter to help other researchers operate and develop dry fractionation processes for other materials.

Key words Air classification, Milling, Protein concentrates, Red lentils, Dry fractionation

1 Introduction

The growing concerns about the environmental, health, and resource implications of animal-based proteins have driven an increased interest in the development of edible plant-based proteins, which offer long-term economic and environmental benefits [1]. The quality of plant proteins is influenced by various factors, including the source, composition, structure, and separation process. Wet fractionation, a commonly used method, produces high-purity protein isolates but requires water, organic solvents, and extreme pH and temperature conditions, which can significantly alter the structure and native functionality of the proteins [2–4]. In

Yonghui Li (ed.), *Plant-Based Proteins: Production, Physicochemical, Functional, and Sensory Properties*, Methods and Protocols in Food Science, https://doi.org/10.1007/978-1-0716-4272-6_4,

contrast, dry separation is a less energy-intensive and more environmentally friendly method that preserves the structure and native functional properties of protein concentrates [5–7].

Pretreatments of pulses and seeds before milling, such as tempering, drying, dehulling, soaking, and defatting, can improve protein separation efficiency during air classification [8]. Air classification is a size fractionation technique that employs an airstream to divide a mixed particle size flour into two distinct size fractions based on a predetermined cut point. The cut point represents the particle size at which there is an equal likelihood of a particle being classified as either coarse or fine [9].

The separation of pulse flours via air classification is feasible due to the substantial size and density difference between starch granules and protein bodies in pulse flours. The size of starch granules serves as a reference point for determining the cut point. Starch granules from ordinary peas (smooth peas) exhibit variability in size, ranging from 2 to 40 micrometers (μm). Bean starch granules have a diameter of 20–40 μm [10], lentil starch granules have a diameter of 22.0 ± 3.0 μm, chickpea starch granules have a diameter of 17.1 ± 2.9 μm [11], and pea starch granules have a diameter of 5–30 μm [10]. Consequently, the optimal cut point for separating protein from starch typically falls within the 15–22 μm range, which is below the size of most pulse starch granules. Particles with a size exceeding 22 μm will invariably be classified into the coarse fraction. Below the cut point, the likelihood of a particle ending up in the fine fraction should consistently increase as particle size decreases [9, 12]. While a lower cut point may enhance the purity of the protein fraction, it comes at the cost of a reduced yield.

1.1 Milling and Air Classification Principle

In the present chapter, Prater's CLM-18 (Prater-Sterling, Bolingbrook, IL) was used to mill and air-classify dehulled or hull-less red lentils. Prater's classifying mill (CLM-18) employs a three-stage grinding process to produce ultra-fine flours with particle sizes (D_{90}) ranging from 35 to 40 microns, suitable for air classification. In the initial grinding stage, conveyed air and metered product enter behind the rotor, where grinding blades impact and propel particles outward. These particles collide with jaws and screens until reduced to a size that enables passage through apertures. Subsequently, in the classifier phase, particles from the first grinding stage are circulated outside the grinding chamber. A secondary air intake is employed to draft pneumatic air, facilitating fluidization and cooling. The material is directed toward the classifying rotor for size separation. On-size particles proceed to the cyclone chamber, where they are collected in the unit, while oversize particles are rejected for second-stage grinding in front of the rotor. The classifier rotor can be independently controlled, allowing precise control over particle size by adjusting the RPM using a variable frequency drive (VFD) [13].

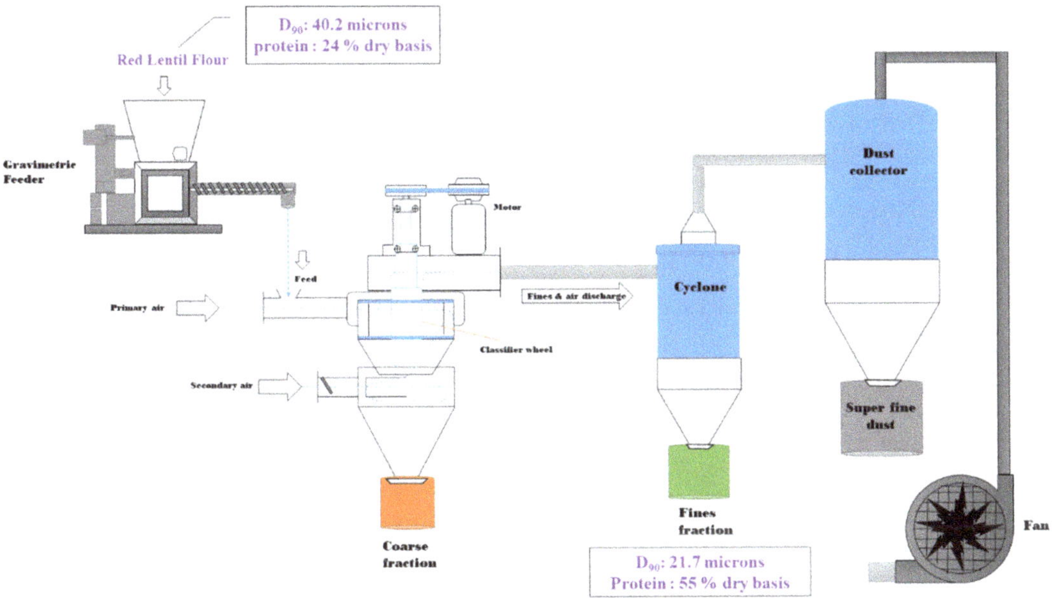

Fig. 1 Schematic of air classification process for red lentil flours

After fine grinding using the CLM-18 Mill, the fine flour will be separated into fine and coarse fractions based on particle size and density differences using an air classifier (Minisplit, Prater-Sterling, Bolingbrook, IL) as shown in Fig. 1. The air classifier consists of a separating rotor (classifier wheel) that is powered by an electric motor with adjustable speed. Upon activation, a spiraling air system is created, which initiates the classification of the parent flour into two size fractions. Flour particles are subjected to a centrifugal force, countered by the centripetal friction drag force, until they reach the cut point. The cut point can be adjusted by changing the air flow (fan speed) and classifier wheel speed.

Fine particles have a tendency to agglomerate and form larger particles, which can be classified as oversized or rejected particles. The separation becomes more difficult as the cut size decreases. To overcome this issue, the CLM-18 Classifier is equipped with a secondary air system that increases the residence time of particles in the separating zone of the cyclone housing and sifts the rejected material as it descends [14]. The optimal classifier speeds and feed rates required to achieve high-efficiency cut points (~ 20–22 μm) for red lentil flours were determined through preliminary trials.

This chapter outlines a detailed procedure for milling and air classification to separate protein-rich fractions from de-husked or hull-less red lentils. It also presents particle size analysis results from different fractions, such as flour, fines, and overs, as well as yield and recovery calculations, to provide a comprehensive understanding of the process.

2 Materials and Equipment

1. De-husked or hull-less red lentils.
2. Prater's CLM-18 Mill and minisplit classifier system.
3. Airtight re-sealable bags.
4. Particle size analyzer (Malvern's).
5. Weighing scale.
6. LECO protein analyzer.

3 Methods

3.1 System Startup

1. Connecting quick disconnect plugs:
   (a) Identify the corresponding connectors on the control panel for the equipment to be used as shown in Fig. 2.
   (b) Insert all quick disconnect plugs securely into their respective connectors.
2. Powering the control panel:

Fig. 2 Panel showing minisplit connection

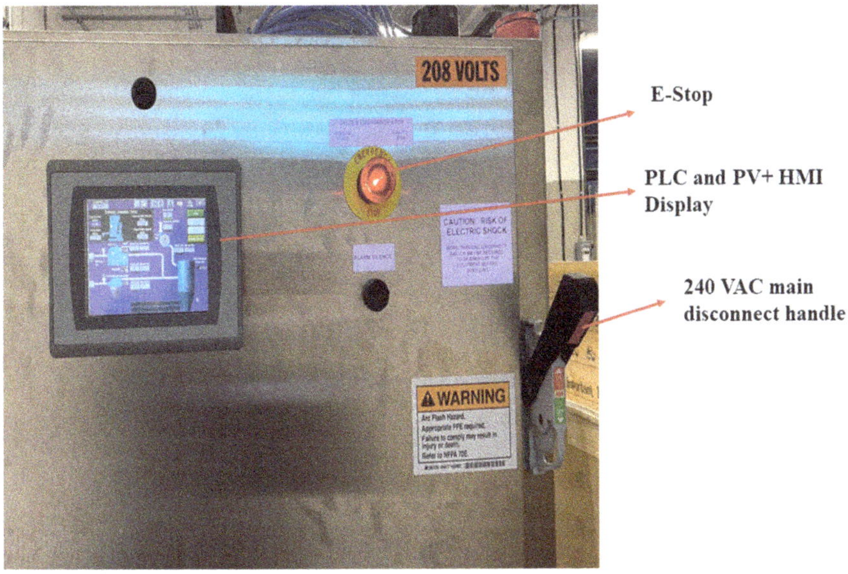

Fig. 3 PLC display view for starting up the system

   (a) Connect the blue Hubbell inlet connector to the appropriate power source.
   (b) Lift the 240VAC main disconnect handle to supply power to the control panel (Fig. 3).
3. Verifying circuit breakers:
   (a) Inspect the PLC panel to ensure all relevant circuit breakers are switched to the "on" position.
4. System boot and initialization:
   (a) Monitor the PV+ terminal as it boots and initializes the interface screen (Fig. 3).
5. Activating e-stop switch:
   (a) Pull out the momentary e-stop switch (Fig. 3) to apply control voltage to all field equipment.
6. Home or login screen:
   (a) Verify that the home or login screen appears, indicating that the system is ready for operation.
   (b) Log in using the user ID and password provided for accessing the process screen, as shown in Figs. 4 and 5.

3.2 Milling Process

1. Verify that the CLM-18 Mill Classifier is properly connected to the system and its power cable is securely plugged into the panel during the startup procedure outlined above.
2. Before starting the mill, ensure the secondary air inlet is fully open (100%).

Fig. 4 Home screen after powering up

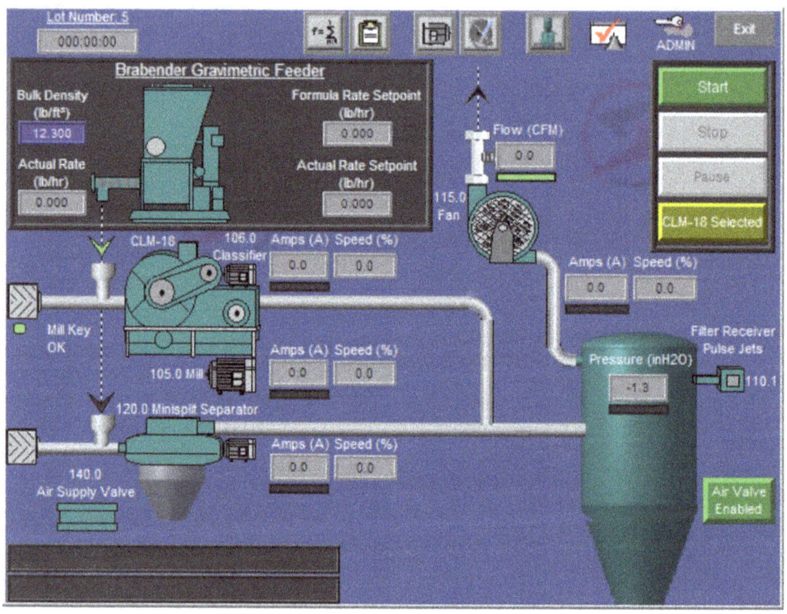

Fig. 5 The main process screen for milling and classifying the material

3. The main toolbar, shown in Fig. 6, is always visible at the top of the screen. Utilize the Formula Editor to create, store, and modify up to ten recipes/formulas for the milling and classifying process.
4. To establish suitable settings for red lentil milling, a new formula was created using the Formula Editor, as depicted in Fig. 7.

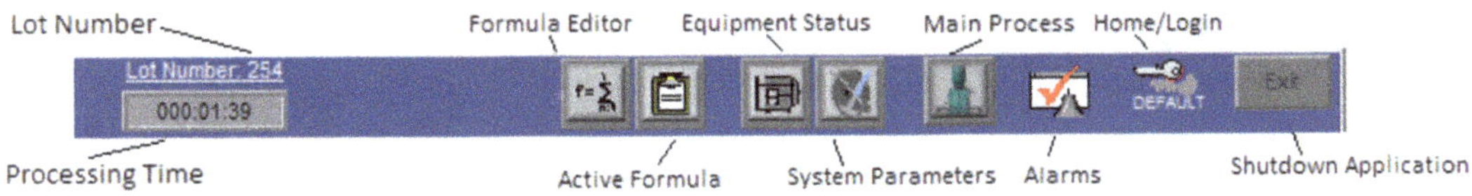

Fig. 6 Main toolbar for operating the equipment

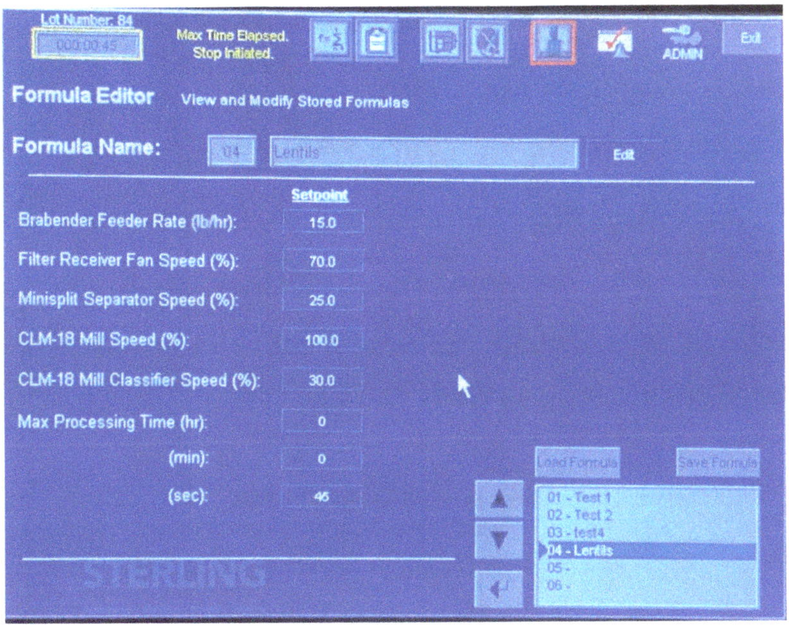

Fig. 7 Formula Editor view with reference set points for milling process

5. The set points (*see* **Note 7**) employed for the red lentil milling formula are as follows:
   (a) Brabender Feeder Rate: 15 lb/h.
   (b) Fan Speed (%): 70.
   (c) Minisplit Separator Speed (%): Not applicable.
   (d) CLM-18 Mill Speed (%): 100. (*see* **Note 3**).
   (e) CLM-18 Mill Classifier Speed (%): 30.
6. Add the grain into the feeder (*see* **Notes 2** and **8**), and set the grain's bulk density. In this case, cleaned dehulled or hull-less red lentils were added to the feeder. The bulk density was set to 53.321 lb/ft^3 by clicking the blue box in the main menu before initiating the process.
7. Use the active formula button (Fig. 8) to select the formula created in **steps 3** and **4** to activate the formula for milling process.

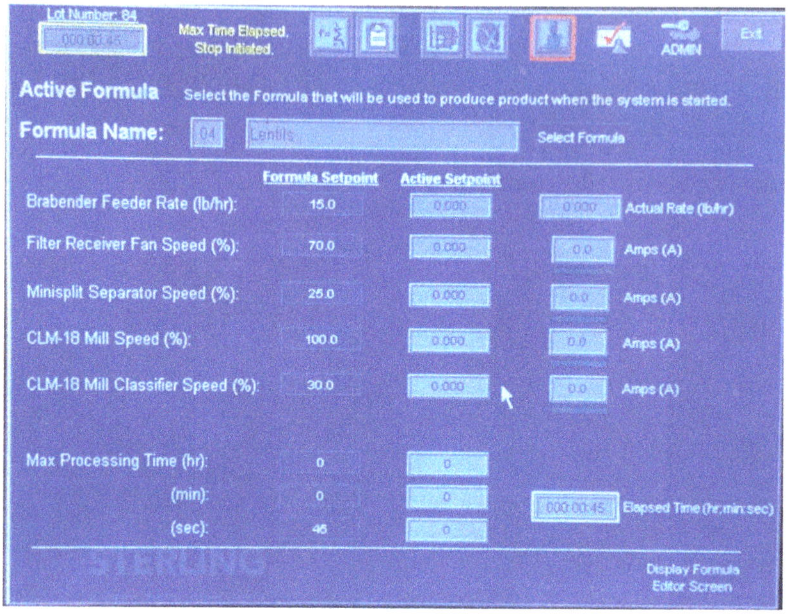

Fig. 8 Active formula view for selecting the formula

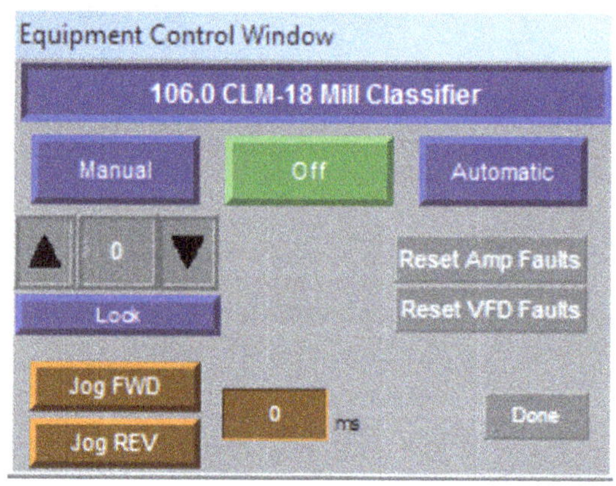

Fig. 9 Equipment control screen selecting automatic mode

8. From the main menu view, individually select each piece of equipment, and transition them all into "Automatic" mode (*see* **Note 1**), as shown in Fig. 9.

   (or)

   All the equipment can be set to automatic mode at a time by clicking the equipment status icon from the main toolbar (Fig. 6) using one button (SET ALL EQUIPMENT TO AUTOMATIC MODE) as shown in Fig. 10.

9. From the main menu, select the CLM-18 for the milling process, and press the SYSTEM START button. The unit will start each equipment sequentially with a programmable delay.

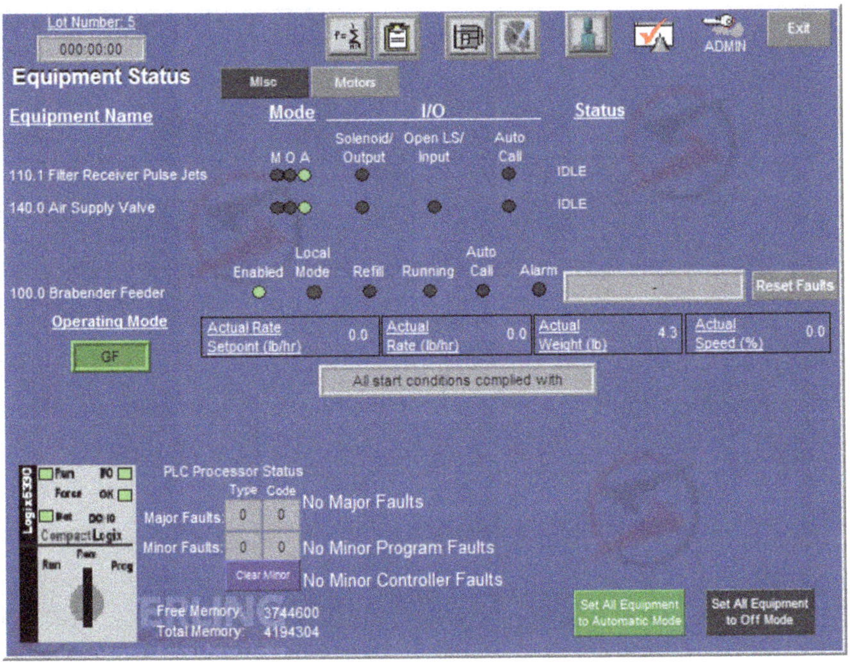

Fig. 10 Equipment status screen for selecting automatic mode

10. Initially, the Fan and Mill Classifier will activate at a programmable speed. After a brief delay, the mill motor will start, followed by the Brabender feeder.
11. An active milling process without any warnings or alarms is shown in Fig. 11.
12. The set feed rate, actual feed rate, and sample weight in the Brabender gravimetric feeder can be monitored from a separate HMI attached to the feeder, as shown in Fig. 12.
13. After reaching the maximum process time or achieving the target amount of milled sample (*see* **Note 6**), press the SYSTEM STOP button.
14. The system will shut down each piece of equipment in the following order after a programmable delay: Brabender gravimetric feeder, CLM-18 Mill, Mill Classifier, Fan.
15. The filter receiver pulse jets will continue to cycle until a programmable delay timer expires, clearing the filter cyclone and collecting the super fine dust into the collector, as shown in Fig. 13.
16. Once the system stopping indicator stops flashing and displays the SYSTEM IDEL sign, gather the milled flour from the container attached to the cyclone system.
17. Before proceeding to the classifying process, analyze the collected flour for particle size distribution (Fig. 16).

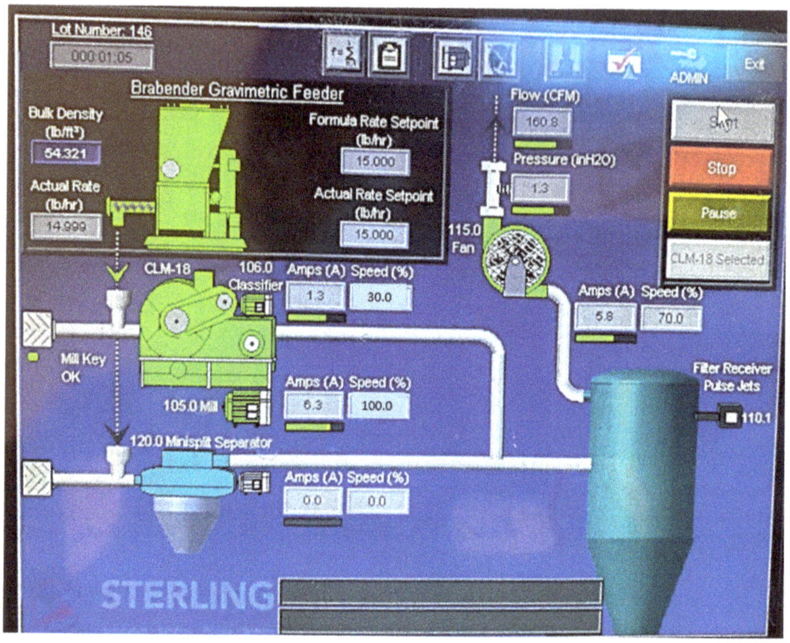

Fig. 11 Active main menu view during feeding and milling process

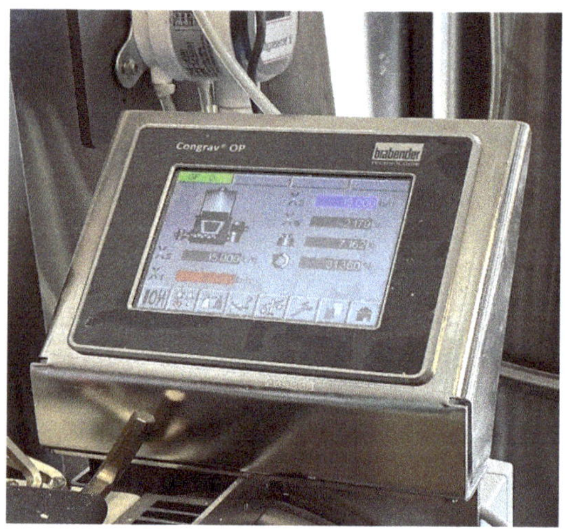

Fig. 12 HMI display for monitoring feeding status

18. In this instance, the red lentil flour was targeted to achieve a D_{90} of 38–40 μm for an optimal breakdown of the starch and protein matrix.

3.3 Classifying Process

1. Verify that the minisplit separator is properly connected to the system, and its power cable is securely plugged into the panel during the startup procedure outlined above.

Fig. 13 Air-pulsed blowback control system and dust collector

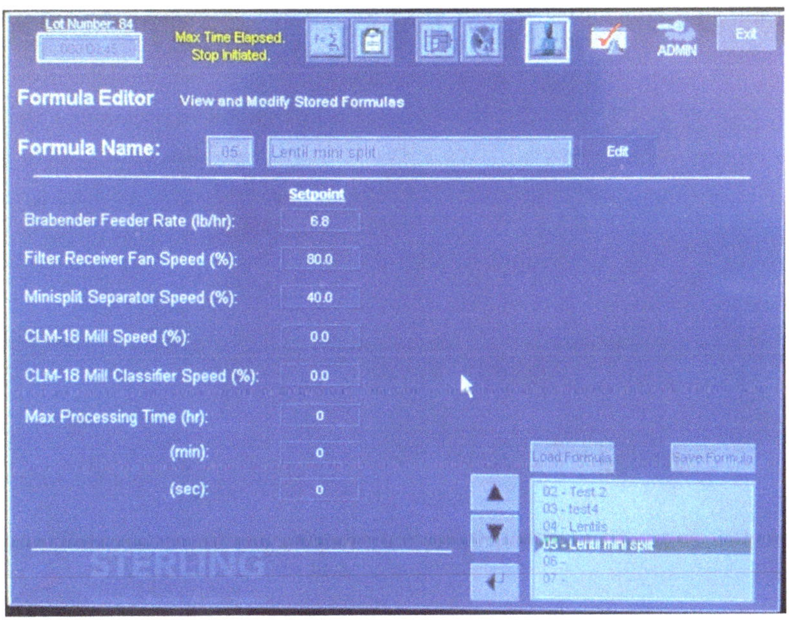

Fig. 14 Formula Editor view with reference set points for air classification process

2. A secondary air opening of 25% was employed for red lentil flour classification to further enhance the separation process (*see* **Note 5**).
3. Figure 1 provides a schematic representation of the process flow during air classification for easy comprehension.
4. Select the Formula Editor, and create a formula specifically designed to separate flour into protein-rich fine fractions and starch-rich overs (Fig. 14).

5. For red lentil flour classification, the following set points (*see* **Notes 4** and 7) were utilized for recipe (Fig. 14):
   (a) Brabender Feeder Rate: 6.8 lb/h.
   (b) Fan Speed (%): 80.
   (c) Minisplit Separator Speed (%): 40.
   (d) CLM-18 Mill Speed (%): Not applicable.
   (e) CLM-18 Mill Classifier Speed (%): Not applicable.
6. Add the milled flour into the feeder (*see* **Notes 2** and **8**), and set the flour's bulk density. In this case, red lentil flour was added to the Brabender gravimetric feeder. The flour,s bulk density was set to 31.6 lb/ft^3 by clicking the blue box in the main menu before initiating the process.
7. Utilize the active formula button to select the formula created in **step 2**. This will direct to the main menu view for the classifying process.
8. From the main menu, choose the minisplit for the classifying process, and press the SYSTEM START button for each piece of equipment. The equipment will start sequentially with a programmable delay.
9. Initially, the fan and minisplit will activate at a programmable speed. After a brief delay, the Brabender feeder will begin feeding the flour into the system.
10. Figure 15 illustrates an active classifying process without any warnings or alarms.

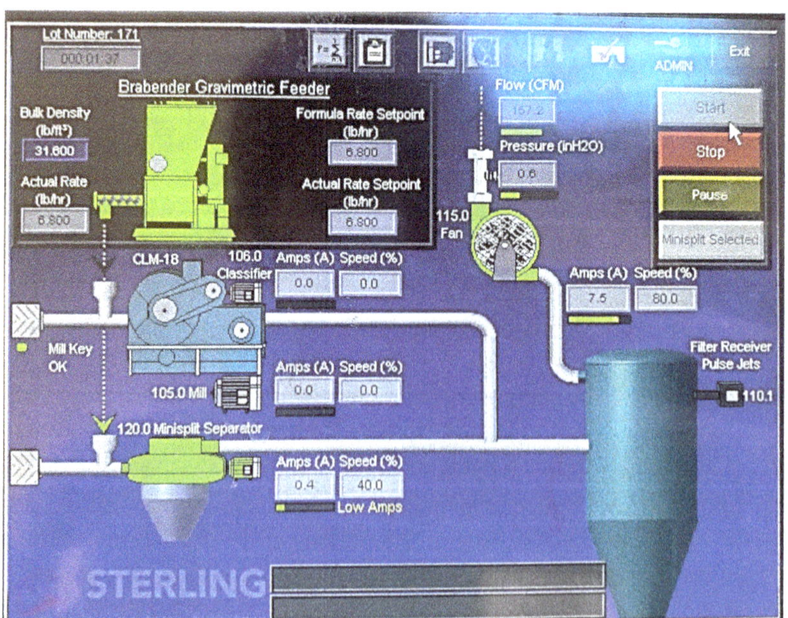

Fig. 15 Active main menu view during the feeding and classifying process

11. Once the maximum process time is reached or the target amount of sample has been classified, press the SYSTEM STOP button.
12. The system will shut down each equipment in the following order after a programmable delay: Brabender gravimetric feeder, minisplit, and fan.
13. The filter receiver pulse jets will continue to cycle until a programmable delay timer expires, clearing the filter cyclone and collecting the super fine dust into the collector.
14. Once the system stopping indicator ceases flashing and displays the SYSTEM IDEL sign, gather the fines and overs from their respective containers, as shown in Fig. 1, and measure their weights.
15. The separation yields of fines and overs can be determined using Eq. 1:

$$\text{yield (\%)} = \frac{\text{weight of fines/overs obtained (g)}}{\text{weight of parent flour (g)}} * 100 \quad (1)$$

16. The collected flour fractions (fines and overs) can be analyzed for particle size distribution (Fig. 16) and protein content analysis to evaluate the separation efficiency.
17. In this case, the red lentil flour was classified to achieve a fine fraction with a D_{90} of 20–22 μm for optimal protein enrichment.
18. After obtaining the protein values of the flour fractions, the total protein recovery (Table 1) can be calculated using Eqs. 2 and 3:

$$\text{Protein recovery or yield (\%)} = \frac{\text{Protein mass offine fraction}}{\text{Protein content of parent flour}} \times 100 \quad (2)$$

Whereas:

$$\text{Protein mass of fine fraction} = \frac{\text{Protein content of fine fraction} \times \text{Mass of fine fraction}}{100} \quad (3)$$

4 Notes

1. The main process screen (Fig. 5) provides a graphical representation of the process parameters and an outline of equipment connections. Individual equipment status can be determined from the main process screen using the following color coding:
   - Automatic mode: light green.
   - Manual mode: light blue.
   - Turned off: dark gray.
   - Faulted: flash from red to white.

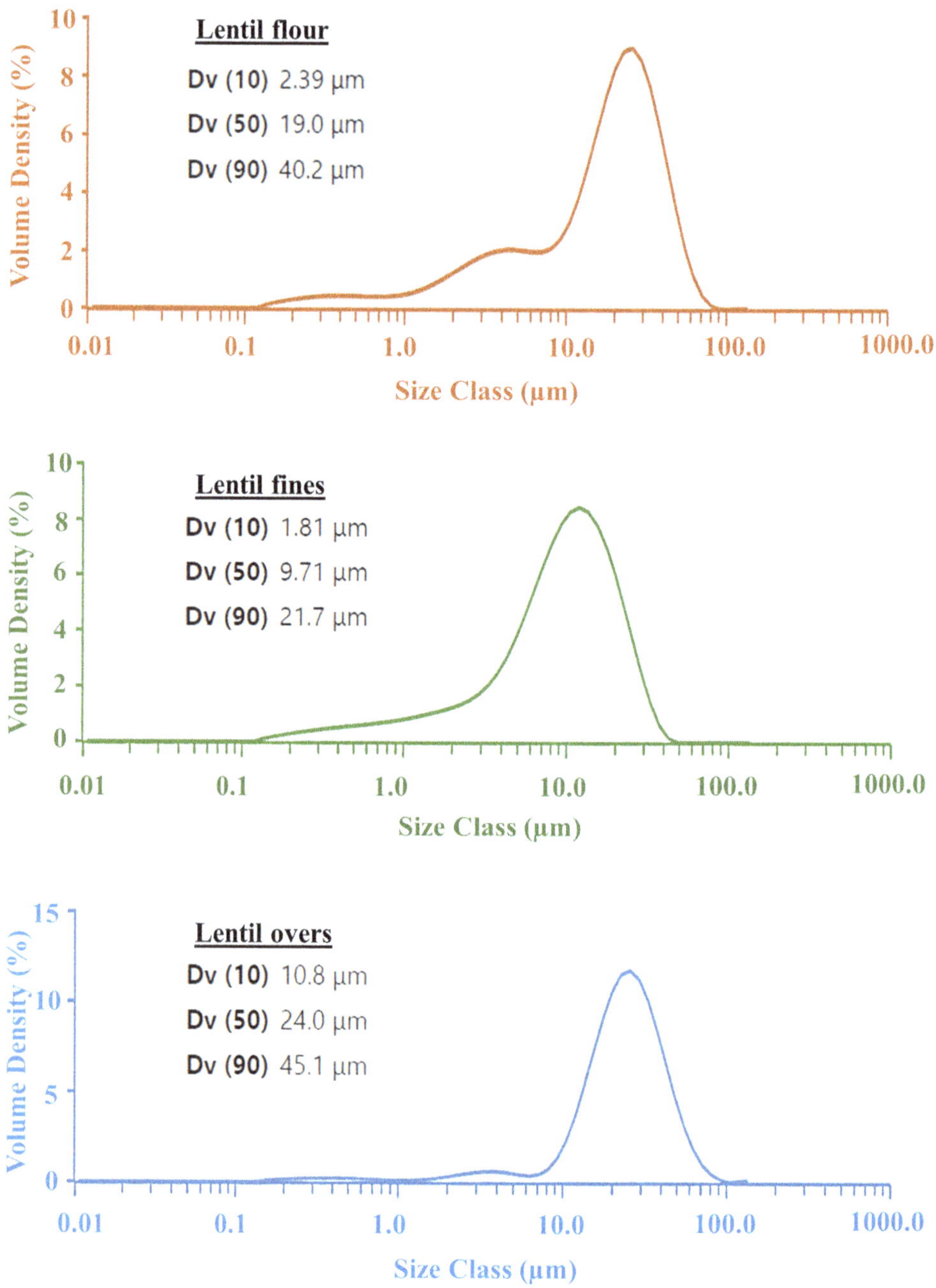

Fig. 16 Particle size distribution of lentil flour, fines, and overs

2. Materials with high fat and moisture content are not suitable for processing using this unit. Pretreatments such as defatting and drying should be performed before milling/air classification.
3. For milling, it is recommended to operate the fan and CLM-18 Mill at full speed (100%) and adjust only the CLM-18 classifier speed and feed rate.

Table 1
Air classification results for red lentils

	Separation yield (%)		Protein content (% dry basis)			Protein recovery (%)
S. no.	Fines	Overs	Parent flour	Fines	Overs	
1	18.8%	81.2%	25.58	51.72	18.33	38.01

4. For air classification, it is recommended to operate the fan at full speed (100%) and adjust only the feed rate and minisplit speed.
5. The secondary air for the minisplit during air classification can be completely closed during initial trials. After optimizing settings for a particular pulse flour, the secondary air opening can be adjusted to further improve separation efficiency.
6. It is advisable to take short breaks every 2–3 h of milling to prevent equipment overheating and potential damage to the mill and product.
7. The speeds of the CLM-18 Mill, CLM-18 classifier, and minisplit are represented as percentages in this lab system. These percentages can be converted to RPM using the following native speeds:
   - CLM-18 Mill rotor at 60HZ: 10,219 RPM—100% speed.
   - CLM-18 classifier rotor at 60HZ: 10,219 RPM—100% speed.
   - Minisplit classifier rotor at 60HZ: 16,320 RPM—100% speed.
8. Clean the entire unit thoroughly before and after every sample change.

Acknowledgments

The authors extend appreciation to Darrell Malczewski, at Prater Industries, Inc., Bolingbrook, IL, for his valuable insights and assistance in elucidating the intricacies of the equipment discussed in this chapter. His expertise significantly contributed to the clarity and depth of the content.

Appendix (Fig. 17)

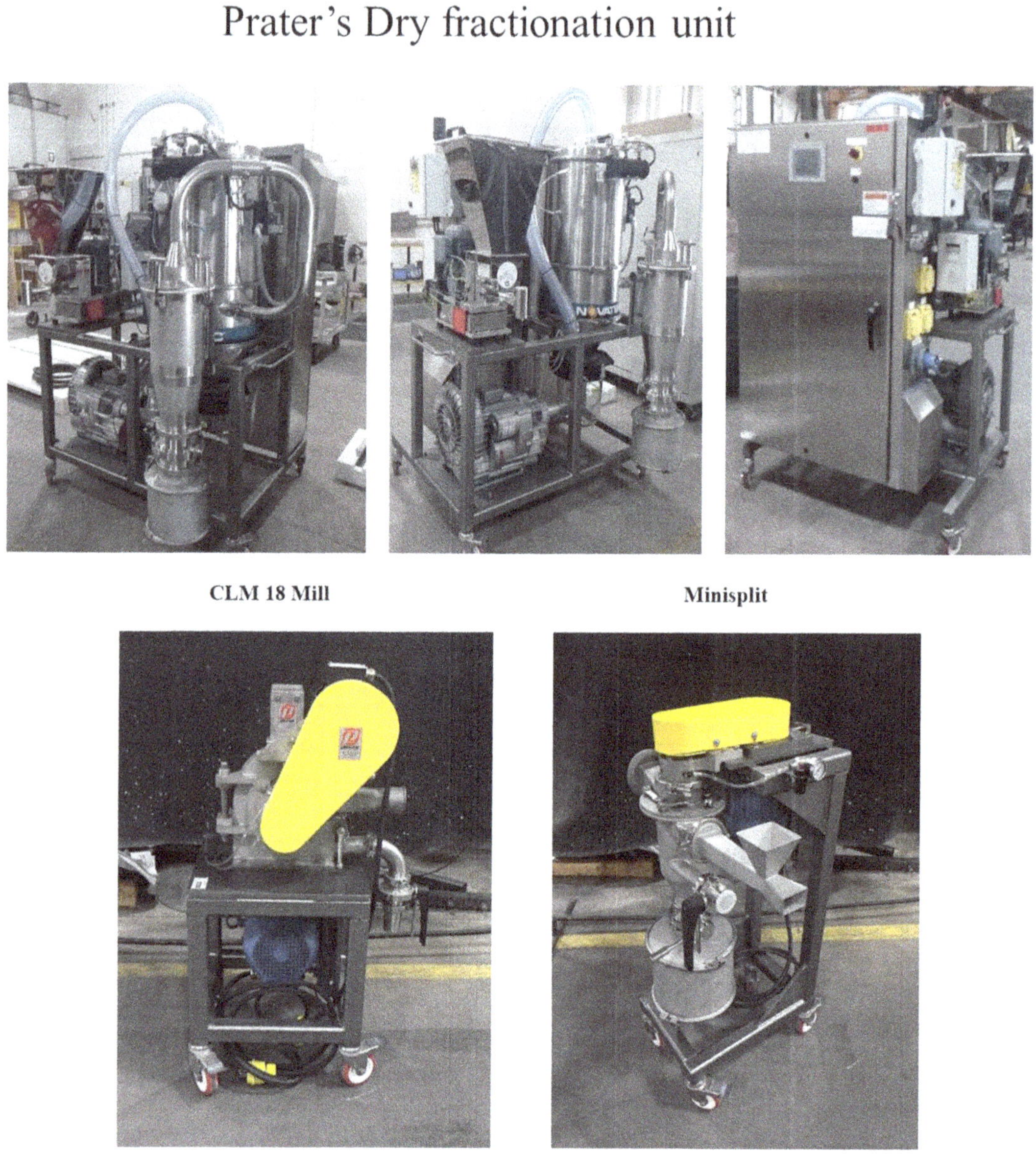

Fig. 17 Prater's dry fractionation unit with CLM-18 Mill and minisplit

References

1. Boye J, Zare F, Pletch A (2010) Pulse proteins: processing, characterization, functional properties and applications in food and feed. Food Res Int 43(2):414–431
2. Tabtabaei S, Konakbayeva D, Rajabzadeh AR, Legge RL (2019) Functional properties of navy bean (Phaseolus vulgaris) protein concentrates obtained by pneumatic tribo-electrostatic separation. Food Chem 283: 101–110
3. Jafari M, Rajabzadeh AR, Tabtabaei S, Marsolais F, Legge RL (2016) Physicochemical characterization of a navy bean (Phaseolus vulgaris) protein fraction produced using a solvent-free method. Food Chem 208:35–41
4. Schutyser MAI, Van der Goot AJ (2011) The potential of dry fractionation processes for sustainable plant protein production. Trends Food Sci Technol 22(4):154–164
5. Wang N, Maximiuk L (2019) Effect of air classification processing variables on yield, composition, and certain antinutrients of air-classified fractions from field peas by response surface methodology. J Food Process Preserv 43(7): e13999
6. Coelho MS, de las Mercedes Salas-Mellado, M. (2018) How extraction method affects the physicochemical and functional properties of chia proteins. LWT 96:26–33
7. Zhu HG, Tang HQ, Cheng YQ, Li ZG, Tong LT (2021) Electrostatic separation technology for obtaining plant protein concentrates: a review. Trends Food Sci Technol 113:66–76
8. Pulivarthi MK, Buenavista RM, Bangar SP, Li Y, Pordesimo LO, Bean SR, Siliveru K (2023) Dry fractionation process operations in the production of protein concentrates: a review. Compr Rev Food Sci Food Saf 22:4670
9. Dijkink BH, Speranza L, Paltsidis D, Vereijken JM (2007) Air dispersion of starch–protein mixtures: a predictive tool for air classification performance. Powder Technol 172(2): 113–119
10. Aguilera JM, Lusas EW, Uebersax MA, Zabik ME (1982) Development of food ingredients from navy beans (Phaseolus vulgaris) by roasting, pin milling, and air classification. J Food Sci 47(4):1151–1154
11. Aguilera Y, Esteban RM, Benitez V, Molla E, Martin-Cabrejas MA (2009) Starch, functional properties, and microstructural characteristics in chickpea and lentil as affected by thermal processing. J Agric Food Chem 57(22): 10682–10688
12. Wu YV, Nichols NN (2005) Fine grinding and air classification of field pea. Cereal Chem 82(3):341–344
13. Prater Industries (n.d.) Classifier mills. Retrieved from https://www.praterindustries.com/products/classifier-mills/. 15th Aug 2023
14. Rempel C, Geng X, Zhang Y (2019) Industrial scale preparation of pea flour fractions with enhanced nutritive composition by dry fractionation. Food Chem 276:119–128

Chapter 5

Wet Extraction of Proteins

Jing Zhao, Changqi Liu, and Selene Alvarado Martinez

Abstract

Proteins could be derived from diverse food sources and have distinct compositions, such as water-soluble albumin, salt-soluble globulin, ethanol-soluble prolamin, and acid- or alkaline-soluble glutelin. The alkaline extraction and isoelectric precipitation method is commonly employed for extracting plant proteins. The pH used for protein solubilization can affect both the protein extraction efficiency and functional properties of the extracted proteins. This chapter outlines a protocol suitable for extracting several major types of plant proteins.

Key words Plant protein, Alkaline extraction, Acid precipitation, Isoelectric point, Centrifugation

1 Introduction

The protein content in various plant sources can vary significantly. In raw fruits and vegetables, the protein content typically falls within the range of less than 1% to approximately 6.9% [1]. Cereals and pseudocereals exhibit a broad spectrum, with some like corn containing as little as about 3% protein, while others like whole wheat, oat, and quinoa can reach levels of 10–15%. Oilseeds and pulses stand out as protein-rich sources, boasting protein contents ranging from approximately 20% in certain dry beans to as high as 36% in raw mature soybeans [2, 3]. To concentrate protein by separating it from other food components, methods such as dry fractionation (*see* Chapter 4) or wet extraction can be employed.

Plant seed proteins, specifically storage proteins found in the endosperm, which are also known as reserve proteins, can be fractionated based on their solubility. This separation involves the classic Osborne extraction method, which employs sequential extractions using water, saline solution, 70% ethanol, and diluted acid or alkaline solutions to isolate albumin, globulin, prolamin,

Yonghui Li (ed.), *Plant-Based Proteins: Production, Physicochemical, Functional, and Sensory Properties*, Methods and Protocols in Food Science, https://doi.org/10.1007/978-1-0716-4272-6_5,

and glutelin, respectively (*see* Chapter 4) [4]. For total protein extraction, a variety of wet extraction techniques are available. The conventional protein extraction methods utilize water, acid, alkali, salt, detergent, or organic solvents for the extraction. Emerging protein extraction techniques apply cell disruption methods such as enzymes, ultrasound, microwave, pulsed electric field, and high pressure, during the wet extraction process to enhance extraction efficiency. This enhancement is typically measured by improved yield (percentage of extracted material weight relative to the starting material weight), protein purity (percentage of protein in the extracted material), and overall protein recovery (percentage of the total protein content in the extracted material relative to the starting material). A comprehensive review of these protein extraction methods was provided by Kumar et al. [5]. Moreover, innovative technologies such as reverse micelle extraction, subcritical water extraction, and aqueous two-phase system extraction have been developed, and a review on this subject is available in the work of Pojic et al. [6]. Following extraction, proteins are often concentrated or precipitated by acid (isoelectric point precipitation using citric acid, trichloroacetic acid, or hydrochloric acid), salt (ammonium sulfate), or organic solvents (ethanol, methanol, or acetone).

This chapter aims to provide a protocol for extracting plant proteins using the alkaline extraction and isoelectric precipitation method. A flow chart of the process is shown in Fig. 1. This protocol is described based on an example of soy protein extraction. The alkaline extraction and acid (isoelectric point) precipitation method represents a conventional and widely employed technique for protein extraction. It serves as the foundation for protocols using various assisting techniques (e.g., salt, enzymes, and physical methods). This method capitalizes on the high solubility of plant proteins at alkaline pHs and the diminished solubility near their isoelectric points, which are typically around pH 4.5 for most plant proteins (Table 1). The alkaline environment maximizes protein solubilization and facilitates the separation of proteins from insoluble components (e.g., insoluble fiber) through centrifugation. Subsequently, the pH of the protein-containing supernatant is adjusted to the isoelectric point of the target proteins, causing the proteins to precipitate out of the solution and thus separating them from soluble non-protein components. Numerous studies have underscored the significant impact of solubilization and precipitation pHs on protein yield and functionalities [7–10]. The judicious selection of pH values is paramount to enhancing both protein yield and purity. The described protocol can be adapted for other plant proteins by adjusting the solubilization and precipitation pHs to values that are optimal for the specific target proteins.

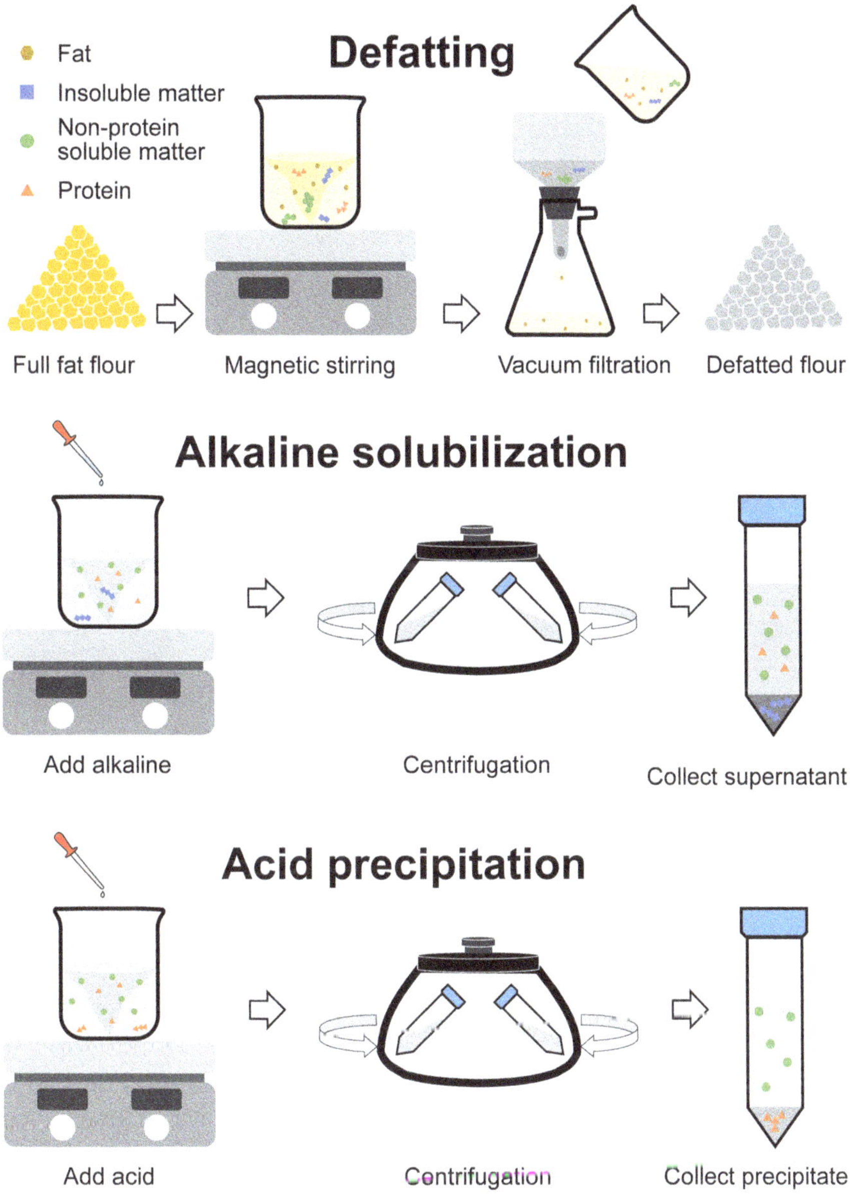

Fig. 1 Procedures of protein extraction by alkaline solubilization and acid precipitation

2 Materials

2.1 Materials for Seed Drying and Milling

The materials needed for seed drying and milling are soybean seeds, aluminum baking tray, oven, blender, blower, and No. 40 sieve.

Table 1
Extraction conditions for various plant proteins

Plant source	Extraction condition	Precipitation condition	Protein recovery (purity)	References
Soybean	Flour: Water = 1:20 (w/v) pH: 9.0 Temperature: RT Time: 1 h	pH 3.5, 4.5, or 5.5	50.8–57.8% (86.5–95.2%)	[11]
Pea	Flour: Water = 1:15 (w/v) pH: 8.5, 9.0, 9.5, 10.0 Temperature: RT Time: 1 h	pH 4.5	NA (80.36–85.72%)	[9]
Pea	Flour: Water = 1:10 (w/v) pH: 9.0 Temperature: RT Time: 2 h	pH 4.5	NA (88.83%)	[10]
Mung bean	Flour: Water = 1:10 (w/v) pH: 9 Temperature: 30 °C Time: 30 min	pH 4.6	72–76% (NA)	[12]
Fava bean	Flour: Water = 1:10 (w/v) pH: 8 Temperature: RT Time: 2 h	pH 4.0	61.7% (81.7%)	
Rice dreg	Flour: Water = 1:12.5 (w/v) pH: 7.8 Temperature: 50 °C Time: 3 h Enzyme: 1% [E/S] α-amylase Ultrasound: Power density of 100 W/L, pulsed on-time and off-time of 6/6 s	pH 4.0	Around 80%[a] (92.99%)	[13]
Extruded soybean powder	Flour: Water = 1:6 (w/v) pH: 8.0 Temperature: 55 °C Time: 2 h Enzyme: 0.5% v/w, Protex 6 L	None	77.32% (76.9%)	[14]

NA not available, *RT* room temperature
[a]The number is estimated based on the graph. The exact values were not available

2.2 Materials for Defatting

The materials needed for defatting are *n*-hexane, ethanol, 1 L filtering flask, porcelain Buchner funnel, grade 4 filter paper, analytical balance or scale, magnetic stirrer, beaker, and aluminum foil.

2.3 Materials for Protein Extraction

The materials needed for protein extraction are sodium hydroxide and hydrochloric acid for pH adjustment, magnetic stirrer, analytical balance, 1 L graduated cylinder, 2 L beaker, centrifuge, and pH meter.

3 Methods

3.1 Seed Drying, Dehulling, and Milling

1. Spread soybean seeds as a thin layer in an aluminum baking tray.
2. Dry the seeds in an oven at 30 °C for 12 h.
3. Use a blender to shatter the seeds (*see* **Note 1**).
4. Return the shattered seeds to the tray. In a fume hood and with a trash bag positioned on one side of the tray, use a blower to remove the husks.
5. Grind the seeds with a blender to pass through a No. 40 sieve to obtain soybean flour (*see* **Note 2**).

3.2 Defatting

1. Weigh 100 g of soybean flour.
2. Mix the soybean flour with 200 mL of *n*-hexane and 50 mL of ethanol (w/v/v = 2:4:1) (*see* **Note 3**).
3. Cover the beakers with aluminum foil, and stir the mixture for 30 min using a magnetic stirrer (*see* **Note 4**).
4. Place a No. 4 filter paper within a porcelain Buchner funnel, which is positioned on top of a 1 L filtering flask. Connect the filtering flask to a vacuum source, and apply vacuum filtration to separate defatted soybean flour from the solvent (*see* **Note 5**).
5. Repeat **steps 2–4** two more times.
6. Place the defatted soybean flour under a fume hood overnight to allow the solvents to evaporate.
7. Refrigerate the defatted soybean flour for future use.

3.3 Protein Extraction

3.3.1 Alkaline Extraction

1. Weigh 100 g of defatted soybean flour into a 2 L beaker.
2. Add 1 L of deionized water.
3. Add a large magnetic stir bar into the beaker, and place it on a magnetic stir plate. Adjust the stir speed to maintain a gentle stirring of the soybean flour suspension.
4. Adjust the pH to 9.0 using 2 M NaOH. Stir the mixture for 2 h at room temperature. The pH may change during stirring. Adjust pH back to 9.00 with NaOH as needed (*see* **Note 6**).
5. Divide the extract among six 250 mL centrifuge tubes. Ensure the centrifuge tubes are balanced in weight before loading them into the centrifuge. Centrifuge at 4 °C and 10,000 *g* for 20 min (*see* **Note 7**).
6. Filter the supernatant through eight layers of cheesecloth into a beaker.

3.3.2 Isoelectric Point (Acid) Precipitation

1. Place the beaker containing the supernatant on a magnetic stirrer plate. Adjust the pH to 4.5 using 2 M HCl. Stir at 350 rpm and room temperature for 30 min.
2. Transfer the pH-adjusted protein suspension into six 250 mL centrifuge tubes. Centrifuge at 4 °C and 2,000 × *g* for 20 min (*see* **Note 8**).
3. Discard the supernatant and collect the precipitate.
4. Weigh the precipitate, and resuspend it in five volumes of distilled water in a beaker (*see* **Note 9**). Adjust the pH to 7.00 ± 0.10 using 1.0 N NaOH.
5. Stir the mixture for 30 min to facilitate the resuspension of proteins. If the pH decreases, bring it back to 7.00 (*see* **Note 10**).
6. Freeze-dry the isolated protein samples.

4 Notes

1. Fill the blender to approximately one third of its capacity. Blend 5–10 s each time for multiple times until all whole seeds are shattered.
2. Avoid excessive heating of the seeds by processing them in small sample batches with short blending time, e.g., 30 s per round, and repeat as necessary. Any seed flour that fails to pass through the sieve can be combined and subjected to additional blending.
3. The flour can also be defatted using four volumes of *n*-hexane [11].
4. The flour may congregate at the bottom of the beaker, hindering the initial stirring process. Manually stir the solids at the bottom of the solvent to facilitate the stirring process. Alternatively, pour the hexane into the beaker, start the stirring first, and gradually introduce the soybean flour into the hexane. Continuously monitor the stirring to ensure it remains uninterrupted.
5. The solvent (hexane/ethanol 4:1 mixture) should be properly disposed of as hazardous waste.
6. The extraction pH, temperature, and time can be optimized for different plant proteins and for different applications. This is also the step where enzymes and physical methods can be applied to achieve improved extractions. When enzymes are used, the optimal temperature and pH of the enzymes need to be used for the extraction. A few examples of reported protein extraction conditions are shown in Table 1.

7. During this step, the insoluble fiber may not distinctly separate from the supernatant. Use a centrifuge speed just below the maximum allowable speed of the centrifuge. Make sure that the centrifuge tubes loaded opposite to each other have the same weight (accurate to 0.01 g), and adhere to the correct centrifuge operation protocols.
8. A higher centrifugation speed can be used in this step but is not necessary. Proteins at its isoelectric point are easy to separate by centrifugation.
9. For example, if the protein pellet weighs 20 g, resuspend it 100 mL of distilled water. Use a spatula to assist in the resuspension of the precipitated protein pellet.
10. As the protein pellet gets resuspended, the pH of the solution may drop. Verify the pH once the protein pellet is fully resuspended. The resolubilization process may take longer than 30 min. If necessary, continue stirring for an extended period or overnight. Keep the protocol consistent between replications.

References

1. Slavin JL, Lloyd B (2012) Health benefits of fruits and vegetables. Adv Nutr 3:506–516
2. Agriculture USDO (2019) FoodData Central. In: Service AR (ed) fdc.nal.usda.gov
3. Division IP (2022) How protein sources measure up. Food Technology Magazine, p 76
4. Osborne TB (1924) The vegetable proteins. Longmans, Green and Company
5. Kumar M, Tomar M, Potkule J et al (2021) Advances in the plant protein extraction: mechanism and recommendations. Food Hydrocoll 115:106595
6. Pojić M, Mišan A, Tiwari B (2018) Eco-innovative technologies for extraction of proteins for human consumption from renewable protein sources of plant origin. Trends Food Sci Technol 75:93–104
7. Potin F, Lubbers S, Husson F et al (2019) Hemp (Cannabis sativa L.) protein extraction conditions affect extraction yield and protein quality. J Food Sci 84:3682–3690
8. Gerzhova A, Mondor M, Benali M et al (2016) Study of total dry matter and protein extraction from canola meal as affected by the pH, salt addition and use of zeta-potential/turbidimetry analysis to optimize the extraction conditions. Food Chem 201:243–252
9. Cui L, Bandillo N, Wang Y et al (2020) Functionality and structure of yellow pea protein isolate as affected by cultivars and extraction pH. Food Hydrocoll 108:106008
10. Yang J, Zamani S, Liang L et al (2021) Extraction methods significantly impact pea protein composition, structure and gelling properties. Food Hydrocoll 117:106678
11. Verfaillie D, Janssen F, Van Royen G et al (2023) A systematic study of the impact of the isoelectric precipitation process on the physical properties and protein composition of soy protein isolates. Food Res Int 163:112177
12. Wang M, Jiang L, Li Y et al (2011) Optimization of extraction process of protein isolate from mung bean. Procedia Eng 15:5250–5258
13. Yang X, Li Y, Li S et al (2018) Effects of ultrasound-assisted α-amylase degradation treatment with multiple modes on the extraction of rice protein. Ultrason Sonochem 40:890–899
14. Tong X, Lian Z, Miao L et al (2020) An innovative two-step enzyme-assisted aqueous extraction for the production of reduced bitterness soybean protein hydrolysates with high nutritional value. LWT 134:110151

Chapter 6

Protein Fractionation Using Osborne Sequential Extraction

Gengjun Chen, Michael Stump, and Yonghui Li

Abstract

Field plants such as legumes and cereal are known as primary sources of essential nutrients such as protein. Food products enriched with plant proteins have been considered a vital resource to meet the nutritional requirements of the increasing global population. Plant proteins consist of polypeptide chains comprising different three-dimensional structures of protein fractions. These supramolecular structures determine the highly specific activities and techno-functionality possessed by proteins. Characterization of the protein fraction profile is essential for evaluating structural, functional, and nutritional values of plant proteins. These characteristics are also important for food formulations in the food industry. Plant proteins are classified based on their solubility in aqueous-based solutions. This system divides proteins into generalized categories called Osborne fractions. Albumin is soluble in water, globulin in dilute saline solutions, prolamin in aqueous ethanol solutions, and glutelin in acidic or alkaline solutions. This chapter outlines a clear procedure for Osborne fractionation as a promising approach to the characterization of plant proteins.

Key words Plant protein, Osborne fractions, Albumin, Globulin, Prolamin, Glutelin

1 Introduction

Field plants are considered as valuable sources of dietary protein. In recent years, plant protein has received significant attention to supplement animal-derived protein due to its better sustainability and high nutritional value [1]. It has been well established that legumes and cereals are able to meet the nutritional requirements of humans and animals [2]. For example, approximately 33% of dietary protein produced globally was provided by legumes and related products, and daily consumption is about 21 g per person [3, 4]. To meet the increasing requirements of their applications in food formulations, knowledge of the composition of plant proteins has been expanded via different extraction procedures. Protein cataloguing and proteomic characterization require the use of efficient fractionation technology. Thus, it is necessary to obtain effective approaches for the determination of plant protein fractions.

Yonghui Li (ed.), *Plant-Based Proteins: Production, Physicochemical, Functional, and Sensory Properties*, Methods and Protocols in Food Science, https://doi.org/10.1007/978-1-0716-4272-6_6,

Protein compositions vary significantly with plant types, varieties, and cropping practices. A classification system for seed proteins was established to group them into four distinct categories based on their solubility in various solutions [5]. Among these protein classes, the categories are albumin, which are water soluble and typically consist of enzymes: globulin, which are insoluble in water but are soluble in dilute salt solutions; prolamin, which are soluble in aqueous ethanol; and glutelin, which are soluble in dilute acid or base solutions such as sodium hydroxide. These protein classes are also applicable to major seed storage proteins. It's worth mentioning that the storage protein fractions are not uniformly present in the seeds of all plant species though. For instance, prolamin and glutelin are predominantly found in cereals as their primary storage proteins, whereas sugar beet seeds usually contain less prolamin compared to the fractions of albumin, globulin, and glutelin. Moreover, each of these protein fractions may display polymorphism across genotypes even within the same species [3]. Protein fractions with varied purity and functionality can be generated by modifying protein solubility, which may also alter their hydrodynamic properties [6].

It is crucial to evaluate the different fractions of plant proteins as suitable extraction technology can aid in increasing their overall yield and nutritional and functional qualities. The objective of this book chapter is to provide a thorough and detailed description of the Osborne fractionation in selected plant protein samples.

2 Materials

1. Prepare all solutions using ultrapure deionized water and/or analytical grade reagents. Prepare and store all reagents at room temperature (unless indicated otherwise).

   Albumin extraction solvent: Deionized water prepared by Direct-Pure Water Systems.

   Globulin extraction solvent: Sodium chloride solution (1 N)—Dissolve 58.44 g of sodium chloride in deionized water, and dilute to 1 L.

   Prolamin extraction solvent: Prepare the reagent as 70% (v/v) ethanol in deionized water.

   Glutelin extraction solvent: Sodium hydroxide (0.1 N)—Dissolve 4.00 g of sodium hydroxide in deionized water, and dilute to 1 L.

2. Commercial wheat flour (protein content of 10.5%, flour weight basis (fwb)) and pea flour (protein content 21.3%, fwb).
3. Refrigerated centrifuge Model Z-366 K (HERMLE Benchmark).
4. Freeze dryers Model FreeZone 4.5 L (Labconco).

5. Platform shaker Model Innova 2000 (New Brunswick Scientific).
6. UP Ultrapure and RO Lab Water Systems (Direct-Pure).
7. pH meter (Fisherbrand).
8. N-EVAP Evaporator (Organomation).
9. Dialysis Tubing 3.5 K (SnakeSkin).
10. Nitrogen Determinator 628 SERIES (LECO).

3 Methods

Carry out all procedures at room temperature unless otherwise specified (*see* **Note 1**).

3.1 Sample Defatting

Flour sample is defatted with hexane (1:10, w/v) and shaken (250 rpm) at room temperature for 3 h. Centrifuge at 6000 × g for 15 min (4 °C), and decant the supernatant. Defatted sample is dried under hood and stored in 4 °C until used.

3.2 Osborne Sequential Extraction

Figure 1 shows a simplified and schematic flowchart of Osborne sequential extraction steps.

1. Selected wheat or pea flour (10.0 g) is dispersed in deionized water (1:10, w/v), and shake the mixture (250 rpm) at room

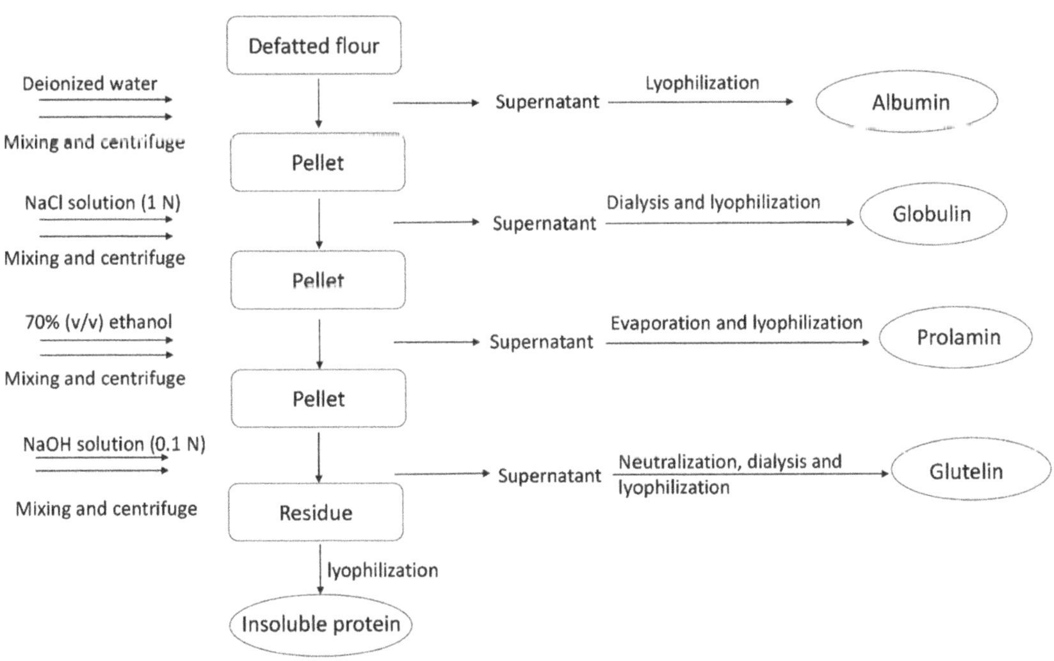

Fig. 1 Schematic and simplified representation of the steps of the Osborne sequential extraction procedure

temperature for 1 h. Centrifuge at 6000 × g for 15 min (4 °C), and decant the supernatant as the albumin fraction. Repeat the step one more time. The extract is freeze-dried and stored at 4 °C.

2. The above pellet is extracted with 1 M NaCl (1:10, w/v), and shake the mixture (250 rpm) at room temperature for 1 h. Centrifuge at 6000 × g for 15 min (4 °C), and decant the supernatant as the globulin fraction. Repeat the step one more time. The extract is dialyzed with deionized water for 48 h at 4 °C, and then the sample is freeze-dried and stored at 4 °C.
3. The above pellet is extracted with 70% ethanol (1:10, w/v), and shake the mixture (250 rpm) at room temperature for 1 h. Centrifuge at 6000 × g for 15 min (4 °C), and decant the supernatant as the prolamin fraction. Repeat the step one more time. Allow the extract to evaporate on a rotary evaporator (45 °C), and freeze-dried and stored at 4 °C.
4. The above pellet is extracted with 0.1 M NaOH (1:10, w/v), and shake the mixture (250 rpm) at room temperature for 1 h. Centrifuge at 6000 × g for 15 min (4 °C), and decant the supernatant as the glutelin fraction. Repeat the step one more time. The extract is neutralized with 1 M HCl and dialyzed with deionized water for 48 h at 4 °C, and then the sample is freeze-dried and stored at 4 °C.

3.3 Protein Content Determination

Protein content in each fraction is determined by modified Dumas combustion method (*see* **Note 2**) using 5.7 (wheat) or 6.25 (pea) as the conversion factor (*see* **Note 3**). The successive extraction with different solvents yields albumin, globulin, prolamin, and glutenin fractions in varying proportions, where different fractionations are expressed as percentages of total protein (Fig. 2).

4 Notes

1. Plant proteins including pea and/or wheat protein fractions (albumin, globulin, prolamin, glutelin) are extracted using the Osborne method with some modification [7].
2. Food protein characterization is not commonly a straightforward procedure, and the determination methods are based on different analytical principles. Except for the Dumas combustion method [8] used in this chapter, there are other common protein quantification methods including the Bradford method [9], Kjeldahl method [10], and Lowry measurement analysis [11]. These indirect analysis methods can usually be performed following the determination of the nitrogen content and/or after chemical reactions between the functional groups within the protein.

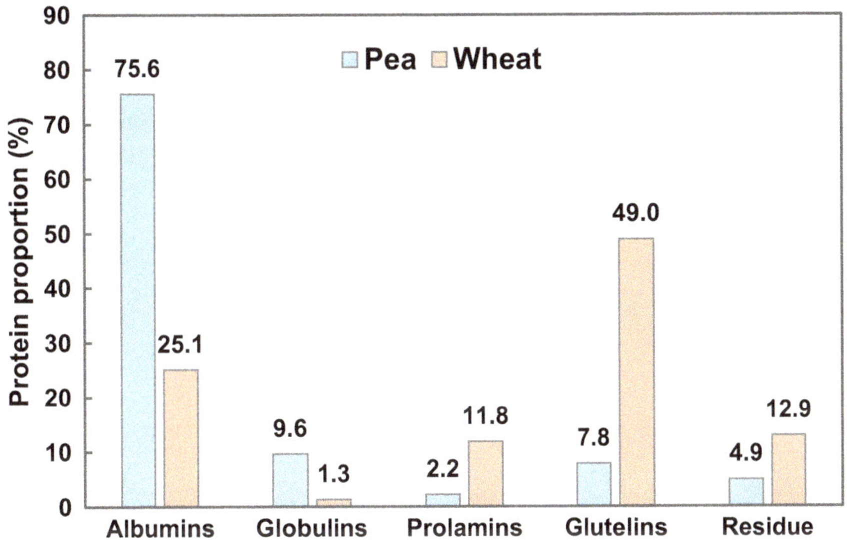

Fig. 2 Proportion of various Osborne sequential fractions in wheat and pea samples

3. A nitrogen-to-protein conversion factor of 5.7 is used for the wheat proteins and 6.25 for pea proteins in the chapter. The nitrogen-to-protein conversion factor of 6.25 is typically used if not specified for a source protein [12].

Acknowledgments

This is contribution no. 24-260-B from the Kansas Agricultural Experimental Station. This work was supported in part by the USDA National Institute of Food and Agriculture Hatch project (Grant Accession No. 7003330).

References

1. Gilbert JA, Bendsen NT, Tremblay A, Astrup A (2011) Effect of proteins from different sources on body composition. Nutr Metab Cardiovasc Dis 21:B16–B31
2. Bennetau-Pelissero C (2019) Plant proteins from legumes. In: Mérillon JM, Ramawat K (eds) Bioactive molecules in food. Reference Series in Phytochemistry. Springer, Cham, pp 223–265
3. Klupšait˙e D, Juodeikien˙e G (2015) Legume: composition, protein extraction and functional properties. A review. Chem Technol 66:5–12
4. Semba RD, Ramsing R, Rahman N, Kraemer K, Bloem MW (2021) Legumes as a sustainable source of protein in human diets. Glob Food Sec 28:100520
5. Chen CH, Bushuk W (1970) Nature of proteins in triticale and its parental species I. Solubility characteristics and amino acid composition of endosperm proteins. Can J Plant Sci 50:9–14
6. Langyan S, Yadava P, Khan FN, Dar ZA, Singh R, Kumar A (2022) Sustaining protein nutrition through plant-based foods. Front Nutr 8:1237
7. Osborne TB, Voorhees CL (1894) Proteins of the wheat kernel. J Am Chem Soc 16(8): 524–535

8. Dumas JBA (1831) Procedes de l'analyse organique. Ann Chim Phys T47:198–213
9. Bradford M (1976) Anal Biochem 72:248–254
10. Kjeldahl J (1883) Neue Methode zur Bestimmung des Stickstoffs in organischen Körpern. Fresenius J Anal Chem 22:366–382
11. Hartree EF (1972) Determination of protein-modification of Lowry method that gives a linear photometric response. Anal Biochem 48:422–427
12. Angell AR, Mata L, de Nys R, Paul NA (2016) The protein content of seaweeds: a universal nitrogen-to-protein conversion factor of five. J Appl Phycol 28:511–524

Chapter 7

Techno-Economic Analysis of Plant Protein Production

Yanhong He and Qing Jin

Abstract

As the demand for sustainable and environmentally friendly protein sources increases, plant-based proteins have emerged as a promising alternative to animal-derived proteins. The techno-economic analysis (TEA) plays a pivotal role in evaluating the economic feasibility and commercial viability of plant protein production processes. This chapter presents a TEA framework tailored for protein production from brewer's spent grain (BSG), offering a comprehensive evaluation of production costs, market potential, and variables that can affect the product price. Critical technical data is collected from lab experiments, and SuperPro Designer is utilized as the process simulation tool for the techno-economic assessment of the technological configuration.

Key words Plant protein, Production efficiency, SuperPro Designer, Process simulation, Production cost, Minimum selling price, Economic feasibility, Sensitivity analysis

1 Introduction

Animal proteins are widely used as protein sources due to their high digestibility and well-balanced amino acids. However, plant protein has gained significant attention and popularity driven by growing awareness and demand for sustainable, nutritious, and environmentally friendly food sources. With rising concerns about climate change and resource depletion, opting for plant protein can aid in creating a more sustainable food system [1]. Common sources of plant-based protein include legumes, pulses, grains, nuts, and seed [2, 3]. The food processing industry often generates significant byproducts and waste, some are rich in plant protein. For example, brewer's spent grain (BSG), a major byproduct generated by the brewing industry, contains 14–30% of protein and is produced in quantities of 38–40 million tons annually [4, 5]. Utilizing these byproducts or wastes as raw materials for protein production can help the food industry reduce waste and maximize resource utilization.

Yonghui Li (ed.), *Plant-Based Proteins: Production, Physicochemical, Functional, and Sensory Properties*, Methods and Protocols in Food Science, https://doi.org/10.1007/978-1-0716-4272-6_7,

The production of plant protein involves a series of processes aimed at separating and concentrating the protein from the raw materials. The specific processes can vary based on the plant source and the intended end product. Common extraction methods include (1) mechanical methods (which encompass physical processes such as grinding, milling, or pressing), (2) solvent extraction (using water, alcohol, or acids to dissolve and extract proteins), (3) biochemical extraction (enzymes are applied to facilitate protein release), (4) membrane separation (techniques such as ultrafiltration and microfiltration separate proteins based on size and charge). Furthermore, innovative methods such as ultrasound-assisted, pulse electric field-assisted, microwave-assisted, and high-pressure-assisted extraction have been developed [6]. These advancements in extraction technology aim to improve efficiency, reduce energy consumption, minimize waste, improve sustainability, and enhance the quality and functionality of extracted proteins. Processing technologies continuously evolve to meet consumer demands and align with the pursuit of sustainable social development. Nevertheless, attempting to commercialize advanced protein extraction technologies without comprehensive economic evaluation would be impractical. A thorough techno-economic analysis (TEA) provides a detailed assessment of the capital and operating costs and offers insights into financial metrics and the profitability factors of the technology or process. Understanding the cost structure empowers us to optimize resource allocation, identify potential cost-saving measures, and make well-informed decisions regarding the scalability and commercialization of the technology.

TEA is a systematic evaluation methodology used to assess the economic viability and feasibility of a technology, process, or project. It serves as a bridge between research and development, engineering, and business domains, integrating technical and economic considerations to determine the economic viability and commercial feasibility of a specific technology or process. The primary objectives of TEA include:

1. Cost evaluation: TEA examines the cost associated with implementing and operating a technology or process, encompassing capital costs (e.g., equipment, infrastructure), operating costs (e.g., raw materials, utilities, labor), and other relevant expenses (e.g., contingency).
2. Revenue estimation: It assesses the potential revenue generated from the sale of products.
3. Sensitivity analysis: This examines how sensitive the project's economics are to variable parameters, such as raw material prices, plant capacity, product selling prices, and process efficiency.

4. Comparison of alternatives: TEA enables comparison among different technologies, processes, or project scenarios to identify the most economically favorable option.

The common steps to conduct TEA involve defining the scope and objectives, collecting data, developing a simulation process model, performing financial analysis, conducting sensitivity analysis, and preparing the final report. The initial step typically involves identifying the technology or process for evaluation and analyzing key parameters. Creating a process flow diagram to illustrate key steps and components aids in understanding the process and identifying cost drivers. Data on process parameters, raw material price, labor expenses, utilities, equipment costs, and other relevant economic factors are gathered from various sources such as publications, equipment suppliers, websites, and more. When dealing with developed bench-scale processes, data on mass flow, yield, and efficiency are critical for constructing reliable TEA models. Several software tools are available for conducting TEA, including Aspen Plus, Aspen HYSYS, SuperPro Designer, ProSimPlus, GeteCycle, and Tecnomatix Plant Simulation. While no software tools are widely recognized for TEA in plant protein production, general-purpose process simulation and economic analysis software mentioned above can be adapted for such purposes. For example, SuperPro Designer has been used for TEA of protein production from microalgae and BSG [7, 8].

The aim of this chapter is to use our previously published work [8] as an illustrative example to demonstrate the TEA of plant-based protein production from BSG, encompassing data collection, process simulation and model development, economic analysis, and sensitivity analysis.

2 Materials

The TEA for protein extraction from BSG is conducted using SuperPro Designer 11.0 (Intelligen, Inc., NJ, USA).

3 Methods

3.1 TEA Scope and Objectives

BSG is the major byproduct generated from the brewing industry. An enzymatic wet fractionation process [9] was developed for producing protein-rich product (PP) and fiber-rich product (FP) from BSG. The PP was identified as a substitute for fishmeal in shrimp feed [4]. In this context, TEA is utilized to assess the feasibility of PP production in a commercial scale and determine its minimum selling price (MSP). The objective includes substituting fishmeal with PP and comparing the selling price of the PP with the market price of fishmeal. All economic calculations were performed for the year of 2019. As the focus of this chapter is to illustrate the utilization of TEA in plant protein production, the experimental details of bench-scale production are omitted. Given the specifics of the process differ depending on the selected raw materials and the intended final product, providing a comprehensive experimental description is deemed unnecessary within this context.

3.2 Process Design and Experiment Data Collection

The developed process aims to produce PP with a substantial protein content and a favorable amino acid profile while also ensuring efficient separation of the protein. The simplified process diagram (Fig. 1) shows the production of PP from BSG, with FP as a co-product. A bench-scale experiment has been conducted to assess the viability of the proposed process and optimize its efficiency. Upon successful optimization, a scaled-up process was implemented to collect technical data that closely resembles commercial production, thereby enhancing the reliability of the TEA. The results, including chemical composition of BSG, PP, and FP, as well as the mass flow throughout the process used for TEA, need to be documented (Fig. 2). The weight unit *t* refers to a metric ton in this chapter.

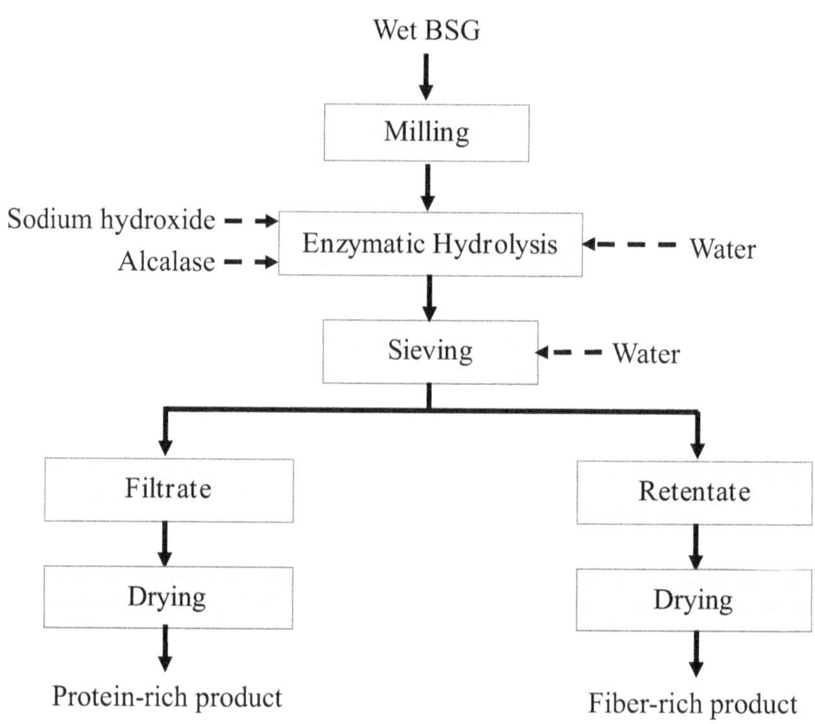

Fig. 1 The wet milling process to produce protein-rich product (PP) and fiber-rich product (FP) from brewer's spent grain (BSG)

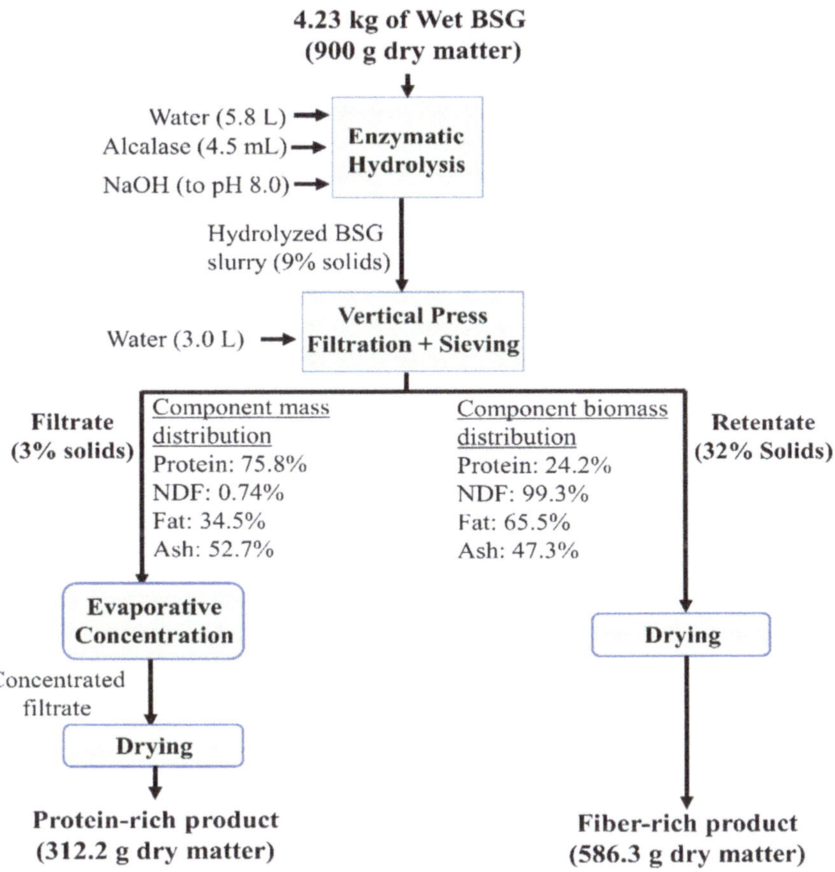

Fig. 2 Process and mass balance of protein-rich and fiber-rich product (PP and FP) production from BSG

3.3 Plant Capacity and Location Determination

Considering the USA has a substantial number of large breweries, which can provide centralized BSG resources, and given the perishable nature and expensive shipping of BSG, it is assumed that the BSG processing plant is co-located at a large brewery (*see* **Note 1**). The co-location strategy offers specific benefits, including the elimination of BSG shipping costs and the prevention of BSG spoilage during transportation. The plant's capacity is designed to accommodate 590 t of raw BSG per day, reflecting the average BSG generation rate observed among large breweries in the USA [10].

3.4 Processing Model Development in the SuperPro Designer

A base model (Fig. 3) for the production of PP and FP from BSG was developed in SuperPro Designer. This model used a sequence of distinct unit operations, each involving input and output of specific materials and chemicals, with users defining specific parameters such as material composition, temperature, hydrolysis time, and more. After that, the software tracks mass flow, establishes mass balance, and computes energy consumption of the whole process. This data is further used to determine equipment size and quantity. The major steps are as follows:

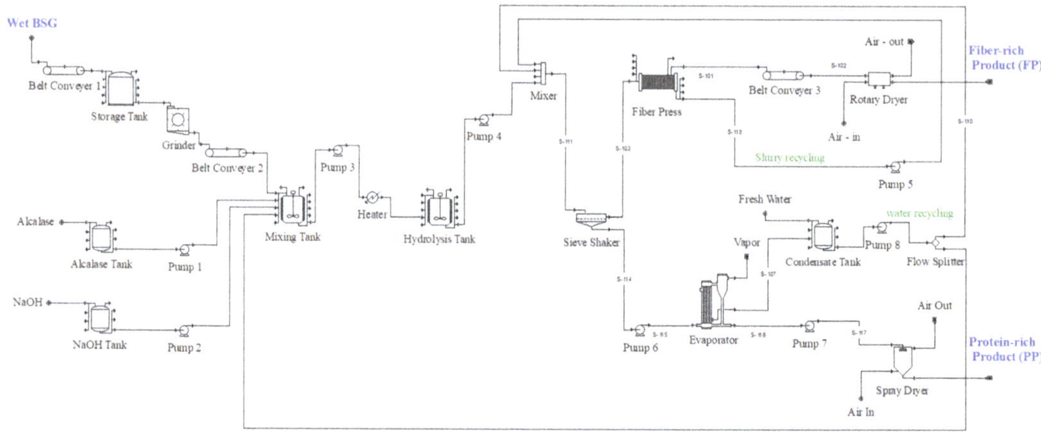

Fig. 3 The process model flow for PP and FP production from BSG using SuperPro Designer

1. Creating a new project: Launch the software and start a New Process File.
2. Selecting operating mode: Choose "Continues" as the process operating mode, with an annual operating time of 330 days. The remaining 35 days are scheduled for equipment maintenance and cleaning. [11]
3. Registering pure components and mixtures: In the Tasks section, register pure components such as ash, fats, fiber, glucose, nitrogen, other solids, oxygen, proteins, sodium hydroxide, and water. Also, register stock mixtures such as air, Alcalase, dry spent grain, and wet spent grain. Enter composition data, which is obtained from experiments, literature, or other references. For instance, given the moisture content of BSG is 78.7% (w/w), the composition of wet BSG comprises 78.7% water and 21.3% dry BSG (*see* **Note 2**).
4. Configuring unit operations: Configure each unit operation within the process which can be found in the Unit Procedures section. Edit the amount of input BSG in the stream information. To set technical parameters, double-click on the equipment. For example, in the hydrolysis tank, set the reaction temperature to 60 °C, retention time to 1 h, and working-to-vessel volume ratio to 90%. The software calculates the working volume accordingly. Similarly, for the sieve shaker, determine the bottom stream component based on the experimental data.
5. Setting equipment cost: Right-click on specific equipment to modify equipment data. Obtain the base cost of each equipment from references, vendor quotes, or the software's database. Use built-in cost models for common equipment such as pumps, heat exchanger, and some tanks. For example, in this model, the currency used is USD, and the equipment costs

from earlier years were adjusted to 2019 using the Chemical Engineering Plant Cost Index, following Eq. 1 (*see* **Note 3**). For equipment of different size costs were adjusted according to Eq. 2, where n is the scaling factor, usually in the range of 0.5–0.8 [12]. The scaling factor of 0.6 is used in this case:

$$2019\ \text{Cost} = (\text{Base Cost})\left(\frac{2019\ \text{Cost Index}}{\text{Base Year Index}}\right) \tag{1}$$

$$2019\ \text{Cost} = (\text{Base Cost})\left(\frac{2019\ \text{size}}{\text{Base size}}\right)^{n} \tag{2}$$

6. Setting material and utility cost: In the Process Explorer, set the costs as shown in Table 1. Determine unit prices for materials and utilities based on their market prices in 2019, using industrial quotes or literature data. The exact market price of wet BSG is complex to obtain. Buffington [13] reported a market price of approximately $44.2/t for wet BSG with 30% dry solids in the USA. Given the lower dry solid content (21%) of the BSG used in the study, a corrected wet BSG price of $31.4/t is used, considering the same dry solid content.
7. Running the simulation: After configuring the model, run the simulation by clicking Calculator. SuperPro Designer will calculate and display the results of mass and energy balances based on the specified inputs and conditions (*see* **Note 4**).

3.5 Economic Analysis

3.5.1 Total Capital Investment (TCI) of the PP Processing Plant (Table 2)

1. Total capital investment (TCI): The sum of fixed capital investment (FCI) and working capital.
2. Fixed capital investment (FCI): Includes total direct cost (TDC) and total indirect cost (TIC).
3. Total direct cost (TDC): Encompasses the expenses associated with equipment installation, warehouse setup, site development, and additional piping. The installed cost of equipment is calculated using Eq. 3: [12]

$$\text{Installed Cost} = (\text{Purchased Equipment Cost}) \times \text{multiplier} \tag{3}$$

4. Total indirect cost (TIC): Includes portable expense, field expense, home office and construction fees, project contingency, and other costs such as start-up expenses, permits, etc.

3.5.2 Minimum Selling Price (MSP) of PP (Table 3)

MSP represents the lowest price at which PP can be sold to cover all associated costs while achieving a target profit margin. This includes various costs such as production costs, operating expenses, taxes, and other relevant expenses. In this example, the MSP aims for a 5% profit return. The main parameters used for determining the minimum selling price of PP are shown in Table 4 (*see* **Note 5**). Here's a step-by-step guide to calculating MSP:

Table 1
Variable and fixed operating costs used in this study

Item	Cost ($)	Unit	References
Variable operating cost items			
Raw material utilities			
Brewer's spent grain	31.4[a]	t	[13]
Alcalase	34.0	kg	Industrial quote[b]
Sodium hydroxide	410.0	t	[15]
Water	0.353	t	[15]
Utilities			
Steam	17.0	t	[15]
Natural gas	188	t	[16]
Electricity	0.0642	KWh	[17]
Co-product credits			
Fiber-rich product	120[c]	t	[18]
Fixed operating cost items			
Total salaries	350,000[d]		
Labor burden (90% of total salaries)	315,000		
Maintenance	3.0% of inside-battery-limit equipment costs [12]		
Property insurance	0.7% of fixed capital investment [12]		
Depreciation	The MACRS[e] 7-year depreciation schedule		

[a]The cost of wet BSG (with 79% moisture content) is calculated to be $31.4/t based on the $44.2/t market price of wet BSG with 70% moisture content [13]
[b]Bulk price quotes from Novozymes Inc.
[c]The selling price of fiber-rich product (FP) is assumed based on the selling price of soy hulls because of the similar chemical composition and applications (animal feed) between FP and soy hulls
[d]Assuming seven employees (one manager, six shift operators) with an average annual salary of $50,000 per employee
[e]MACRS: modified accelerated cost recovery system

1. Identify total annual sales of PP and FP: Begin by estimating the unit price for PP while setting the selling price of FP at $120/t. The annual production volumes of PP and FP are output from the SuperPro Designer.
2. Determine the annual production cost: This includes the total annual costs of feedstock (wet BSG, Alcalase, sodium hydroxide, and water), utilities (electricity, natural gas, and steam), labor cost, and additional overheads:
   - Labor cost: Accounts for total salaries and additional labor burdens, such as engineering, payroll overheads (including benefits), safety, plant security, utilities, and plant

Table 2
Capital investment for the processing plant with an annual capacity of 194,700 t wet BSG

Item	Cost ($, thousands)
Process equipment	
BSG storage tank	127.5
BSG disc mill	114.9
Alcalase storage tank	35.6
NaOH storage tank	72.4
Slurry mixing tank	49.4
Slurry heater	19.5
Hydrolysis tank	113.7
Vibrating screen	36.8
Fiber-rich product filter press	228.6
Fiber-rich product rotary dryer	734.1
Protein-rich product evaporator	409.5
Process condensate tank	85.0
Spray dryer	1031.7
Belt conveyer [3]	164.2
Pump [8]	180.8
Total equipment cost (TEC)	3403.7
Total installed equipment cost[a]	6137.8
Warehouse, 4% TEC	136.1
Site development, 9% TEC	306.3
Additional piping, 4.5% TEC	91.5
Total direct cost (TDC)	6671.8
Proratable expense, 10% TDC	667.2
Field expense, 10% TDC	667.2
Home office and construction expense, 20% TDC	1334.4
Project contingency, 10% TDC	667.2
Other costs (start-up, permits), 10% TDC	667.2
Total indirect cost (TIC)	4003.1
Fixed capital investment (FCI), TDC + TIC	10,674.9
Working capital (WC), 5% FCI	533.7
Total capital investment, FCI + WC	11,208.6

[a]Total installed equipment cost = cost of equipment purchased × multiplier. The multiplier varies from 1.5 to 2.3 depending on different equipment [12]

Table 3
Annual production cost of 16,847 t of PP from BSG

Description	Annual use (thousands)	Annual cost ($, thousands)	Unit cost ($/t)
Raw materials (t)			
BSG	194.7	6113.6	
Enzyme	0.24	8296.0	
Sodium hydroxide	0.53	216.5	
Water	6.93	2.6	
Subtotal raw materials		*14,628.7*	*868.3*
Utilities			
Electricity (kWh)	8297.8	532.7	
Natural gas (t)	4.5	837.9	
Steam (t)	142.3	2419.0	
Subtotal utilities		*3789.6*	*224.9*
Labor			
Total salaries		350.0	
Labor burden		315.0	
Subtotal labor		*665.0*	*39.5*
Other overhead			
Maintenance		286.2	
Property insurance		74.7	
Subtotal overhead		*361.0*	*21.4*
Co-product (FP) credit		*−3579.0*	*−212.5*
Capital depreciation		355.8	21.1
Average income tax		279.3	16.6
Average return on investment		1079.0	64.1
Net operating costs			1043.5

communications, calculated at 90% of the total salaries. The average salary is set at $50,000/year, assuming a requirement for seven employees (*see* **Note 6**).

- Overhead costs: Includes maintenance costs, which are 3% of the sum of total equipment costs and equipment installation expenses, while property insurance is 0.7% of the fixed capital cost.

Table 4
Main parameters used for determining the minimum selling price (MSP) of PP

Parameter	Value
Project lifetime	30 years
Income tax	35%
Working capital	5% of fixed capital costs
Construction period	1 year
Start-up time	3 months
Revenues during start-up	50%
Variable costs incurred during start-up	75%
Fixed costs incurred start-up	100%

3. Depreciation charges: These are derived by multiplying the fixed capital investment (FCI) by the annual depreciation rate.
4. Income tax: The annual income tax, set at 35%, is computed based on the net annual revenue, which is the total annual sales minus production costs and depreciation charges.
5. Cash income: The annual cash income is the total annual sales minus the sum of annual production costs and income tax.
6. Present value: The annual present value is the annual cash income multiplied by the discount factor.
7. Net present value (NPV): NPV is the difference between the annual present value and the sum of total capital investment and interest. By adjusting the unit price of PP and monitoring the NPV change, the unit price of PP is set as the MSP when NPV is positive and close to zero. In this case, the MSP of PP is determined as \$1044/t to achieve a 5% return. This price can be used for comparison with the fishmeal prices.

3.6 Sensitivity Analysis

Upon determining the MSP, a single-point sensitivity analysis is conducted to assess variables that are uncertain or significantly affect MSP. Seven variables are chosen for the analysis, including BSG price, enzyme cost, steam price, solid loading of hydrolysis, protein separation efficiency, and daily input of BSG. The variables are adjusted by ±20% from the base model to establish their low and high range values (*see* **Note** 7). For example, the base model price of Alcalase is \$34/kg, and the adjusted low and high values would be \$27.2/kg and \$40.8/kg, respectively. Each parameter alteration leads to a recalculated MSP. The new MSPs alongside the base model MSP are used to create a tornado chart (Fig. 4). The length of the bars indicates the magnitude of the impact, and the

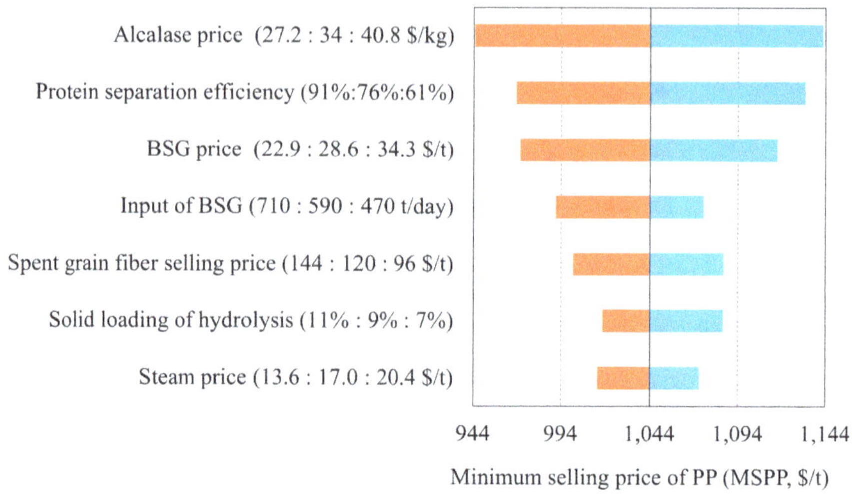

Fig. 4 Sensitivity of PP production cost to various parameters. The numbers in parentheses on the Y-axis represent the potential low, base, and high values of each parameter

position of the bars indicates the range of the output variable. The sensitivity analysis shows that Alcalase price is the most influential to the MSP of PP, and the steam has the least affect compared to other variables.

4 Notes

1. When selecting a plant location, two major factors should be considered: (1) proximity to raw materials and (2) transportation and logistics. If a centralized location is chosen for the plant, then the radius of raw materials collection, transportation method, diesel/gas price, driver wages and benefits, etc. need to be calculated accordingly. Further details can be drawn from a previous study on a TEA of a grape pomace biorefinery [14].
2. To establish a comprehensive composition input for stock mixtures, it is advisable to begin by registering pure components first and then stock mixtures. Pure components serve as the foundation for the composition data used in stock mixtures.
3. If the economic calculations are conducted for a year other than 2019, then the new size and new year cost index should be used for that specific year.
4. The economic report can be generated by the SuperPro Designer or via alternative software which can be used for handling the economic aspects.
5. The selling price of PP can be manually adjusted in the SuperPro Designer to make the NPV close or equal to zero.

6. The salary considerations should be aligned with real-world location-specific values to enhance the accuracy of the TEA. The number of employees may also be affected by factors like the shifts per day, which can be estimated through labor modeling.
7. The scope and percentage adjustments for variables are adaptable and can be referenced from various sources.

Acknowledgments

This work was supported by the Virginia Tech John Lee Pratt Animal Nutrition Program, the Southern Sustainable Agriculture Research and Education Graduate Student Grant (project # 00002107), and the Virginia Agricultural Experiment Station. We thank Dr. Haibo Huang for his assistance in the TEA.

References

1. Lonnie M, Johnstone AM (2020) The public health rationale for promoting plant protein as an important part of a sustainable and healthy diet. Nutr Bull 45(3):281–293. https://doi.org/10.1111/nbu.12453
2. Semba RD, Ramsing R, Rahman N, Kraemer K, Bloem MW (2021) Legumes as a sustainable source of protein in human diets. Glob Food Sec 28:28. https://doi.org/10.1016/j.gfs.2021.100520
3. Poutanen KS, Karlund AO, Gomez-Gallego C, Johansson DP, Scheers NM, Marklinder IM, Eriksen AK, Silventoinen PC, Nordlund E, Sozer N, Hanhineva KJ, Kolehmainen M, Landberg R (2022) Grains—a major source of sustainable protein for health. Nutr Rev 80(6): 1648–1663. https://doi.org/10.1093/nutrit/nuab084
4. He YH, Galagarza OA, Wang HJ, Taylor ZW, Ferguson CS, Ogejo JA, O'Keefe SF, Fraguas CF, Yu DJ, Poe NE, Wiersema BD, Kuhn DD, Huang HB (2020) Protein-rich product recovered from brewer's spent grain can partially replace fishmeal in diets of Pacific white shrimp, Litopenaeus vannamei. Aquac Res 51:3284–3296. https://doi.org/10.1111/are.14664
5. Devnani B, Moran GC, Grossmann L (2023) Extraction, composition, functionality, and utilization of Brewer's spent grain protein in food formulations. Food Secur 12(7):1543. https://doi.org/10.3390/foods12071543
6. Kumar M, Tomar M, Potkule J, Verma R, Punia S, Mahapatra A, Belwal T, Dahuja A, Joshi S, Berwal MK, Satankar V, Bhoite AG, Amarowicz R, Kaur C, Kennedy JF (2021) Advances in the plant protein extraction: mechanism and recommendations. Food Hydrocoll 115:115. https://doi.org/10.1016/j.foodhyd.2021.106595
7. Asiedu A, Ben S, Resurreccion E, Kumar S (2018) Techno-economic analysis of protein concentrate produced by flash hydrolysis of microalgae. Environ Prog & Sustain 37(2): 881–890. https://doi.org/10.1002/ep.12722
8. He Y, Kuhn DD, O'Keefe SF, Ogejo JA, Fraguas CF, Wang H, Huang H (2021) Protein production from brewer's spent grain via wet fractionation: process optimization and techno-economic analysis. Food Bioprod Process 126:234–244. https://doi.org/10.1016/j.fbp.2021.01.005
9. He YH, Kuhn DD, Ogejo JA, O'Keefe SF, Fraguas CF, Wiersema BD, Jin Q, Yu DJ, Huang HB (2019) Wet fractionation process to produce high protein and high fiber products from brewer's spent grain. Food Bioprod Process 117:266–274. https://doi.org/10.1016/j.fbp.2019.07.011
10. TTB, Beer Statistics U.S. Department of the Treasury. 2019. https://www.ttb.gov/beer/statistics. Accessed 12 Mar 2020
11. Davis R, Aden A, Pienkos PT (2011) Techno-economic analysis of autotrophic microalgae for fuel production. Appl Energ 88(10): 3524–3531. https://doi.org/10.1016/j.apenergy.2011.04.018

12. Humbird D, Davis R, Tao L, Kinchin C, Hsu D, Aden A, Schoen P, Lukas J, Olthof B, Worley M (2011) Process design and economics for biochemical conversion of lignocellulosic biomass to ethanol: dilute-acid pretreatment and enzymatic hydrolysis of corn stover; National Renewable Energy Lab. (NREL). Golden, CO
13. Buffington J (2014) The economic potential of brewer's spent grain (BSG) as a biomass feedstock. Adv Chem Engineer Sci 4:308–318. https://doi.org/10.4236/aces.2014.43034
14. Jin Q, O'Keefe SF, Stewart AC, Neilson AP, Kim Y-T, Huang H (2021) Techno-economic analysis of a grape pomace biorefinery: production of seed oil, polyphenols, and biochar. Food Bioprod Process 127:139–151. https://doi.org/10.1016/j.fbp.2021.02.002
15. Huang HB, Long S, Singh V (2016) Techno-economic analysis of biodiesel and ethanol co-production from lipid-producing sugarcane. Biofuels Bioprod Biorefin 10(3): 299–315. https://doi.org/10.1002/bbb.1640
16. EIA Natural gas prices (annual). 2019. https://www.eia.gov/dnav/ng/ngpri sum dcu nus a.htm. Accessed Feb 12
17. EIA Electric power annual. 2019. https://www.eia.gov/electricity/annual/. Accessed Oct 21
18. Anon (2020) Extension, University of Missouri, Price of by-products, http://agebb.missouri.edu/dairy/byprod/allcompanies.asp. Accessed 29 July 2020

Chapter 8

Microbial Analysis

Cangliang Shen, Wentao Jiang, Corey Coe, and Carly Long

Abstract

First in this chapter, we will enumerate overall microbial populations of plant protein using "Beyond Burger" as an example to use 3M petrifilm including aerobic plate counts (APCs), total coliform counts (TCCs), and generic *Escherichia coli* (ECCs). APCs, TCCs, and generic *E. coli* are recognized as indicator microorganisms, which are usually suggested to be a surrogate (non-pathogenic) to evaluate the presence of pathogens in that environment. APCs usually indicate the total aerobic microorganisms. TCCs are from the intestine of warm-blooded animals; are gram-negative, non-endospore-forming, facultative, fermentation lactose-generating acid and gas; and belong to the *Enterobacteriaceae* spp. Finally, generic ECC is a member of TCCs, and its presence indicates fecal contamination. 3M petrifilm plate is called a "paper medium" with a foam barrier accommodating a 20 cm^2 circular plating surface covered with a top plastic film. It also has a cold water-soluble gelling agent containing dehydrated nutrients in the middle of the film for plating, incubating, and numerating indicator microorganisms. Second in this chapter, we will isolate *Salmonella* spp. and *Listeria monocytogenes*, and using a selective, enriched medium, immunology assay and multi-channel biochemistry test to confirm the presence/absence of these two foodborne pathogens.

Key words Petrifilm, Aerobic plate counts, Coliforms, Generic *Escherichia coli*, Plant protein meat, *Listeria monocytogenes*, *Salmonella*, Analytical profile index 20E test, 12L *Listeria* test, Latex agglutination test

1 Introduction

Plant protein has been utilized as a substitute protein resource since the tenth century. As technology advanced, texturized vegetable proteins (TVP) emerged as potential replacements for conventional meat in modern times [1]. Recently, the global food landscape has shifted due to consumer demand for sustainable and environmentally friendly plant protein alternatives. Commercial plant-based meat companies like Impossible Foods, Beyond Meat, and Gardein have risen as prominent players in this domain [2].

The growing interest in new plant protein sources is driven by various factors, including environmental concerns, animal welfare, and the health implications of excessive red meat consumption

Yonghui Li (ed.), *Plant-Based Proteins: Production, Physicochemical, Functional, and Sensory Properties*, Methods and Protocols in Food Science, https://doi.org/10.1007/978-1-0716-4272-6_8,

[2]. To address these challenges, innovative food technology companies have leveraged cutting-edge research to create plant-based meat alternatives that replicate the taste, texture, and nutritional profile of traditional meat products [3].

However, the widespread acceptance of plant protein meat products necessitates a comprehensive assessment of their microbial safety. Microorganisms throughout the food production chain can pose potential hazards. Contamination of plant protein with harmful microorganisms, such as bacteria, viruses, and fungi, can result in foodborne illnesses and outbreaks if not effectively managed [4].

To ensure the microbial safety of plant protein meat, a multifaceted approach is required, including rigorous testing, process validation, and compliance with strict food safety regulations [5]. Furthermore, understanding the challenges of processing plant-based proteins and potential microorganism interactions with these novel substrates is crucial [6].

This book chapter aims to explore the complex relationship between microbial safety to plant protein, understand the correlation between novel plant protein products and microorganisms, investigate potential microbial safety concerns, and seize the implications for the future development of plant-based proteins in food safety. Approaching from various perspectives to elaborate and ensure the safe consumption of plant protein and contribute to a sustainable and healthier future for all. By critically analyzing microbial population especially foodborne pathogens, which will provide a comprehensive resource for researchers, food technologists, health professionals, and policymakers interested in alternative protein sources.

2 Materials

2.1 Materials for Enumeration of Aerobic Plate Counts (APCs), Total Coliform Counts (TCCs), and Generic *Escherichia coli* (ECCs) of Plant Protein on 3M Petrifilm

1. 25 g of freshly thawed plant protein meat.
2. 3 M APC and ECC/TCC petrifilm.
3. 0.1% of sterilized buffered peptone water.
4. Multiple 15 mL plastic dilution tubes.
5. Sterilized 200 μL and/or 1000 μL of pipette tips with pipette guns.
6. Incubator for 30–35 °C.

2.2 Materials for Isolation/Identification of *Salmonella* and *Listeria monocytogenes* in Plant Protein Meat

1. 25 g of freshly thawed Impossible plant protein meat.
2. Tryptic soy broth.
3. Buffered *Listeria* enrichment broth (LEB).
4. Whirl-Pak food sampling bags.
5. Stomacher.

6. Modified Oxford Agar.
7. Blood Agar.
8. H_2O_2 reagent.
9. 12L Listeria test kit.
10. Rappaport-Vassiliadis (RV) medium.
11. Tetrathionate Broth (TT).
12. XLT-4 agar.
13. API 20E biochemistry test kits.
14. Mineral oil.
15. Kovac Reagent.
16. 10% of ferric chloride.
17. 40% of KOH.
18. 6% of α-naphthol.
19. Latex agglutination test kit.

3 Methods

3.1 Procedures for Enumeration of APCs, TCCs, and ECCs of Plant Protein on 3M Petrifilm

1. As shown in Fig. 1: dilution factor for APCs, 10^1 and 10^3; coliform/*E.coli*, 0 and 10^2; total coliform, 0, 10^1, and 10^3.
2. The basic dilution schedule is shown in Fig. 2.
3. The colonies of APCs and TCC/ECCs are shown in Fig. 3.
4. The whole protocol is shown in Fig. 4.
5. Weigh 25 g of the plant protein products on the scale, and add into a filtered Whirl-Pak food sample bag.
6. Add 100 mL of pre-sterilized tryptic soy broth into the food sample bag.
7. Stomach for 60 s (1.0 min).
8. Label at the top of each petrifilm including sample name, date, and dilution factor.
9. Absorb 1000 μL of the "stomached" solution from the filtered side of the sample bag.
10. Prepare the diluted solution using 0.1% of BPW based on the dilution schedule as shown in Fig. 1. Because 1000 μL (1.0 mL) of the final diluted solution will be added onto the petrifilm, the final dilution is the same as tube dilution factor, which is different from our microbial spread-plating on agars.
11. Gently pull up the film followed by adding 1000 μL of the diluted solutions from **step 2** onto the center of petrifilm.
12. Gently roll down the film for preventing the air bubbles.

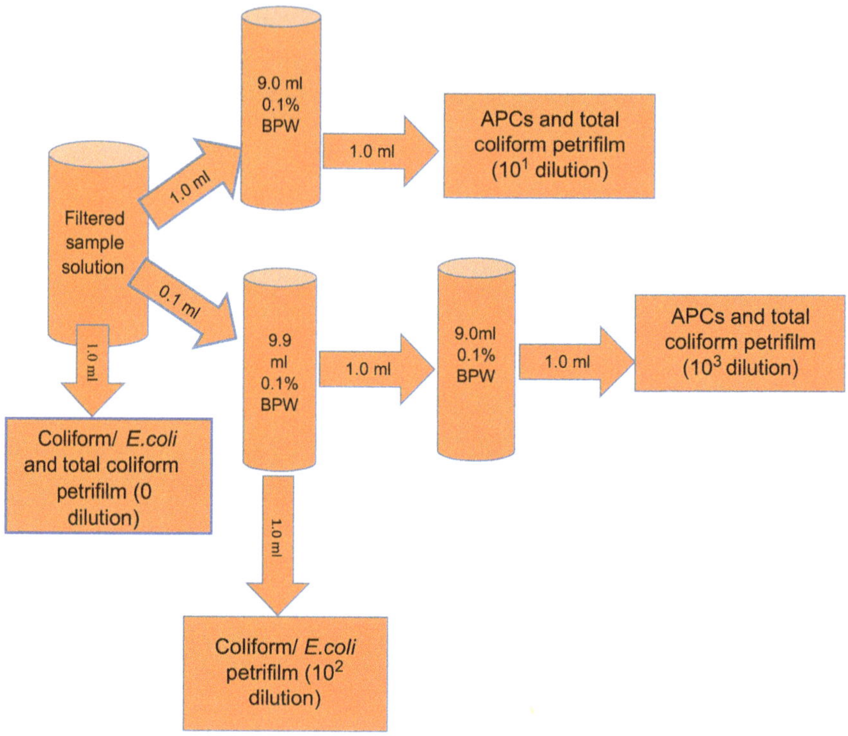

Fig. 1 Dilution protocol for analyzing aerobic plate counts (APCs), total coliform counts (TCCs), and generic *Escherichia coli* (ECCs) in the plant protein products

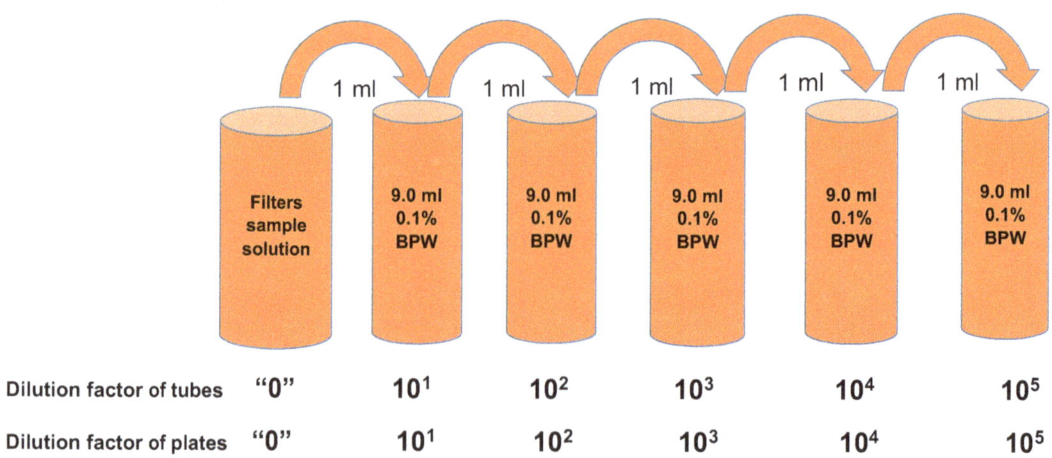

Make sure you vortexing each tube for 15 s after adding previous diluted solution.

Fig. 2 Dilution technique figure (tenfold serial dilution, 1.0 mL of each dilution tube adding onto petrifilm)

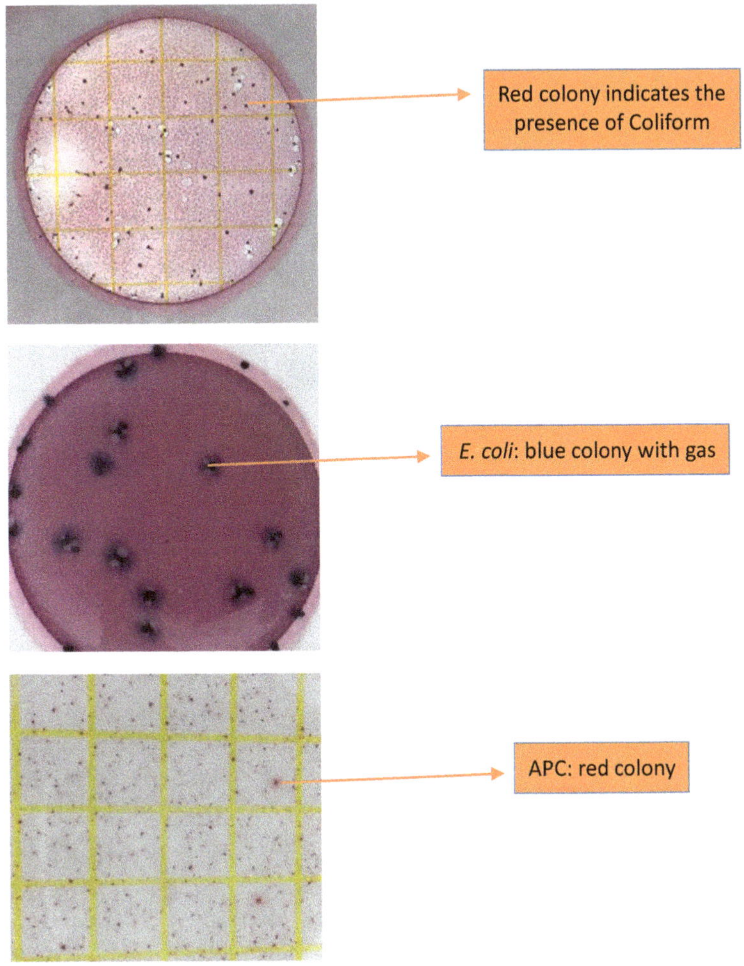

Fig. 3 Results for APCs, total coliform counts (TCCs), and generic *Escherichia coli* (ECCs) on petrifilms

13. Gently press the sample using a plastic template to fill the area.

    Note: For petrifilms of TCCs and ECCs, please skip this step.

14. For APCs, incubate the petrifilms at 25 °C or room temperature for 72 h.
15. For TCC/ECCs, incubate the petrifilms at 35 °C for 24–48 h.
16. Petrifilms cannot be stacked more than 20 pieces.
17. Manually count the colonies, and record into the data sheet.
18. As shown in Fig. 3, APCs, red colony; coliform, pink colony; *E. coli*, blue colony with gas. TCCs = numbers of coliform + numbers of *E. coli*.

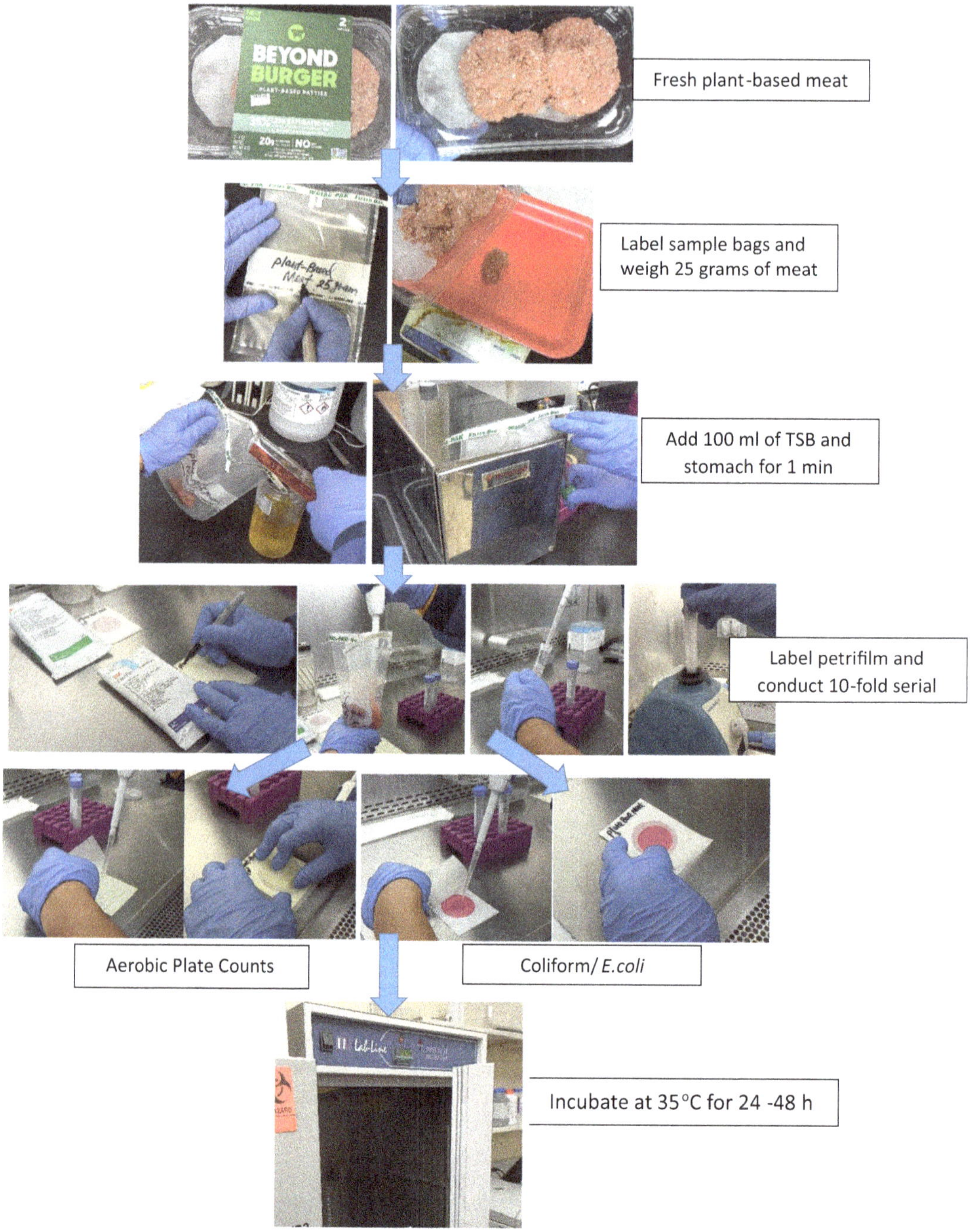

Fig. 4 Procedure for testing APCs, total coliform counts (TCCs), and generic *Escherichia coli* (ECCs) from plant protein meat on petrifilms

3.2 Procedures of Isolation of Salmonella

1. The whole protocol is shown in Fig. 5.
2. Weigh 25 g of the plant protein products on the scale, and add into a filtered Whirl-Pak food sample bag.
3. Add 100 mL of pre-sterilized tryptic soy broth into the food sample bag.
4. Stomach for 60 s (1.0 min).
5. Incubate at 35 °C for 24 h (pre-enriched solution).
6. Subculture 0.1 mL (100 μL) of the pre-enriched solution into Rappaport-Vassiliadis (RV) enrichment medium, and incubate at 35 °C for 24 h (subculture solution).
7. Subculture 1.0 mL (1000 μL) of the pre-enriched solution into Tetrathionate Broth (TT) enrichment medium, and incubate at 35 °C for 24 h (subculture solution).
8. Streak-plating above subcultured solutions onto XLT-4 Agar.
9. Incubate at 35 °C for 24–48 h.
10. Presumptive positive colonies (red color with black center) will be conducted latex agglutination test and multi-channel biochemical tests (API 20E).

3.3 Procedure for Latex Agglutination Test

1. The whole protocol is shown in Fig. 6.
2. Shake latex solution bottle very well.
3. Place one drop of latex antibody solution onto two wells of the test card, one for the control and the other one for the tested sample.
4. Use a prepared sterile plastic toothpick to mix the presumptive positive colonies from XLT-4 with the latex antibody solution; do this step for both control and texted sample colonies.
5. Slowly rock the test card with approximately 200 rpm for 1 min.
6. Observe the clumping for both control and tested samples.

3.4 Procedures for Multi-channel Biochemical Tests (API 20E)

1. The whole protocol is shown in Fig. 7.
2. Pick two to three presumptive positive *Salmonella* colonies from XLT-4 agar, and suspend into a prepared sterilized saline solution (5 mL).
3. Use a 1000 μL pipette to add 100 μL saline suspension into the cupules of the strip.
4. Mineral oil is filled for underlined test *ADH*, *LDC*, and *URE*, to create anaerobic environment.
5. Both the tube and cupule should be filled for boxed three tests CIT, VP, and GEL.

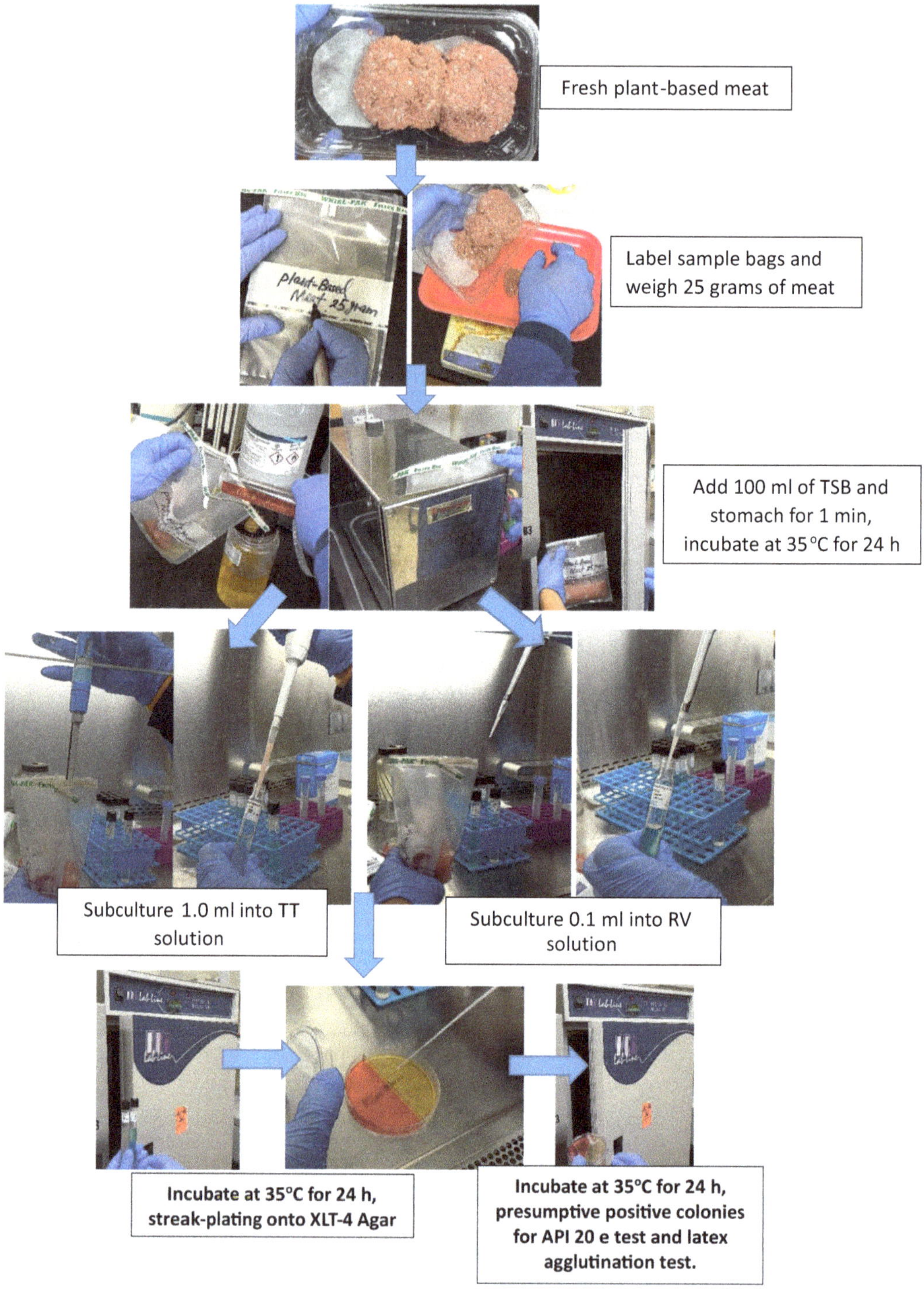

Fig. 5 Procedure for isolation of *Salmonella* spp. from plant protein meat

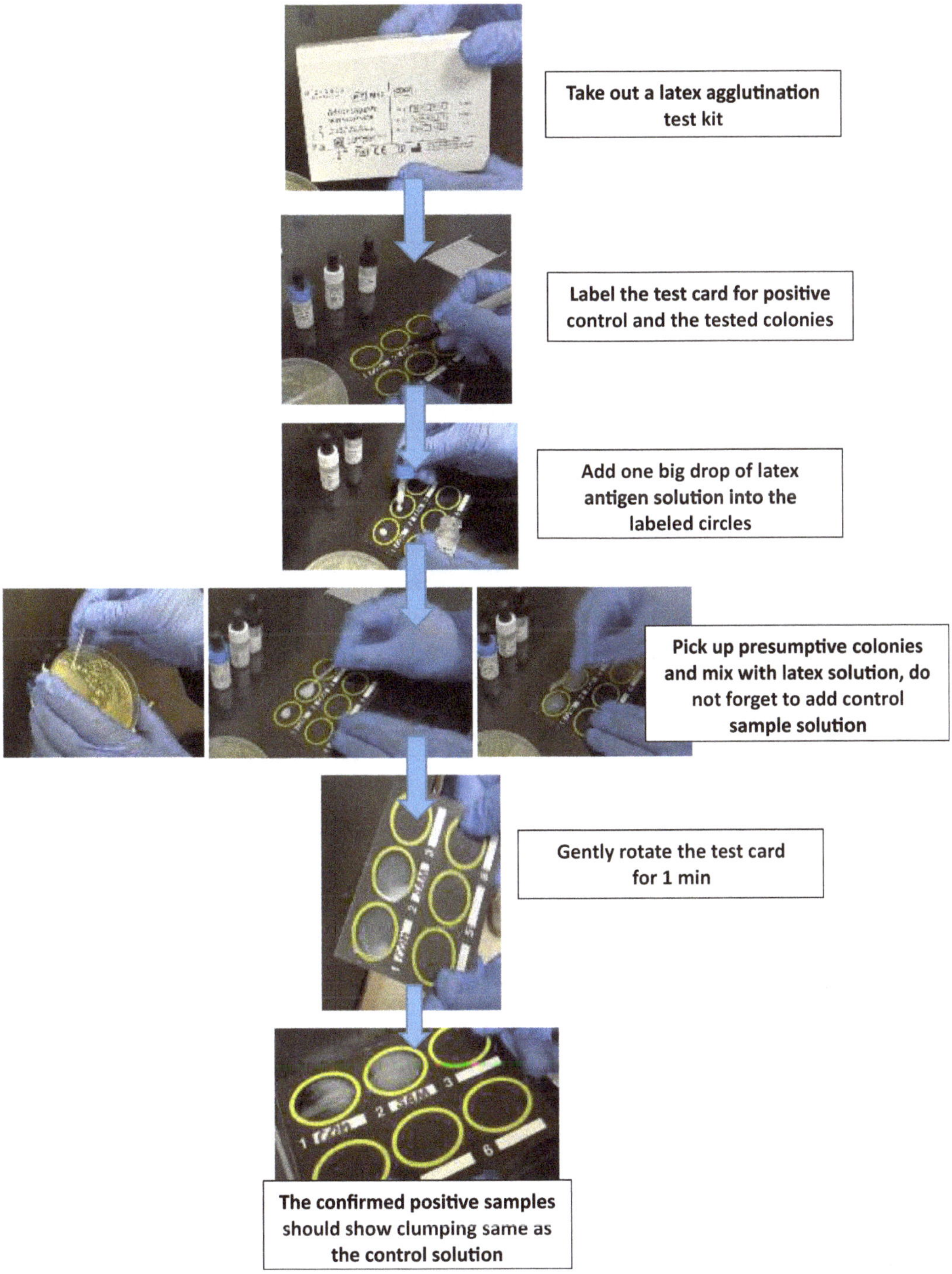

Fig. 6 Procedure for conducting latex agglutination test

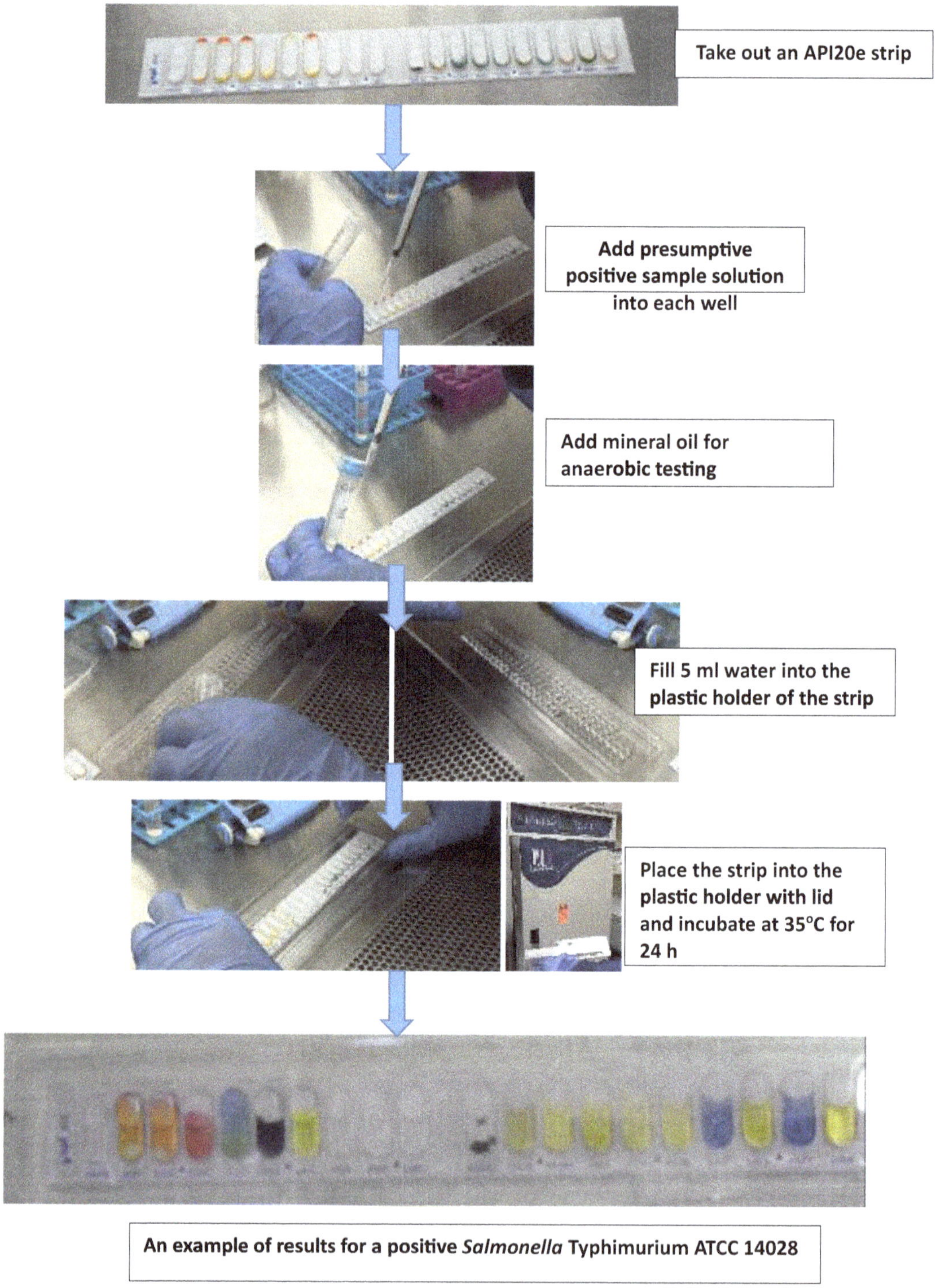

Fig. 7 Procedure for conducting API 20E test

Table 1
Result explanation for the positive and negative of API 20E test

Test items	Positive results	Negative results
ONPG	Yellow	Colorless
ADH	Red or orange	Yellow
LDC	Red or orange	Yellow
ODC	Red or orange	Yellow
CIT	Turquoise/dark blue	Light green/yellow
H_2S	Black deposit	No black deposit
URE	Red or orange	Yellow
TDA	Golden brown/red	Yellow
IND	Pink-red ring	Yellow
VP	Pink-red	Colorless
GEL	Diffusion	No diffusion
GLU	Yellow or gray	Blue to green
MAN, INO, SOR, RHA, SAC, MEL, AMY, ARA	Yellow-yellow green	Blue to blue-green

6. Add 5–8 mL of water into the plastic base, and place strip in the base to avoid drying during incubation.
7. Incubate the whole strip for at 35 °C for 24 h.
8. After 24 h of incubation, add one drop of Kovac Reagent to record results in 2 min for IND.
9. Immediately after 24 h of incubation, add one drop of 10% ferric chloride for TDA.
10. After 24 h of incubation, add one drop of 40% KOH and then one drop of 6% α-naphthol in 10 min for VP.
11. Record the positive (+) or negative (−) results on the worksheet as shown in Table 1.
12. Record the API number, add number up for the positive reaction, and search the code in the API Reference Book.
13. A possible *Salmonella* spp. code is 6704752 and positive results are in Table 2.

3.5 Procedures of Isolation of L. monocytogenes

1. Part of the protocol is shown in Fig. 8.
2. Weigh 25 g of the plant protein products on the scale, and add into a filtered Whirl-Pak food sample bag.

Table 2
API 20E test results for *Salmonella* Typhimurium

Test items	Results
ONPG	Colorless
ADH	Orange
LDC	Orange
ODC	Orange
CIT	Dark blue
H_2S	Black deposit
URE	Yellow
TDA	Yellow
IND	Yellow
VP	Colorless
GEL	Diffusion
GLU	Yellow
MAN	Yellow
INO	Yellow
SOR	Yellow
RHA	Yellow
SAC	Blue
MEL	Yellow
AMY	Blue
ARA	Yellow

3. Add 100 mL of pre-sterilized tryptic soy broth into the food sample bag.
4. Stomach for 60 s (1.0 min).
5. Incubate at 35 °C for 24 h (pre-enriched solution).
6. Subculture 1 mL of the pre-enriched solution into buffered *Listeria* enrichment UVM broth, and incubate at 35 °C for 24 h (subculture solution).
7. Streak-plating above subcultured solution onto Modified Oxford Agar.
8. Incubate at 35 °C for 48 h.

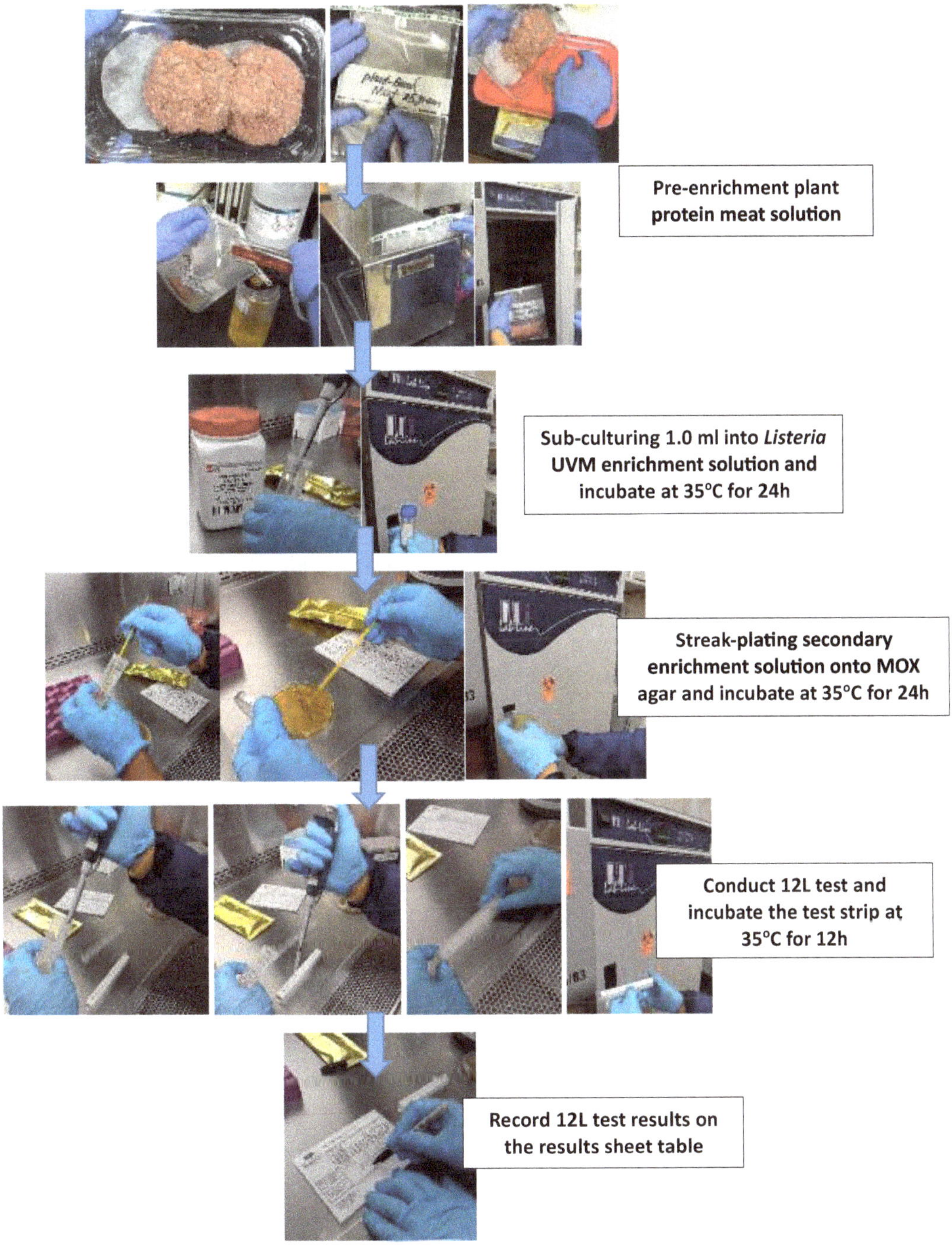

Fig. 8 Procedure for isolation of *Listeria monocytogenes* from plant protein meat and 12L biochemistry test

9. Typical presumptive colonies (black color) will be conducted catalase test (bubbling).
10. Positive catalase test colonies will conduct multi-channel biochemical tests (12L tests).
11. Positive catalase test colonies will be streak-plated onto blood agar for observing β-hemolytic *L. monocytogenes* colonies (transparent zone).

3.6 Catalase Test Procedures

1. Place one drop of H_2O_2 on a clean glass slide.
2. Use sterilized loop pick presumptive colony from MOX Agar, and slowly immerse with cells into H_2O_2.
3. A positive test is read immediately from the slide, showing oxygen gas bubbles.

   Note: It can also be conducted by adding one tiny drop of H_2O_2 directly onto a typical colony, and observe for bubbling or not.

3.7 Procedures for Multi-channel Biochemical Tests (12L Tests) Procedures

1. Part of the protocol is shown in Fig. 8.
2. Pick two to three presumptive colonies from MOX agar.
3. Suspend the colonies into the prepared "*Listeria* Suspending Medium."
4. Place 12L test strip in the "holding tray," and remove the plastic lid.
5. Add 100 uL of bacterial suspension to each well from wells 1 to 11.
6. Add one drop of prepared hemolysis to well 12. This step can be done on the blood agar.
7. Cover the 12L test strip with the lid.
8. Incubate it at 35 °C for 24 h.
9. Record results on notebook or record sheet, and interpret results using the color sheet.
10. The explanation of 12L test results is shown in Table 3.
11. A sample of *L. monocytogenes* is shown in Table 4.

Acknowledgments

The development of this book chapter was supported by the US Department of Agriculture, the National Institute of Food and Agriculture—Research and Extension Experiences for Undergraduates (REEU) program (Award #2020-68018-30657), and the West Virginia Agricultural and Forestry Experiment Station Hatch Grant # WVA00684 and WVA00736.

Table 3
Result explanation for the positive and negative of 12L test

Number of wells	Substrate of biochemical tests	Colors of negative results	Colors of positive results
1	Esculin (ESC)	Pink or brown	Black
2	Mannitol (MAN)	Purple	Yellow, brown, or straw
3	Xylose (XYL)	Purple	Yellow, brown, or straw
4	Arabitol (ARL)	Purple	Yellow, brown, or straw
5	Ribose (RIB)	Purple	Yellow, brown, or straw
6	Rhamnose (RHA)	Purple	Yellow, brown, or straw
7	Trehalose (TRE)	Purple	Yellow, brown, or straw
8	Tagatose (TAG)	Purple	Yellow, brown, or straw
9	Glucose-1-phosphate (G1P)	Purple	Yellow, brown, or straw
10	Methyl-D-glucose (MDG)	Purple	Yellow, brown, or straw
11	Methyl-D-mannose (MDM)	Purple	Yellow, brown, or straw
12	Hemolysin (HEM)	Red cell deposit	Partial or complete red cell lysis

Table 4
12L test results for *Listeria monocytogenes*

1ESC	2MAN	3XYL	4ARL	5RIB	6RHA	7TRE	8TAG	9G1P	10MDG	11MDM	12HEM
+	–	–	+	–	+	+	–	–	+	+	+

Note: All *Listeria* spp. test positive for esculin (well 1), arabitol (well 4), and trehalose (well 7)

References

1. Joshi V, Kumar S (2015) Meat analogues: plant based alternatives to meat products—a review. Int J Food Ferment Technol 5:107
2. Zhang L, Hu YY, Badar IH, Xia XF, Kong BH, Chen Q (2021) Prospects of artificial meat: opportunities and challenges around consumer acceptance. Trends Food Sci Technol 116:434–444. https://doi.org/10.1016/j.tifs.2021.07.010
3. Velasco C, Barbosa Escobar F, Petit O, Wang QJ (2021) Impossible (food) experiences in extended reality. Front Comput Sci 3:716846. https://doi.org/10.3389/fcomp.2021.716846
4. Hadi J, Brightwell G (2021) Safety of alternative proteins: technological, environmental and regulatory aspects of cultured meat, plant-based meat, insect protein and single-cell protein. Food Secur 10.1226. https://doi.org/10.3390/foods10061226
5. Draft Guidance for Industry: Hazard Analysis and Risk-Based Preventive Controls for Human Food (2023) U.S. Food and Drug Administration (FDA). https://www.fda.gov/regulatory-information/search-fda-guidance-documents/draft-guidance-industry-hazard-analysis-and-risk-based-preventive-controls-human-food. Accessed 12 Aug 2023
6. Ahmad M, Qureshi S, Akbar MH, Siddiqui SA, Gani A, Mushtaq M (2022) Plant-based meat alternatives: compositional analysis, current development and challenges. Appl Food Res 2:100154. https://doi.org/10.1016/j.afres.2022.100154

Part III

Biochemical and Nutritional Analysis of Plant-Based Proteins

Chapter 9

Quantification of Crude and Soluble Protein Content

Changqi Liu, Jing Zhao, Sahil Gupta, and Marta Carrillo

Abstract

Reliable quantitative techniques for assessing protein content in food products and ingredients are crucial for ensuring food quality and safety. A variety of protein quantification methods have been developed, each grounded in distinct principles and necessitating unique laboratory setups. This chapter describes the protocols for quantifying the crude protein content in plant-based samples based on their nitrogen content using Kjeldahl and Dumas methods, as well as the analysis of soluble proteins using bicinchoninic acid, biuret, Bradford, Lowry, and ultraviolet absorption methods.

Key words Protein analysis, Nitrogen analysis, Soluble protein

1 Introduction

Proteins constitute an essential macronutrient in human diets. However, the consumption of animal proteins is associated with climate change, freshwater depletion, biodiversity loss, and adverse health outcomes [1]. Plant-based proteins are progressively gaining traction as a cost-effective, sustainable replacement for animal proteins in human nutrition and as functional ingredients in product formulation [2]. Accurate protein quantification holds significance in identifying plant sources rich in proteins, isolating protein ingredients, assessing their functional properties and bioactivities, creating nutrition labels, and determining pricing [3]. Various analytical methods exist for protein quantification. The fundamental principles underlying these techniques involve analyzing nitrogen content, peptide bonds, aromatic amino acids, dye-binding properties, and ultraviolet absorption of proteins [3].

The crude protein content of a food can be determined using either the Kjeldahl method or the Dumas method. In the Kjeldahl method, nitrogen within proteins is converted into ammonia through digestion, neutralization, and distillation. The nitrogen content is measured by titrating the resulting ammonia. In the

Yonghui Li (ed.), *Plant-Based Proteins: Production, Physicochemical, Functional, and Sensory Properties*, Methods and Protocols in Food Science, https://doi.org/10.1007/978-1-0716-4272-6_9,

Dumas method, proteins are subjected to combustion, resulting in the formation of nitrogen gas and nitrogen oxides. The nitrogen oxides are subsequently converted into nitrogen gas, and the combined amount of nitrogen gas is measured using a thermal conductivity detector. The calculated nitrogen content is converted into the crude protein content using a conversion factor [4].

Various methods for quantifying soluble proteins rely on the formation of colored complexes between cupric ions and peptide bonds, including the biuret method [5], Lowry method [6], and bicinchoninic acid (BCA) method [7]. The Lowry method enhances the biuret method by adding the Folin-Ciocalteu reagent, which reacts with tyrosine and tryptophan residues. Similarly, the BCA method combines the biuret method with BCA, leading to a reaction with cuprous ions. The Bradford method is based on the capacity of proteins to bind with Coomassie Brilliant Blue G-250 dye and elicit a color change [8]. Meanwhile, the ultraviolet absorption method relies on the absorption of ultraviolet light at 280 nm by protein [4].

In this chapter, we provide protocols for quantifying plant proteins using the Kjeldahl [9], Dumas [10], BCA, biuret [11], Bradford [12], Lowry [13], and ultraviolet absorption [14] methods. These protocols stem from scientific literature and our practical familiarity with these techniques. The Kjeldahl protocol is conducted using a Labconco Rapid Digestor and distillation unit, while the Dumas protocol is based on the operation of an Elementar rapid N exceed nitrogen analyzer.

2 Materials

2.1 Materials for Crude Protein Analysis

2.1.1 Materials for Sample Preparation

1. Equipment and supplies: analytical balance, grinder, sieve, oven (optional), and fat extraction system (optional).
2. Materials: plant or protein samples.

2.1.2 Materials for the Kjeldahl Method

1. Equipment and supplies: digestor, distillation unit, analytical balance, digestion tubes (250 mL), flasks (500 mL), graduated cylinders, burette, and glass beads.
2. Chemicals and reagents: boric acid, cupric sulfate, ethanol, methyl red, methylene blue, potassium sulfate, sodium hydroxide, and sulfuric acid (concentrated, 36 N) (*see* **Note 1**).
3. 60% *w/v* sodium hydroxide solution (*see* **Note 2**).
4. Boric acid indicator solution: Dissolve 200 mg of methyl red in 100 mL of ethanol, and dissolve 100 mg of methylene blue in 50 mL of ethanol. Mix two volumes of methyl red with one volume of methylene blue. Dissolve 20 g of boric acid in

800 mL of deionized water. Add 10 mL of the methyl red/-methylene blue indicator solution. Dilute the mixture to a total volume of 1 L with deionized water (*see* **Note 3**).

5. 0.1 N sulfuric acid solution.

2.1.3 Materials for the Dumas Method

1. Equipment and supplies: nitrogen analyzer, analytical balance, tin foil squares, cone-forming tool, manual press, multi-well plates, tweezers, and instrument consumables.
2. Chemicals and reagents: aspartic acid, carbon dioxide gas, oxygen gas, and silicone grease.

2.2 Materials for Soluble Protein Analysis

1. Equipment and supplies: analytical balance, vortex mixer, magnetic stirrer, shaker, spectrophotometer, microplate reader, heat block, pipettes, microplates, cuvettes, volumetric flasks, graduated cylinders, beakers, test tubes, and microcentrifuge tubes.

2.2.1 Materials for Sample Preparation

1. Equipment and supplies: analytical balance, grinder, sieve, oven (optional), fat extraction system (optional), vortex mixer, magnetic stirrer, shaker, pipettes, graduated cylinders, beakers, and microcentrifuge tubes.
2. Chemicals and reagents: plant or protein samples and solvent of choice.

2.2.2 Materials for Preparing Bovine Serum Albumin (BSA) Stock Solution

1. Equipment and supplies: analytical balance, spectrophotometer, magnetic stirrer, vortex mixer, pipettes, membrane filter (0.22 μm), beakers, microcentrifuge tubes, and cuvettes.
2. Chemicals and reagents: BSA.

2.2.3 Materials for BCA Method

1. Equipment and supplies: analytical balance, pH meter, pipettes, vortex mixer, heat block, microplate reader, microplates, volumetric flasks, graduated cylinders, beakers, and microcentrifuge tubes.
2. Chemicals and reagents: bicinchoninic acid (BCA), cupric sulfate, sodium carbonate monohydrate, sodium hydroxide, and sodium tartrate.
3. Reagent A: Dissolve 8.0 g of cupric sulfate and 1.6 g of sodium tartrate in deionized water. Adjust the pH to 11.25. Bring the volume to 100 mL with deionized water.
4. Reagent B: Dissolve 4 g of BCA in deionized water, and bring the volume to 100 mL.
5. Reagent C: Dissolve 0.4 g of cupric sulfate in water, and bring the volume to 10 mL.
6. Reagent D: Mix one part of reagent C and 25 parts of reagent B. Subsequently, mix one part of the mixture with one part of reagent A.

2.2.4 Materials for Biuret Method

1. Equipment and supplies: analytical balance, spectrophotometer, vortex mixer, magnetic stirrer, pipettes, volumetric flasks, graduated cylinders, beakers, test tubes, and cuvettes.
2. Chemicals and reagents: cupric sulfate, sodium hydroxide, and sodium potassium tartrate.
3. Biuret reagent: Dissolve 1.5 g of cupric sulfate and 6.0 g of sodium potassium tartrate in 500 mL of deionized water in a 1 L beaker. Dissolve 30 g of sodium hydroxide in 300 mL of deionized water, and add it to the 1 L beaker containing the cupric sulfate and sodium potassium tartrate solution. Stir the solution constantly. Bring the volume in a volumetric flask to 1 L (*see* **Note 4**).

2.2.5 Materials for Bradford Method

1. Equipment and supplies: analytical balance, pipettes, multichannel pipettes, vortex mixer, microplate reader, microplates, volumetric flasks, graduated cylinders, beakers, microcentrifuge tubes, and reagent reservoir.
2. Chemicals and reagents: Coomassie Brilliant Blue G-250, ethanol, and *o*-phosphoric acid.
3. Bradford reagent stock solution: Dissolve 0.25 g of Coomassie Brilliant Blue G-250 in 125 mL of ethanol, and add 250 mL of *o*-phosphoric acid. Bring the volume to 500 mL with deionized water (*see* **Note 5**).
4. Bradford reagent: Dilute the Bradford reagent stock solution five times with deionized water.

2.2.6 Materials for Lowry Method

1. Equipment and supplies: analytical balance, spectrophotometer, vortex mixer, magnetic stirrer, pipettes, volumetric flasks, graduated cylinders, beakers, test tubes, test tube stand, and cuvettes.
2. Chemicals: cupric sulfate, Folin-Ciocalteu's phenol reagent (2 N), potassium tartrate, sodium carbonate, and sodium hydroxide.
3. Reagent A: Dissolve 20 g of sodium carbonate, 1.6 g of potassium tartrate, and 4 g of sodium hydroxide in deionized water. Bring the volume to 1 L with deionized water.
4. Reagent B: Dissolve 4 g of cupric sulfate pentahydrate in deionized water. Bring the volume to 100 mL with deionized water.
5. Reagent C: Mix 100 parts of reagent A with 1 part reagent B while stirring constantly (*see* **Note 6**).
6. Folin-Ciocalteu's phenol reagent (1 N): Dilute Folin-Ciocalteu's phenol reagent (2 N) 1:1 *v/v* with deionized water (*see* **Note 6**).

2.2.7 Materials for Ultraviolent Absorption Method

1. Equipment and supplies: spectrophotometer, vortex mixer, pipettes, microcentrifuge tubes, and ultraviolet clear cuvettes.
2. Chemicals and reagents: solvent of choice.

3 Methods

3.1 Crude Protein Analysis

3.1.1 Sample Preparation

1. Grind plant samples in a grinder.
2. Dehydrate or defat the samples as needed (*see* **Note 7**).

3.1.2 Kjeldahl Method

Sample Digestion

1. Into each 250 mL digestion tube, place 0.1 g of sample (*see* **Note 8**), 0.42 g of cupric sulfate (*see* **Note 9**), 9 g of potassium sulfate, and three to four glass beads (to prevent sample bumping).
2. Prepare at least one tube for a blank containing all components except the sample.
3. Document the sequence in which you place the tubes into the digestion rack (*see* **Note 10**).
4. Working under the hood, add 15 mL of concentrated sulfuric acid to each tube.
5. Place the digestion rack into the digestor. Turn on the heating block, and set the temperature to 410 °C. Place the fume removal system above the tubes. Turn on the water flow to create suction for effective fume removal.
6. Digest the samples for 45–75 min until a clear green solution devoid of any black residues is obtained, guaranteeing the complete oxidation of all organic matter.
7. After the digestion is completed, lift the rack while keeping the fume removal system attached to the upper section of the digestor. Allow the tubes to cool for a minimum of 10 min.
8. Tilt the digestion tube, and gradually pour 50 mL of deionized water down the side of the tube (*see* **Note 11**).
9. Place the bottom of the digestion tube under running water to facilitate cooling.

Neutralization and Distillation

1. During digestion, prepare the distillation unit by flushing it with the 60% *w/v* sodium hydroxide solution. Configure the unit to dispense 45 mL of the sodium hydroxide solution.
2. Set up a 500 mL flask for every tube in the digestor, each holding 100 mL of the boric acid indicator solution.
3. Turn on the water flow and the heating element of the distillation unit.

4. Insert the distillation outlet tube into the flask containing the boric acid indicator solution (*see* **Note 12**).
5. Attach the cooled digestion tube to the distillation head.
6. Introduce 45 mL of 60% *w/v* sodium hydroxide solution, allowing the dispenser to release it gradually.
7. Once the sodium hydroxide solution has been introduced into the tube, initiate the distillation procedure by shifting the lever from standby to distill. Maintain distillation for a duration of 5 min, a period sufficient to collect approximately 125 mL of distillate in the receiving flask.
8. Upon completion of distillation, revert the lever to the standby position.
9. Remove the digestion tube carefully as it will be very hot. Take out the receiving flask, and rinse the distillation outlet tube with deionized water.
10. Check the water level in the steam generator between samples. Never let it run dry. Do not refill while a digestion tube or receiving flask is in place.

Titration

1. Set up the burette with 0.1 N sulfuric acid solution, and perform a verification against a standard alkali solution.
2. Titrate the receiving solution using 0.1 N sulfuric acid solution to a purple end point. Perform the blank titration first.
3. Calculate the nitrogen content using the following equation:

$$\text{Nitrogen content } (\%) = c \times \frac{(V_S - V_b) \times 14}{m \times 1000} \times 100\% \quad (1)$$

   where c is the normality of sulfuric acid (N), V_s is the sulfuric acid volume for sample titration (mL), V_b is the sulfuric acid volume for blank titration (mL), m is the mass of sample (g).
4. After acquiring the nitrogen content, move on to Subheading 3.1.4 for the conversion into crude protein content.

Shutdown and Cleanup

1. Turn off the heating block of the digestor and water flow of the fume removal system.
2. Turn off the heating element and water flow of the distillation unit.
3. Rinse the sodium hydroxide dispenser and tubing by priming at least 2 L of deionized water through it. Subsequently, drain and disconnect the dispenser. Wash it with water and allow it to dry.
4. Wipe down the digestor and the distillation units with water.

3.1.3 Dumas Method

Sample Preparation

1. Transform tin foil squares into tin foil cones using a cone-forming tool (*see* **Note 13**).
2. Weigh 250 mg of aspartic acid or 100–400 mg of samples into each tin foil cone (*see* **Note 14**).
3. Encase the sample with the foil, and utilize a manual press to compact it into a pellet with a uniform diameter.
4. Place the pellets into a sample holder such as a multiple-well plate.

Sample Running

1. Turn on the nitrogen analyzer and its furnace (*see* **Note 15**).
2. Turn on the gas flow of carbon dioxide and oxygen.
3. Perform a leak test to ensure the system is airtight (*see* **Note 16**).
4. Input the details for the blank, standard, and sample into the system, and choose the suitable method for each.
5. Load the samples onto the sample carousel.
6. Run at least two blank samples to ensure the thermal conductivity detector signal is at baseline.
7. Run at least three aspartic acid standards to calculate the correction factor.
8. Run each sample in a minimum of three replicates.
9. Export the data.
10. After acquiring the nitrogen content, move on to Subheading 3.1.4 for the conversion into crude protein content.

Shutdown and Maintenance

1. Turn off the furnace and the gas flow of carbon dioxide and oxygen.
2. Turn off the nitrogen analyzer after the furnace is cool.
3. Empty the ash crucible in the combustion tube (*see* **Note 17**).
4. Remove any ashes from the oxygen lance and nickel flap (*see* **Note 18**).
5. Refill the Regainer in the post-combustion tube (*see* **Note 19**).
6. Replace the drying agent in the drying tubes as needed (*see* **Note 19**).
7. Replace the combustion tube, post-combustion tube, and reduction tube as needed (*see* **Note 19**).

3.1.4 Nitrogen-to-Protein Conversion

Upon obtaining the nitrogen content, use a nitrogen-to-protein conversion factor to transform it into crude protein content using the following equation:

$$\%\text{protein} = \%N \times \mathrm{k_p} \quad (2)$$

where % protein is the crude protein content (%), % N is the nitrogen content (%), and k_p is the nitrogen-to-protein conversion factor.

Because of variations in amino acid compositions and the presence of non-protein nitrogen compounds, nitrogen conversion factors vary among various food items. Table 1 provides the nitrogen-to-protein conversion factors for select foods. In cases where the factor is unknown, a commonly employed default value of 6.25 is used, although research suggests that a smaller factor (5.60) may be more appropriate [15].

Table 1
Nitrogen-to-protein conversion factors for select protein-rich plant samples

Food	Ref [16]	Ref [17]	Ref [18]	Ref [19]
Cereals				
Barley	5.83	5.40	5.50	
Corn	6.25		5.65	5.72
Millet, foxtail			5.80	
Millet, pearl		5.63	5.33	
Oat	5.83	5.32	5.36	
Rice	5.95		5.17	5.61
Rye	5.83	5.35	5.33	
Sorghum			5.65	5.93
Triticale		5.62	5.36	
Wheat	5.83	5.49	5.33	5.75
Wheat, bran	6.31	4.71		
Wheat, endosperm	5.70			
Wheat, germ	5.80	4.99		
Legumes				
Adzuki bean	6.25			
Broad bean			5.40	
Dry bean				5.44
French bean			5.33	
Jack bean	6.25			
Lima bean	6.25			
Lupin			5.40	

(continued)

Table 1
(continued)

Food	Ref [16]	Ref [17]	Ref [18]	Ref [19]
Mung bean	6.25			
Navy bean	6.25			
Pea			5.44	5.40
Peanut	5.46			
Soybean	5.71	5.44	5.52	
Velvet bean	6.25			
Oilseeds				
Cantaloupe seed	5.30			
Castor bean	5.30			
Cottonseed	5.30			
Flaxseed	5.30			
Hempseed	5.30			
Mustard seed		5.12		
Pumpkin seed	5.30			
Rapeseed		5.29		
Sesame seed	5.30			
Squash seed	5.30			
Sunflower seed	5.30	5.29		
Pseudocereals				
Buckwheat		5.24		
Tree nuts				
Almond	5.18			
Brazil nut	5.46			
Butternut	5.30			
Coconut	5.30			
Hazelnut	5.30			
Tuber and roots				
Potato				5.18

3.2 Soluble Protein Analysis

3.2.1 Sample Preparation

When handling with isolated plant proteins, disperse the protein in appropriate solvent directly. Centrifuge or filter the samples as needed. If proteins need to be extracted from plant samples, complete Subheading 3.1.1. first, and proceed to the following instructions:

1. Disperse the plant samples in an appropriate amount of solvent.
2. Extract soluble proteins by continuous magnetic stirring, vortex mixing, or shaking (*see* **Note 20**).
3. Centrifuge the mixture and collect the supernatant.

3.2.2 BSA Stock Solution Preparation

1. Dissolve 0.4 g of BSA in 40 mL of deionized water (*see* **Note 21**).
2. Filter the BSA solution using a membrane (0.22 μm).
3. Pipette 100 μL of the BSA solution in a microcentrifuge tube containing 900 μL of deionized water, and vortex mix.
4. Read the absorbance of the diluted solution at 280 nm (*see* **Note 22**).
5. Calculate the concentration of the BSA stock solution using the following equation:

$$c = \frac{A \times d \times 10}{\varepsilon^{\%} \times 1} \quad (3)$$

where c is the concentration (mg/mL), A is the absorbance at 280 nm, d is the dilution factor ($d = 10$ in this case), $\varepsilon^{\%}$ is the percent extinction coefficient (6.6 for 1% BSA solution at 280 nm), and l is the light path (cm) (*see* **Note 23**).

3.2.3 BCA Method

1. Prepare 0, 20, 40, 60, 80, and 100 μg/mL of BSA standards in triplicate (*see* **Note 24**, Table 2).
2. Adjust the concentration of the protein samples to fit within the range covered by the standard curve.

Table 2
Standard preparation for the BCA method

Concentration (μg/mL)	100 μg/mL BSA stock (mL)	Solvent (mL)
0	0.0	2.0
20	0.4	1.6
40	0.8	1.2
60	1.2	0.8
80	1.6	0.4
100	2.0	0.0

Table 3
Standard preparation for the biuret method

Concentration (mg/mL)	10 mg/mL BSA stock (mL)	Solvent (mL)
0.0	0.00	5.00
1.0	0.50	4.50
2.5	1.25	3.75
5.0	2.50	2.50
7.5	3.75	1.25
10.0	5.00	0.00

3. Pipette 500 μL of the standard or sample into a microcentrifuge tube.
4. Add 500 μL of reagent D and vortex mix.
5. Incubate at 60 °C in a heat block for 1 h.
6. Let the samples to cool.
7. Pipette 200 μL of the reaction mix to a microplate.
8. Read the absorbance at 562 nm using a microplate reader (*see* **Note 25**).
9. Calculate the protein concentration from the standard curve.

3.2.4 Biuret Method

1. Prepare 0, 1.0, 2.5, 5.0, 7.5, and 10.0 mg/mL of BSA standards in triplicate (*see* **Note 24**, Table 3).
2. Adjust the concentration of the protein samples to fit within the range covered by the standard curve.
3. Pipette 1.0 mL of the standard or sample into a test tube.
4. Add 5.0 mL of biuret reagent and vortex mix.
5. Incubate at room temperature for 30 min.
6. Read the absorbance at 540 nm using a spectrophotometer (*see* **Note 26**).
7. Calculate the protein concentration from the standard curve.

3.2.5 Bradford Method

1. Prepare 0, 50, 100, 200, 300, 400, 500, and 600 μg/mL of BSA standards in triplicate (*see* **Note 24**, Table 4).
2. Adjust the concentration of the protein samples to fit within the range covered by the standard curve.
3. Pipette 10 μL of the standard or sample into a microplate.
4. Add 200 μL of diluted Bradford reagent, and incubate for 5 min at room temperature.

Table 4
Standard preparation for the Bradford method

Concentration (μg/mL)	600 μg/mL BSA stock (μL)	Solvent (μL)
0	0	60
50	5	55
100	10	50
200	20	40
300	30	30
400	40	20
500	50	10
600	60	0

Table 5
Standard preparation for the Lowry method

Concentration (μg/mL)	200 μg/mL BSA stock (mL)	Solvent (mL)
0	0.0	5.0
20	0.5	4.5
40	1.0	4.0
60	1.5	3.5
80	2.0	3.0
100	2.5	2.5
120	3.0	2.0
140	3.5	1.5
160	4.0	1.0
180	4.5	0.5
200	5.0	0.0

5. Read the absorbance at 595 nm using a microplate reader (*see* **Note 25**).
6. Calculate the protein concentration from the standard curve.

3.2.6 Lowry Method

1. Prepare 0, 20, 40, 60, 80, 100, 120, 140, 160, 180, and 200 μg/mL of BSA standards in triplicate (*see* **Note 24**, Table 5).
2. Adjust the concentration of the protein samples to fit within the range covered by the standard curve.

3. Pipette 1.0 mL of the standard or sample into a test tube.
4. Add 3.0 mL of reagent C and vortex mix.
5. Incubate at room temperature for 60 min.
6. Add 0.3 mL of 1 N Folin-Ciocalteu's phenol reagent and vortex mix immediately.
7. Incubate in dark for 45 min at room temperature.
8. Read the absorbance at 660 nm using a spectrophotometer (*see* **Note 26**).
9. Calculate the protein concentration from the standard curve.

3.2.7 Ultraviolet Absorption Method (See Note 27)

1. Adjust the protein concentration so that it falls within the detectable range of the spectrophotometer.
2. Read the absorbance at 280 nm against a solvent blank using a spectrophotometer.
3. Calculate the protein concentration using the following equation:

$$c = \frac{A \times d}{\varepsilon \times 1} \quad (4)$$

where c is the molar concentration of the protein solution (M), A is the corrected absorbance at 280 nm, d is the dilution factor, ε is the molar extinction coefficient (M^{-1} cm^{-1}), and 1 is the light path (cm).

4 Notes

1. Exercise caution and wear protective gears while handling sodium hydroxide and concentrated sulfuric acid, both of which are highly corrosive.
2. Make sure to prepare the 60% *w/v* sodium hydroxide solution in advance, allowing several hours for it to cool.
3. If the temperature in the laboratory is cool, it is necessary to warm the water to aid in dissolving the boric acid. Failure to do so might result in the formation of a red precipitate upon adding the methyl red/methylene blue indicator solution. The final boric indicator solution should be green. If this is not achieved, it indicates the pH of the water is overly acidic and requires neutralization.
4. Ensure that the solution is cooled when making it to the final volume. The reagent can be stored in a plastic bottle in the refrigerator. Discard the solution if precipitate appears.
5. Store the Bradford reagent stock solution at 4 °C.

6. Reagent A and reagent B can be stored indefinitely at room temperature, whereas reagent C and Folin-Ciocalteu's phenol reagent (1 N) need to be freshly prepared on the day of use.
7. Make sure the samples are homogeneous.
8. Make sure that all samples are at the bottom of the tube without adherence to the walls.
9. Weigh 0.65 g if cupric sulfate pentahydrate is used.
10. Using a diagram of the digestion rack could be useful.
11. While the bottom of the tube will become very hot, there should be no observable formation of steam as you introduce the water. If steam becomes apparent, it indicates that the tube's contents were excessively hot, causing the loss of nitrogen and rendering the tube unusable.
12. Ensure that the tip of the distillation outlet tube is submerged beneath the surface of the boric acid indicator solution.
13. Tin foils have a tendency to adhere together. Avoid creating tin foil cones using multiple layers of foil, as this can lead to incomplete combustion and hasten the buildup of ash.
14. For samples with low protein content, opt for a larger sample weight. Samples abundant in moisture or fat might generate elevated pressure during combustion. Consider using a reduced sample weight or subjecting the samples to dehydration and defatting procedures.
15. Turn on the furnace of the nitrogen analyzer in advance, as it requires time to reach the operational temperature.
16. Only perform the leak test after the furnace reaches the operational temperature. During heating, the changing temperature will interfere with the pressure and thus test outcome. If a leak is detected, follow the diagnostic instructions to identify the leak location. Leaks are usually eliminated by applying a thin layer of silicone grease to the connections and O-rings.
17. When running a large number of samples, it may be necessary to empty the ash crucible while the furnace is hot. Exercise caution and wear protective gear during this procedure.
18. Verify that the oxygen lance is unobstructed. If it is blocked, clear the obstruction or replace the oxygen lance.
19. The software tracks the maintenance schedule of the instrument. Follow the manufacturer's instructions for tube replacement or refill.
20. Protein extraction conditions (e.g., solvent, sample-to-solvent ratio, extraction time, temperature, etc.) need to be optimized.
21. The actual concentration of the BSA stock solution may be slightly lower than 10 mg/mL due to impurities and insolubility of the BSA standard. Therefore, it is important to determine

the concentration of the BSA stock solution. The BSA stock solution can be aliquoted and kept frozen at −20 °C to maintain stability.

22. Ensure that the correct side of the cuvette is oriented toward the light source and is free of fingerprints.
23. It is important to use the correct light path based on the specific cuvette used.
24. Prepare the BSA standards in the same solvent used for the sample protein solution to ensure consistency in the experimental conditions.
25. Before performing measurements, make sure that the microplate reader is adequately warmed up and that the software is loaded in advance to prevent extended incubation periods.
26. Centrifuge the samples prior to reading the absorbance if the reaction mixture is not optically clear or pass the solution through a membrane filter.
27. Proteins strongly absorb ultraviolet light around 280 nm, primarily due to the presence of tryptophan and tyrosine residues. Since different proteins possess distinct aromatic amino acid compositions, this method is more suitable for purified proteins with a known molar extinction coefficient.

Acknowledgments

This project was supported by the Hispanic-Serving Institutions Education Grants Program (Grant No. 2022-77040-37620) and the From Learning to Leading: Cultivating the Next Generation of Diverse Food and Agriculture Professionals Program (Grant No. 2023-70440-40156) from the USDA National Institute of Food and Agriculture and Southern California Institute of Food Technologists Section (SCIFTS) Education and Research Grant (2017).

References

1. Willett W, Rockström J, Loken B et al (2019) Food in the Anthropocene: the EAT–lancet commission on healthy diets from sustainable food systems. Lancet 393:447–492
2. Sá AGA, Moreno YMF, Carciofi BM (2020) Plant proteins as high-quality nutritional source for human diet. Trends Food Sci Technol 97:170–184
3. Chang SKC (2010) Protein analysis. In: Nielsen SS (ed) Food analysis. Springer, New York, pp 133–146
4. Moore JC, Devries JW, Lipp M et al (2010) Total protein methods and their potential utility to reduce the risk of food protein adulteration. Compr Rev Food Sci Food Saf 9:330–357
5. Robinson HW, Hogden CG (1940) The biuret reaction in the determination of serum proteins. 1. A study of the conditions necessary for the production of a stable color which bears a quantitative relationship to the protein concentration. J Biol Chem 135:707–725

6. Lowry O, Rosebrough N, Farr AL et al (1951) Protein measurement with the Folin phenol reagent. J Biol Chem 193:265–275
7. Smith PE, Krohn RI, Hermanson G et al (1985) Measurement of protein using bicinchoninic acid. Anal Biochem 150:76–85
8. Bradford MM (1976) A rapid and sensitive method for the quantitation of microgram quantities of protein utilizing the principle of protein-dye binding. Anal Biochem 72:248–254
9. Joshi AU, Liu C, Sathe SK (2015) Functional properties of select seed flours. LWT-Food Sci Technol 60:325–331
10. Lu S, Pentico D, Castro R et al (2022) Effect of ultraviolet light exposure and compost tea supplementation on growth, antioxidant activities, and microbiome of hydroponically grown mustard greens. ACS Agric Sci Technol 2:521–533
11. Wang Y, Liu C, Ma T et al (2019) Physicochemical and functional properties of γ-aminobutyric acid-treated soy proteins. Food Chem 295:267–273
12. Liu C, Chhabra GS, Sathe SK (2015) Pistachio (*Pistacia vera* L.) detection and quantification using a murine monoclonal antibody-based direct sandwich enzyme-linked immunosorbent assay. J Agric Food Chem 63:9139–9149
13. Gupta S, Sathe SK, Su M et al (2021) Germination reduces black gram (*Vigna mungo*) and mung bean (*Vigna radiata*) vicilin immunoreactivity. LWT-Food Sci Technol 135:110217
14. Noble KA, Liu C, Sathe SK et al (2017) A cherry seed-derived spice, mahleb, is recognized by anti-almond antibodies including almond-allergic patient IgE. J Food Sci 82:1786–1791
15. Mariotti F, Tomé D, Mirand PP (2008) Converting nitrogen into protein - beyond 6.25 and Jones' factors. Crit Rev Food Sci Nutr 48:177–184
16. Jones DB (1931) Factors for converting percentages of nitrogen in foods and feeds into percentages of proteins. US Department of Agriculture
17. Tkachuk R (1969) Nitrogen-to-protein conversion factors for cereals and oilseed meals. Cereal Chem 46:419–424
18. Mosse J (1990) Nitrogen-to-protein conversion factor for ten cereals and six legumes or oilseeds. A reappraisal of its definition and determination. Variation according to species and to seed protein content. J Agric Food Chem 38:18–24
19. Sosulski FW, Imafidon GI (1990) Amino acid composition and nitrogen-to-protein conversion factors for animal and plant foods. J Agric Food Chem 38:1351–1356

Chapter 10

Free Amino and Sulfhydryl Group Content

Ruoshi Xiao, Ruijia Hu, Nandan Kumar, and Yonghui Li

Abstract

The amino (-NH_2) and sulfhydryl (-SH) groups represent two critical functional moieties frequently analyzed to elucidate protein structural alterations. Determining the abundance of free amino groups is imperative for quantifying the extent of hydrolysis, a pivotal process in protein modification. Precise control over the degree of hydrolysis is essential to achieve the desired protein properties. Sulfhydryl groups also play vital structural roles in proteins, particularly when protein characteristics change due to the rearrangement or formation of disulfide bonds resulting from various treatments. Owing to their functional significance in protein structure and their utility in protein modification processes, both amino and thiol groups are acknowledged as indispensable functional entities in numerous food proteins. In this chapter, two methods for quantifying these functional groups are outlined: the use of 2,4-dinitrobenzene-1-sulfonic acid (DNBS) for quantifying free amino groups and the use of 5,5′-dithiobis-(2-nitrobenzoic acid) (DTNB) for measuring free sulfhydryl groups. These methods are specifically designed to enhance the analysis of protein samples, contributing to a deeper understanding of their characteristics.

Key words Free amino groups, Free sulfhydryl groups, TNBS, DTNB

1 Introduction

Proteins, the fundamental building blocks of life, are intricate macromolecules composed of amino acids linked together by peptide bonds. The unique sequence and arrangement of amino acids give rise to diverse protein structures and functions essential for biological processes. Because of this complexity, the determination of free amino and sulfhydryl groups within proteins holds profound significance, resonating across various scientific disciplines and industrial applications.

It is well-established that the 20 distinct amino acids share a common molecular structure, and through a process of dehydration condensation, the peptide bonds are formed, connecting them into peptide chains. These peptide chains then fold into three-dimensional protein structures [1]. Despite the involvement of amino groups in peptide bond formation, proteins retain specific

Yonghui Li (ed.), *Plant-Based Proteins: Production, Physicochemical, Functional, and Sensory Properties*,
Methods and Protocols in Food Science, https://doi.org/10.1007/978-1-0716-4272-6_10,

amino groups capable of reacting with other molecules or undergoing modifications. However, identifying which specific amino groups remain reactive was uncertain until Skraup and Kaas [2] proposed in 1907 that the free amino groups in proteins could originate from the ε-amino group of lysine. Later, Van Slyke and Birchard [3] also suggested that the nitrogen content of free amino groups in native proteins is approximately half that of lysine, indicating that one of the two amino groups in lysine remains free within the protein molecule. Therefore, the determination of free amino groups in native proteins essentially requires quantification of the number of amino groups in lysine.

In industrial settings, proteins find extensive applications in sectors such as food processing, pharmaceuticals, and biotechnology. Protein hydrolysis, a common biochemical process, involves the enzymatic or chemical cleavage of peptide bonds within proteins, is often employed to make specific protein modifications [4]. However, precise control over the extent of protein hydrolysis is necessary depending on the intended purpose. This has led researchers to introduce a critical parameter for monitoring protein hydrolysis reactions, known as the degree of hydrolysis (DH). DH is defined as the percentage of peptide bonds that have been cleaved [5]. When peptide bonds are hydrolyzed, free amino and carboxyl groups are released. Consequently, variations in the concentration of free amino or carboxyl groups can effectively serve as indicators of the progress of protein hydrolysis [6]. Hence, the selection of a method to measure free amino groups in proteins becomes of paramount importance.

Traditionally, the ninhydrin assay was the most widely utilized method for determining the content of free amino groups in proteins. However, this method produced different results for different peptides and was not sensitive to peptides at high concentrations. These differences could be attributed to the numerous side reactions associated with this assay, which compromise result accuracy [7]. In 1960, Satake and Okuyama [8] introduced a method using TNBS (2,4-dinitrobenzene-1-sulfonic acid) for the measurement. In this method, amino groups react with TNBS reagent at pH 8 in mild conditions without interference from unnecessary side reactions. During the reaction, amino groups displace sulfite ions in the TNBS reagent, forming a trinitrophenyl derivative (TNP-NH_2). The reaction is terminated using acid before measuring, and the TNP-NH_2 exhibits the highest molar absorptivity at 340 nm [9]. Therefore, in this chapter, we adopted TNBS reagent to determine the free amino groups in samples according to the method from Nissen [4] with some modifications.

The disulfide bond in proteins is also a critical covalent structure that plays a key role in maintaining the stability of protein structures [10]. When proteins are exposed to abnormal conditions, such as changes in temperature, pH levels, or the introduction of reducing agents, they may partially unfold, leading to a phenomenon known as protein denaturation [11]. This denaturation can occur reversibly or irreversibly. During the unfolding process, disulfide bonds in the protein may be broken or rearranged. However, in most cases, these bonds tend to break, resulting in the formation of free thiol groups [10]. Thus, thiol groups in proteins have been widely used as a measurement index in various studies investigating changes in protein structure under different experimental conditions.

There was a large amount of research conducted in previous studies. Shimade et al. [12] investigated the alterations in sulfhydryl groups and disulfide bonds within soybean protein isolate following heat-induced gelation. Taylor et al. [13] examined the active thiol content and total thiol content in heated skim milk as part of their research on skim milk's antioxidant activity, analyzing the role of thiol in the milk's antioxidant properties. Additionally, Plancken et al. [14] focused on studying the impact of high-temperature and high-pressure conditions on the sulfhydryl content of egg white protein. The measurement of thiol groups in proteins has become increasingly crucial as more researchers employ it to assess the reaction effects, emphasizing its significance in scientific research.

In 1958, researcher Ellman [15] synthesized a novel reagent, 5,5′-dithiobis-(2-nitrobenzoic acid) (DTNB), which is now commercially available. Later scholars widely used this DTNB reagent and named it Ellman's reagent. DTNB is an aromatic disulfide compound that reacts with thiol groups in proteins, forming a mixed disulfide and 2-nitrothiobenzoic acid (TNB) in a 1:1 molar ratio [16]. The nitromercaptobenzoic acid anion exhibits an intense yellow color and has an absorbance at 412 nm, making it useful for measuring thiol groups [17]. Therefore, in this experiment, the DTNB reagent was used to measure the content of thiol groups in proteins. The specific experiment combined the method of Beveridge [18] and Tang [19] with some modifications to achieve the desired results.

2 Materials

All the solutions are prepared using analytical grade reagents and distilled water. These solutions can be stored at room temperature (unless indicated otherwise). Diligently follow all waste disposal regulations when disposing of waste materials. This experiment aims to determine the free amino groups in proteins using the 2,4,6-trinitrobenzene sulfonic acid (TNBS) method, and determine the free and total sulfhydryl groups in proteins using 5,5'-dithiobis-(2-nitrobenzoic acid) (DTNB) method.

2.1 Materials for Determination of Free Amino Groups

1. 1% (w/v) Sodium dodecyl sulfate (SDS).
2. 0.2125 M phosphate buffer.
3. 0.1% (v/v) 2,4,6-trinitrobenzene sulfonic acid (TNBS).
4. 0.1 N HCl.
5. 0–2.4 mM L-leucine solution.
6. Vortex.
7. Water bath shaker.
8. Avanti® J-E high-speed centrifuge (Beckman Coulter Inc., Brea, CA, USA).
9. Double-beam spectrophotometer (VWR, UV-6300PC).
10. Flasks, volumetric flask, and centrifuge tubes for reagent preparation.

2.2 Materials for Determination of Free and Total Sulfhydryl Groups

1. Tris-Gly buffer: 0.086 M Tris(hydroxymethyl)aminomethane (Tris), 0.09 M glycine, 0.004 M ethylenediaminetetraacetic acid (EDTA), and pH 8.0.
2. Tris-Gly buffer containing 8 M urea: 0.086 M Tris, 0.09 M glycine, 0.004 M EDTA, and 8 M urea.
3. Ellman's reagent: Dissolve 4 mg of 5,5′-dithio-bis-2-nitrobenzoic acid in Tris-Gly buffer to make a 4 mg/mL solution. Cover the solution with aluminum foil (*see* **Note 8**).
4. β-Mercaptoethanol (*see* **Note 9**).
5. 12% (w/v) Trichloroacetic acid (TCA): Dissolve TCA in deionized water (*see* **Note 10**).
6. Vortex mixer.
7. Water bath shaker.
8. Platform shaker.
9. Avanti® J-E high-speed centrifuge (Beckman Coulter Inc., Brea, CA, USA).
10. Double-beam spectrophotometer (VWR, UV-6300PC).
11. Flasks, volumetric flasks, and centrifuge tubes for reagent preparation and reaction.

3 Methods

3.1 Method for Determination of Free Amino Groups

3.1.1 Preparation of Reagents

1. *1% (w/v) sodium dodecyl sulfate (SDS):* For 100 mL of 1% (w/v) SDS solution, dissolve 1 g of SDS in deionized water in a 100 mL volumetric flask to make a total 100 mL solution.
2. *0.2125 M phosphate buffer (pH 8.2):* Adding 4.5 mL of 0.2125 M sodium phosphate monobasic anhydrous (NaH_2PO_4) to 100 mL of 0.2125 M sodium phosphate dibasic

Table 1
Preparation of standards for determination of free amino groups

Tubes	2.4 mM of L-leucine standard (mL)	1% of SDS solution (mL)	Resulting L-leucine concentration (mM)
1	2.5	0.5	2.0
2	2.0	1.0	1.6
3	1.5	1.5	1.2
4	1.0	2.0	0.8
5	0.5	2.5	0.4
6	0.0	3.0	0.0

anhydrous (Na_2HPO_4) can get the final mixed phosphate solution with pH 8.20 ± 0.02 (*see* **Note 1**).

3. *0.1% (v/v) 2,4,6-trinitrobenzene sulfonic acid (TNBS):* The purchased TNBS is 5% (v/v) in methanol. Prepare a 0.1% (v/v) TNBS solution by diluting the 5% (v/v) TNBS solution with deionized water (*see* **Notes 2** and **3**).

3.1.2 Preparation of Samples

Add 2 mg of sample (*see* **Note 4**) and 5 mL of 1% (w/v) SDS solution to a 50 mL centrifuge tube to obtain a sample solution at a concentration of 4 mg/mL (*see* **Note 5**). Mix the sample solution by vortex for 5 s at 3000 rpm.

3.1.3 Preparation of Standards

1. Prepare a fresh 2.4 mM L-leucine standard by dissolving 15.74 mg of L-leucine in 50 mL of 1% (w/v) SDS solution and shaking at 150 rpm for 15 min using a platform shaker.
2. Prepare a series of remaining standards according to Table 1, by diluting the 2.4 mM of L-leucine standard with 1% (w/v) SDS solution (*see* **Note 6**).

3.1.4 Procedures

1. To the 50 mL tube containing 0.5 mL of the sample solution, add 4 mL of phosphate buffer and 4 mL of 0.1% TNBS. Vortex the mixture for 5 s at 3000 rpm.
2. Shake the mixture in a 50 °C water bath at 200 rpm for 1 h (in dark).
3. After the 1-h reaction, add 8 mL of 0.1 N HCl to the tube, and vortex for 10 s at 3000 rpm.
4. Store the centrifuge tube in the dark at room temperature for 30 min, and then centrifuge for 10 min at 8000 g under 20 °C (*see* **Note 7**).
5. Measure the absorbance of the supernatant at 340 nm using a spectrophotometer.

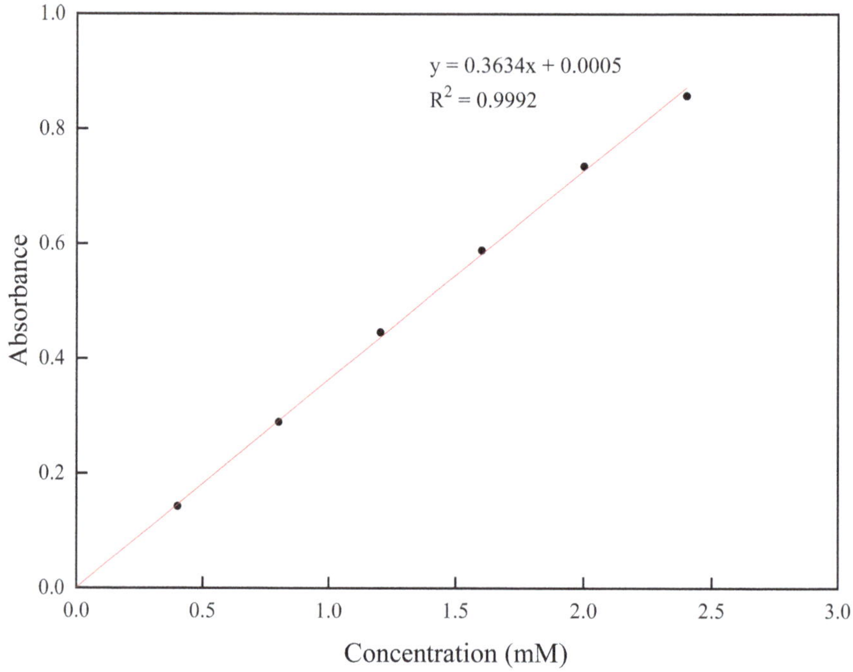

Fig. 1 Standard curve for the determination of free amino groups using the TNBS method

Table 2
Free amino group content (mmol/g protein) of soy protein isolate (SPI), pea protein isolate (PPI), and wheat gluten

	Free amino group (mmol/g protein)
SPI	9.48 ± 0.04
PPI	9.90 ± 0.05
Gluten	5.01 ± 0.03

6. Measure the L-leucine solutions with different concentrations (0–2.4 mM) in the same way as above to establish the standard curve (Fig. 1).

3.1.5 Example of Results

Soy protein isolate (SPI), pea protein isolate (PPI), and wheat gluten were used as examples, the standard curve was shown in Fig. 1, and the free amino group content of testing samples was shown in Table 2.

3.2 Method for Determination of Free and Total Sulfhydryl Groups

3.2.1 Sample Preparation

1. Add 10 mL of Tris-Gly buffer containing 8 M of urea to a 15 mL centrifuge tube.
2. Weigh 75 mg of sample (*see* **Notes 11**, **12**, and **13**), and transfer it to the tube from **step 1**.
3. Vortex the mixture for 10 s at 3000 rpm and then shake it overnight at 200 rpm.

3.2.2 Determination of Free Sulfhydryl Groups

1. Transfer 1 mL of sample suspension to a 15 mL centrifuge tube (*see* **Note 14**).
2. Add 4 mL of Tris-Gly buffer and 0.05 mL of Ellman's reagent to the sample suspension, and vortex the mixture for 5 s.
3. After incubating the mixture in the dark for 15 min, centrifuge it at 8000 g under 4 °C for 10 min.
4. Measure the absorbance of the supernatant at 412 nm using a spectrophotometer.

3.2.3 Determination of Total Sulfhydryl Groups

1. Transfer 1 mL of sample suspension to a 50 mL centrifuge tube (*see* **Note 14**).
2. Add 4 mL of Tris-Gly buffer and 0.05 mL of β-mercaptoethanol to the sample suspension, and vortex the mixture for 10 s at 3000 rpm.
3. Shake the mixture at 200 rpm for 1 h.
4. After shaking the mixture at 200 rpm for 1 h, add 10 mL of 12% (w/v) TCA, and vortex the mixture for 5 s at 3000 rpm. Then, shake the mixture at 200 rpm for another 1 h.
5. Centrifuge the mixture at 8000 g under 4 °C for 10 min to remove the supernatant.
6. Wash the precipitate twice with 5 mL of 12% (w/v) TCA.
7. Add 10 mL of Tris-Gly buffer to the precipitate, and vortex the mixture for 5 s at 3000 rpm. Then, shake the mixture at 200 rpm for 15 min.
8. Transfer 4 mL of the sample solution from **step 7** to a 5 mL centrifuge tube, and add 0.04 mL of Ellman's reagent to the sample solution (*see* **Note 14**).
9. After incubating the mixture in the dark for 15 min, centrifuge it at 8000 g under 4 °C for 10 min.
10. Measure the absorbance of the supernatant at 412 nm using the spectrophotometer.

3.2.4 Calculations of Free and Total Sulfhydryl Groups

Calculation of the concentration of sulfhydryl groups is as follows:

$$\mu\text{mol SH/g} = \frac{73.53 \times A_{412} \times D}{C} \quad (1)$$

$$\text{Disulfide bonds content } (\mu\text{mol/g}) = (\text{Total SH content} - \text{Free SH content})/2 \quad (2)$$

where A_{412} is the absorbance at 412 nm (*see* **Note 15**), C is the sample concentration (mg/mL), D is the dilution factor (5 is used for free sulfhydryl content, and 10 is used for total sulfhydryl content), 73.53 is derived from $10^6/(1.36 \times 10^4)$, 1.36×10^4 is

Table 3
Free SH group (μmol/g protein), total SH group (μmol/g protein), and SS bond contents (μmol/g protein) of soy protein isolate (SPI), pea protein isolate (PPI), and wheat gluten

	Free SH group (μmol/g protein)	Total SH group (μmol/g protein)	SS bond (μmol/g protein)
SPI	4.74 ± 0.15	80.88 ± 1.09	38.07 ± 0.51
PPI	6.26 ± 0.17	63.92 ± 2.32	28.83 ± 1.24
Gluten—Commercial	2.86 ± 0.23	94.71 ± 2.41	45.92 ± 1.09
Gluten—Lab	2.83 ± 0.28	100.75 ± 2.17	48.96 ± 1.11

the molar absorptivity of Ellman' reagent, and 10^6 is for conversions from the molar basis to the μmol/mL basis and from mg solids to g solids.

3.2.5 Example of Results

Soy protein isolate, pea protein isolate, commercial wheat gluten, and wheat gluten from our lab were used as examples. The free SH group content, total SH group content, and SS bond content of testing samples were shown in Table 3.

4 Notes

4.1 Notes for Determination of Free Amino Groups

1. Check the pH of 0.2125 M phosphate buffer after preparation; the pH of the buffer should be 8.20 ± 0.02.
2. The TNBS reagent must be prepared immediately before use and prepared in dark.
3. The TNBS reagent is unstable and toxic and must be handled carefully.
4. Make sure samples are fine powder before use.
5. The sample concentration could be adjusted according to the properties of the user's samples.
6. The diluted standard solutions cannot be stored overnight.
7. Following the addition of TNBS reagent to the sample solution, an orange color will progressively develop and intensify over time. In contrast, the blank solution will transition to a pale-yellow color. After acidification, the sample solution will transition from orange to a stable yellow color, particularly stable in a dark room environment. Simultaneously, the blank solution will become colorless.

4.2 Notes for Determination of Total and Free Sulfhydryl Groups

8. Prepare fresh Ellman' reagent each time before test.
9. β-Mercaptoethanol is a volatile liquid with a strong pungent odor; please add it in hood.
10. TCA is very acidic, so be careful when preparing it.
11. Make sure samples are fine powder before use.
12. The precise concentration of each reagent in the buffer is very important. Errors in these concentrations must be avoided as they can significantly affect the results.
13. Make sure the sample is dispersed well in the buffer after vortex; otherwise, vortex longer to get a homogeneous suspension without any clumps.
14. Vortex the sample solution to get a uniform situation before each transformation or utilization.
15. $A_{412} = A_1 – A_2 – A_3$: A_1, the sample absorbance measured at 412 nm; A_2, the absorbance of reagent blank, which is used in which the same volume of standard buffer instead of protein solution; and A_3, the absorbance of protein blank, which is used in which the same volume of standard buffer replaced Ellman's reagent solution.

Acknowledgments

This is contribution no. 24-262-B from the Kansas Agricultural Experimental Station. This work was supported in part by the USDA National Institute of Food and Agriculture Hatch project (Grant Accession No. 7003330).

References

1. Branden CI, Tooze J (1998) Introduction to protein structure. Garland Science, New York
2. Skraup ZH, Kaas K (1907) Ueber die Einwirkung von salpetriger Säure auf Ovalbumin. Justus Liebigs Ann Chem 351:379–389. https://doi.org/10.1002/jlac.19073510130
3. Van Slyke DD, Birchard FJ (1914) The nature of the free amino groups in proteins. Proc Soc Exp Bio Med 16:539–547. https://doi.org/10.1016/S0021-9258(18)88462-9
4. Adler-Nissen J (1979) Determination of the degree of hydrolysis of food protein hydrolysates by trinitrobenzenesulfonic acid. J Agric Food Chem 27:1256–1262. https://doi.org/10.1021/jf60226a042
5. Nielsen PM, Petersen D, Dambmann C (2001) Improved method for determining food protein degree of hydrolysis. J Food Sci 66:642–646. https://doi.org/10.1111/j.1365-2621.2001.tb04614.x
6. Navarrete del Toro MA, García-Carreño FL (2003) Evaluation of the progress of protein hydrolysis. Curr Protoc Food Anal Chem 10: B2.2.1–B2.2.14. https://doi.org/10.1002/0471142913.fab0202s10
7. Satake K, Okuyama T, Ohashi M et al (1960) The spectrophotometric determination of amine, amino acid and peptide with 2,4,6- trinitrobenzene 1-sulfonic acid. J Biochem 47: 654–660. https://doi.org/10.1093/oxfordjournals.jbchem.a127107
8. Okuyama T, Satake K (1960) On the preparation and properties of 2, 4, 6-trinitrophenyl-amino acids and-peptides. J Biochem 47:454–466. https://doi.org/10.1093/oxfordjournals.jbchem.a127083
9. Fields R (1972) The rapid determination of amino groups with TNBS. In: Methods in enzymology. Elsevier, pp 464–468

10. Wedemeyer WJ, Welker E, Narayan M et al (2000) Disulfide bonds and protein folding. Biochemist 39:4207–4216. https://doi.org/10.1021/bi992922o
11. Kosuri P, Alegre-Cebollada J, Feng J et al (2012) Protein folding drives disulfide formation. Cell 151:794–806. https://doi.org/10.1016/j.cell.2012.09.036
12. Shimada K, Cheftel JC (1988) Determination of sulfhydryl groups and disulfide bonds in heat-induced gels of soy protein isolate. J Agric Food Chem 36:147–153. https://doi.org/10.1021/jf00079a038
13. Taylor MJ, Richardson T (1980) Antioxidant activity of skim milk: effect of heat and resultant sulfhydryl groups. J Dairy Sci 63:1783–1795. https://doi.org/10.3168/jds.S0022-0302(80)83140-7
14. Van Der Plancken I, Van Loey A, Hendrickx MEG (2005) Changes in sulfhydryl content of egg white proteins due to heat and pressure treatment. J Agric Food Chem 53:5726–5733. https://doi.org/10.1021/jf050289+
15. Ellman GL (1959) Tissue sulfhydryl groups. Arch Biochem Biophys 82:70–77. https://doi.org/10.1016/0003-9861(59)90090-6
16. Habeeb AF (1972) Reaction of protein sulfhydryl groups with Ellman's reagent. In: Methods in enzymology. Academic Press, pp 457–464
17. Sedlak J, Lindsay RH (1968) Estimation of total, protein-bound, and nonprotein sulfhydryl groups in tissue with Ellman's reagent. Anal Biochem 25:192–205. https://doi.org/10.1016/0003-2697(68)90092-4
18. Beveridge T, Toma SJ, Nakai S (1974) Determination of SH- and SS-groups in some food proteins using Ellman's reagent. J Food Sci 39:49–51. https://doi.org/10.1111/j.1365-2621.1974.tb00984.x
19. Tang C-H, Sun X (2010) Physicochemical and structural properties of 8S and/or 11S globulins from mungbean [*Vigna radiata* (L.) Wilczek] with various polypeptide constituents. J Agric Food Chem 58:6395–6402. https://doi.org/10.1021/jf904254f

Chapter 11

Amino Acid Composition

Liuyi Chang and Bingcan Chen

Abstract

Amino acids are crucial constituents of plant proteins as they serve as the fundamental building blocks. Obtaining the amino acid profile is essential for evaluating the nutritional value of plant protein. This chapter outlines a clear procedure for accurately quantifying the absolute content of 17 amino acids in plant protein. The process involves hydrolyzing the solid protein sample with a hydrochloric acid solution, followed by derivatizing the amino acids using 6-aminoquinolyl-N-hydroxysuccinimidyl carbamate. The derivatized amino acids are then separated using reversed-phase high-performance liquid chromatography with an ultraviolet detector. The absolute quantity of each amino acid is calculated based on the internal standard and amino acid standard curves. This procedure utilizes common instruments and chemicals, and it involves a straightforward operation, providing high accuracy and good repeatability. Detailed explanations of experimental steps and calculation methods enable easy replication of this procedure by other researchers.

Key words Hydrolyzed amino acids, Amino acid profile, Internal standard, Absolute quantity, Derivation

1 Introduction

Amino acids are fundamental molecules of plant protein. They are organic compounds consisting of an amino group, a carboxyl group, and a unique side chain. In total, there are 20 different amino acids involved in protein synthesis, with nine of them being essential amino acids that human beings must obtain through diet [1]. These amino acids are assembled and connected through peptide bonds to form polypeptides, which serve as the primary structure of plant protein [2].

In recent years, plant protein has gained popularity as a competitive alternative to animal-derived protein, thanks to its affordability, high protein content, environmental sustainability, and valuable nutritional profiles [3, 4]. However, the nutritional aspect of plant protein warrants careful consideration, as its amino acid composition differs from that of animal proteins, affecting its

Yonghui Li (ed.), *Plant-Based Proteins: Production, Physicochemical, Functional, and Sensory Properties*, Methods and Protocols in Food Science, https://doi.org/10.1007/978-1-0716-4272-6_11,

overall nutritional value [5]. For example, proteins derived from peas, lupins, and soy are limited in methionine and cysteine, whereas other plant proteins found in corn, rice, and wheat may be limited in lysine [5–7]. To assess protein quality, the Food and Agriculture Organization (FAO) has developed the indispensable amino acid content (IAA) index, which compares essential amino acid content to a reference pattern [8]. Additionally, various other indexes, such as predicted biological value (p-BV), essential amino acid index (EAAI), and digestible indispensable amino acid score (DIAAS), have been created to evaluate the nutritional quality of plant proteins from different perspectives [9–12].

Hence, obtaining the amino acid composition profile is crucial in assessing the nutritionally quality of plant protein. All the above-mentioned indexes rely on the amino acid composition profile for their calculations [5, 9, 10, 13, 14]. Therefore, the objective of this book chapter is to provide a thorough and detailed description of an analytical procedure for the quantitative determination of hydrolyzed amino acids in plant protein samples. The procedure encompasses three essential stages: protein hydrolysis, amino acid derivatization, and high-performance liquid chromatography (HPLC) analysis. Firstly, the protein is hydrolyzed in 6 N HCl at 105 °C for 24 h. Next, the amino acids are derivatized using 6-aminoquinolyl-N-hydroxysuccinimidyl carbamate (AQC). The derivatized amino acids are subsequently separated using reversed-phase HPLC with ultraviolet (UV) detection at 260 nm.

It is important to note that in this method, glutamine and asparagine are converted to glutamic acid and aspartic acid, respectively. Therefore, the glutamine value represents the combined values of glutamine and glutamic acid, while the asparagine value represents the combined values of asparagine and aspartic acid. Additionally, since tryptophan undergoes hydrolysis during the acid hydrolysis step, this method enables the analysis of 17 amino acids, including 9 essential amino acids and 8 non-essential amino acids [2, 15].

2 Materials

All solutions should be prepared using deionized water and/or analytical grade reagents, ensuring adherence to waste disposal regulations when disposing of waste materials. All reagents are store at room temperature unless specified otherwise.

2.1 Protein Hydrolyzation

1. Hydrochloric acid solution (6 N): Mix 497 mL of HCl (37%) with 503 mL of deionized water.

2. Hydrochloric acid solution (0.1 N): Measure out 8.3 mL of HCl (37%), and slowly pour it into 991.7 mL of deionized water in a beaker.
3. Phenol: Heat the solid phenol to 60 °C until it liquefies.
4. Internal Standard A—25 mmol/L aminobutyric acid (AABA) solution: Measure 64.5 mg AABA and bring it to 25 mL 0.1 N HCl.
5. Internal Standard B—2.5 mmol/L AABA solution: Ten times dilute the Internal Standard A (25 mmol/L AABA solution) with 0.1 N HCl.

2.2 Amino Acid Standard Curve

1. Amino Acid Standard A—2.5 μmol/mL: Buy the Amino Acid Standard H from Thermo Fisher Scientific (Catalog #: 20088), which contains 2.5 μmol/mL of 17 amino acids, except cystine with a concentration of 1.25 μmol/mL.
2. Amino Acid Standard B—0.5 μmol/mL: Five times dilute the Amino Acid Standard A (2.5 μmol/mL amino acid solution except cystine) with 0.1 N HCl.
3. Amino Acid Standard C—0.05 μmol/mL: Ten times dilute the Amino Acid Standard B (0.5 μmol/mL amino acid solution except cystine) with 0.1 N HCl.
4. Prepare a series of concentrations of the mixed amino acid standard to generate the standard curve, as outlined in Table 1 (*see* **Notes 1** and **2**).
5. The AABA concentration in each standard curve sample is 250 μmol/L.

2.3 Derivation

1. AccQ-Tag Ultra Derivatization Kit (SKU#: 186003836): The commercial derivation kit supplied by Waters includes AccQ-Fluor Reagent Powder, AccQ-Fluor Reagent Diluent, and AccQ-Fluor Reagent Borate Buffer.
2. AccQ-Fluor Reagent: Add 3 mL of AccQ-Fluor Reagent Diluent in the small glass bottle containing AccQ-Fluor Reagent Powder. Seal the bottle with the original cap, and then vortex for 10 s. Place the bottle in a water bath at 55 °C, and heat for 5 min or until the powder is fully dissolved. The prepared AccQ-Fluor Reagent can be stored in a refrigerator at 4 °C for up to 2 weeks.

2.4 HPLC Mobile Phases

1. Eluent A: Ten times dilute the Waters AccQ-Tag Eluent A Concentrate (SKU#: WAT052890) with deionized water. The Eluent A should be stored in a refrigerator at 4 °C before using.
2. Eluent B: Mix 999 mL of HPLC-grade acetonitrile with 1 mL of HPLC-grade formic acid.

Table 1
Volume (μL) of the amino acid standards, internal standard, and 0.1 N HCl for the elaboration of the different points of the standard curve (final volume 500 μL)

Standard curve sample number	Amino acid standard A	Amino acid standard B	Amino acid standard C	Internal standard B	HCl solution of 0.1 N	The concentrations of each amino acid except cystine (μmol/L)	The concentrations of cystine (μmol/L)
1	0	0	50	50	400	5	2.5
2	0	0	100	50	350	10	5
3	0	20	0	50	430	20	10
4	0	30	0	50	420	30	15
5	0	50	0	50	400	50	25
6	0	100	0	50	350	100	50
7	40	0	0	50	410	200	100
8	60	0	0	50	390	300	150
9	80	0	0	50	370	400	200
10	120	0	0	50	330	600	300
11	160	0	0	50	290	800	400
12	200	0	0	50	250	1000	500

3 Methods

3.1 Protein Hydrolyzation

1. Weigh 20 mg of plant protein powder sample into a 20 mL glass tube, ensuring that each sample contains 10–20 mg of protein. Take note of the protein weight, and determine the protein content using the Dumas combustion method. Then, add 10 mL of 6 N HCl to the tube. Next, add 200 μL of internal standard A and 100 μL of liquid phenol to the tube. Purge the tube with nitrogen and immediately seal it tightly (*see* **Note 3**).
2. Place the tube in a 105 °C oven, and allow it to undergo hydrolysis for 24 h.
3. Once the hydrolyzed solution is cooled to room temperature, filter 2 mL of the solution through a 0.45 μm Cellulose Acetate (CA) Syringe Filter (*see* **Note 4**).

4. Transfer 500 μL of the filtered solution to an Eppendorf tube, and place it in a freeze-dryer for approximately 3 h until complete dryness (*see* **Note 5**).
5. Add 1 mL of 0.1 N HCl to the Eppendorf tube, and vortex until fully dissolved (*see* **Note 6**).

3.2 Amino Acid Derivation

1. Ten μL of the standard curve solution or hydrolyzed sample solution is added into 70 μL of AccQ-Fluor Reagent Borate Buffer and 20 μL of AccQ-Fluor Reagent in a glass HPLC Vail Insert (Agilent, ref. 5183-2085). Mix them immediately for 10 s with vortex (*see* **Note 7**).
2. Add 1.0 mL of deionized water to the brown HPLC Screw Top Vial (Agilent, ref. #: 5183-2085). Insert the glass HPLC Vial Insert into the HPLC Screw Top Vial, and securely tighten the cap.
3. Place the HPLC Screw Top Vial in a water bath set at 70 °C, and allow it to heat for 30 min (*see* **Note 8**).

3.3 HPLC

1. Inject either a prepared sample or a standard of 1 μL into the AccQ-Tag amino acid column Nova-Pak C18, 4 μm (150 × 3.9 mm) from Waters (SKU #: WAT086344RF). Ensure that the column is thermostatically controlled at 37 ° C (*see* **Note 9**).
2. Run each sample for a duration of 60 min, following the HPLC settings specified in Table 2. The settings include the flow rate and solvent gradient to be used.

3.4 Data Analysis

1. The peak identification and the retention time are shown in Fig. 1 (*see* **Note 11**).
2. Integrate these 18 identified amino acids peaks using Agilent OpenLab Software, and get the absolute peak area (*see* **Note 12**).
3. Standard curve calculation.

   The response concentration of each amino acid in the amino acid standard can be calculated by Eq. 1, where Response C_{AA} (μmol/L) represents the response concentration of each individual amino acid (μmol/L), A_{AA} is the peak area of each individual amino acid, A_{AABA} refers to the peak area of AABA (peak 10), and C_{AABA} is the concentration of AABA (250 μmol/L):

$$\text{Response } C_{AA} = \frac{A_{AA}}{A_{AABA}} \times C_{AABA} \tag{1}$$

   Once the series of response concentrations for each individual amino acid is obtained, construct the standard curves for

Table 2
The HPLC conditions used for amino acid analysis

Parameters	Values			
HPLC system	Agilent 1260 HPLC system			
Injection volume	1 μL			
Oven temperature	37 °C			
Detector	UV 260 nm			
Gradient elution	Time	Eluent A (%)	Eluent B (%)	Flow rate (mL/min)
	0.0	99.9	0.1	0.4
	17.0	95.0	5.0	–[a]
	24.0	91.0	9.0	–
	34.0	89.2	10.8	–
	35.0	89.2	10.8	0.1
	46.0	86.2	13.8	0.1
	46.1	852	14.8	–
	56.0	61.0	39.0	–
	58.0	99.9	0.1	0.4

[a] "–" represents "not setting" (*see* **Note 10**)

the 17 amino acids. The X-axis represents the real amino acid concentration (μmol/L), while the Y-axis represents the response amino acid concentration (μmol/L) (*see* **Note 13**).

4. The calculation of each amino acid concentration in a liquid hydrolyzed sample solution.

   The response concentration of each amino acid in the hydrolyzed sample solution can be calculated by Eq. 2, where Response C_{AA} (μmol/L) represents the response concentration of each individual amino acid, A_{AA} is the peak area of each individual amino acid, A_{AABA} refers to the peak area of AABA (peak 10), and C_{AABA} is the concentration of AABA (250 μmol/L):

$$\text{Response } C_{AA} = \frac{A_{AA}}{A_{AABA}} \times C_{AABA} \qquad (2)$$

   Once the response concentration for each amino acid is obtained, refer to the standard curve to calculate the actual concentration of each individual amino acid.

5. The calculation of each amino acid concentration in the original protein sample.

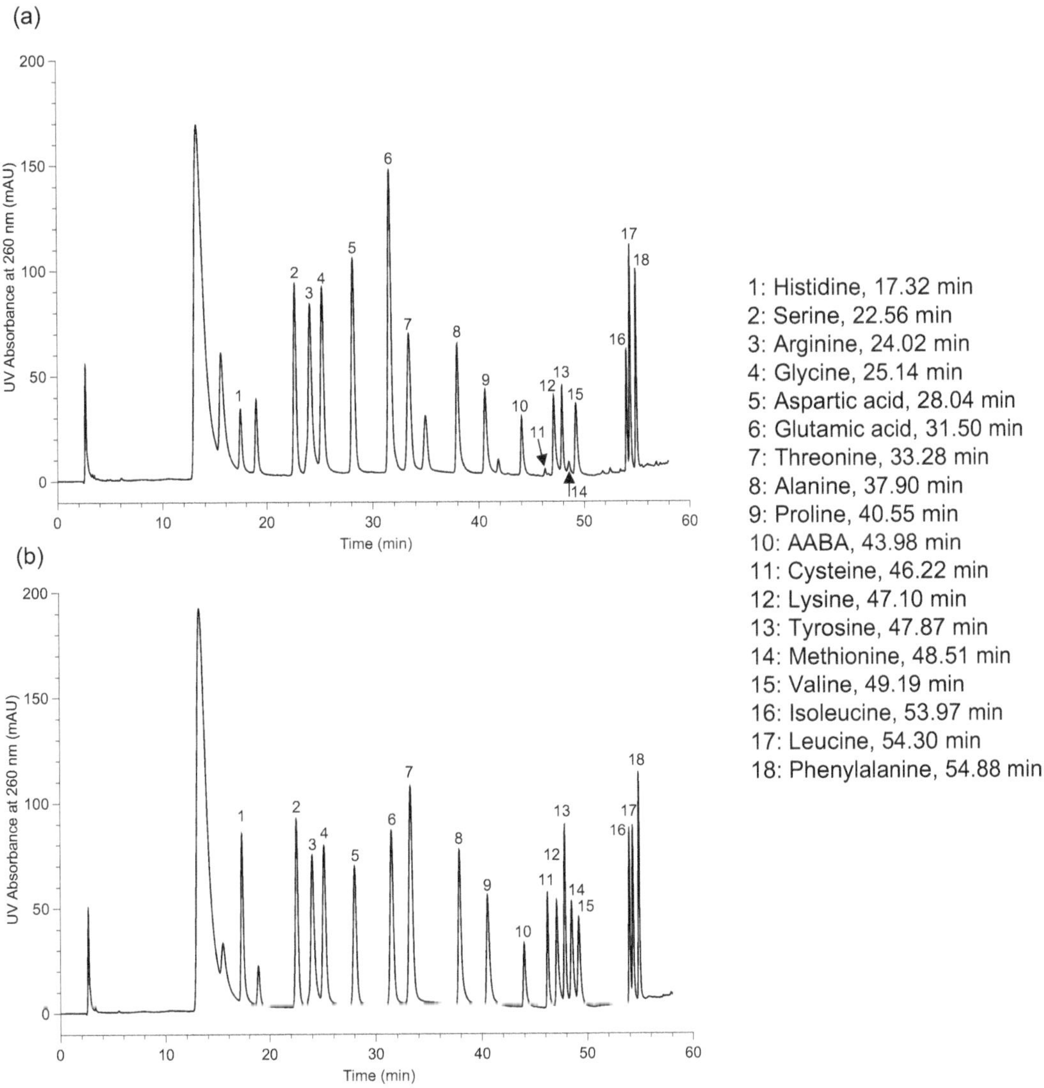

Fig. 1 The peak identification of hydrolyzed lentil protein sample (**a**) and amino acid standard (**b**)

Use Eq. 3 to determine the amino acid concentration ($C_{\mathrm{AA\ original}}$) (g/100 g) in the original protein sample, where $C_{\mathrm{AA\ H}}$ (μmol/L) is the actual amino acid concentration in the hydrolyzed liquid solution, Mw_{AA} (g/μmol) is the weight of the individual amino acid, V_{H} (L) is the volume of the hydrolyzed liquid solution, W_{protein} (g) is the protein weight used for hydrolysis, and C_{protein} (%) is the protein content tested by Dumas combustion:

$$C_{\mathrm{AA\ original}} = \frac{C_{\mathrm{AA\ H}} \times Mw_{\mathrm{AA}} \times V_{\mathrm{H}} \times 2}{W_{\mathrm{protein}} \times C_{\mathrm{protein}} \times 100} \tag{3}$$

4 Notes

1. Prepare 12 gradient concentrations of mixed amino acids for the standard curve, as each amino acid may have its optimal concentration range. Choose at least five consecutive points for each amino acid to create the standard curve, based on the actual conditions.
2. The amino acid standard curve can be stored at −20 °C for future use.
3. The tube must be completely sealed and free of leaks when exposed to a temperature of 105 °C. Alternatively, a glass ampoule can be used as a substitute for the tube.
4. Ensure the use of a CA Syringe Filter in this step. If a Nylon Syringe Filter is used, precipitation may occur after filtering.
5. The solution contains 6.0 N HCl, which means it cannot be frozen at −20 °C. Therefore, simply put the liquid solution in the freeze dryer without the need for prior freezing.
6. This step ensures that the AABA concentrations in the sample solution and standard solutions are maintained at 250 μmol/mL. Avoid diluting the solution, as it will affect the AABA concentration. If adjustments to the amino acid concentration in the sample solution are needed, modify the protein content used for hydrolyzation, and ensure that the AABA concentration aligns with that of the standard curve solutions.
7. To prevent sample contamination, it is recommended to add AccQ-Fluor Reagent Borate Buffer and AccQ-Fluor Reagent, followed by sample solution.
8. Using a foam float is recommended to ensure even heating of the HPLC Screw Top Vial.
9. Three blank samples (acetonitrile) should be injected before the standard and amino acid samples. After running the replicates of each sample, an additional blank sample should be injected before proceeding to the next sample.
10. It is recommended to ensure a smooth gradient change during the HPLC run. For example, if the flow rate is not adjusted at 17.0 min, 24.0 min, and 34.0 min, the flow rate will automatically change from 0.4 mL/min to 0.1 mL/min between 0.0 min and 35.0 min. This gradual change in flow rate has been found to improve the separation efficiency of amino acids compared to abrupt flow rate changes.
11. There may be additional noise peaks in the samples (Fig. 1), particularly in the plant protein sample (Fig. 1a), that are not identified as amino acids due to the complex composition of the samples.

12. Avoid integrating these noise peaks, and focus only on the peaks corresponding to amino acids.
13. Ensure that the coefficient of determination (R^2) falls within the range of 0.99 to 1.00.

Acknowledgments

This work is supported by ND Specialty Crop Block Grant Program.

References

1. Dolowy M, Pyka A (2014) Application of TLC, HPLC and GC methods to the study of amino acid and peptide enantiomers: a review. Biomed Chromatogr 28:84–101
2. Chang L, Lan Y, Bandillo N et al (2022) Plant proteins from green pea and chickpea: extraction, fractionation, structural characterization and functional properties. Food Hydrocoll 123:107165. https://doi.org/10.1016/j.foodhyd.2021.107165
3. Iqbal A, Khalil IA, Ateeq N, Sayyar Khan M (2006) Nutritional quality of important food legumes. Food Chem 97:331–335. https://doi.org/10.1016/j.foodchem.2005.05.011
4. Hall C, Hillen C, Robinson JG (2017) Composition, nutritional value, and health benefits of pulses. Cereal Chem 94:11–31. https://doi.org/10.1094/CCHEM-03-16-0069-FI
5. Herreman L, Nommensen P, Pennings B, Laus MC (2020) Comprehensive overview of the quality of plant- and animal-sourced proteins based on the digestible indispensable amino acid score. Food Sci Nutr 8:5379–5391. https://doi.org/10.1002/fsn3.1809
6. Cervantes-Pahm SK, Liu Y, Stein HH (2014) Digestible indispensable amino acid score and digestible amino acids in eight cereal grains. Br J Nutr 111:1663–1672. https://doi.org/10.1017/S0007114513004273
7. Boye J, Wijesinha-Bettoni R, Burlingame B (2012) Protein quality evaluation twenty years after the introduction of the protein digestibility corrected amino acid score method. Br J Nutr 108:S183–S211. https://doi.org/10.1017/S0007114512002309
8. FAO/WHO (2013) Dietary protein quality evaluation in human nutrition Report of an FAO Expert Consultation
9. Shen P, Gao Z, Xu M et al (2020) Physicochemical and structural properties of proteins extracted from dehulled industrial hempseeds: role of defatting process and precipitation pH. Food Hydrocoll 108:106065. https://doi.org/10.1016/j.foodhyd.2020.106065
10. Chang L, Gu Z, Bandillo N et al (2023) Fractionation, structural characteristics, functionality, aromatic profile, and in vitro digestibility of lentil (*Lens culinaris*) proteins. ACS Food Sci Technol 3:478. https://doi.org/10.1021/acsfoodscitech.2c00429
11. Mathai JK, Liu Y, Stein HH (2017) Values for digestible indispensable amino acid scores (DIAAS) for some dairy and plant proteins may better describe protein quality than values calculated using the concept for protein digestibility-corrected amino acid scores (PDCAAS). Br J Nutr 117:490–499. https://doi.org/10.1017/S0007114517000125
12. Fanelli NS, Bailey HM, Guardiola LV, Stein HH (2021) Values for digestible indispensable amino acid score (DIAAS) determined in pigs are greater for milk than for breakfast cereals, but DIAAS values for individual ingredients are additive in combined meals. J Nutr 151:540–547. https://doi.org/10.1093/jn/nxaa398
13. Rutherfurd SM, Fanning AC, Miller BJ, Moughan PJ (2015) Protein digestibility-corrected amino acid scores and digestible indispensable amino acid scores differentially describe protein quality in growing male rats. J Nutr 145:372–379. https://doi.org/10.3945/jn.114.195438
14. Yang F, Huang X, Zhang C et al (2018) Amino acid composition and nutritional value evaluation of Chinese chestnut (: Castanea mollissima Blume) and its protein subunit. RSC Adv 8: 2653–2659. https://doi.org/10.1039/c7ra13007d
15. Yue J, Gu Z, Zhu Z et al (2021) Impact of defatting treatment and oat varieties on structural, functional properties, and aromatic profile of oat protein. Food Hydrocoll 112: 106368. https://doi.org/10.1016/j.foodhyd.2020.106368

Chapter 12

Protein Digestibility Through In Vitro Gastrointestinal Digestion

Guo Huang, Qin Zhang, Rongxue Zhou, Yan Zhang, and Xiaonan Sui

Abstract

The process of food digestion is a multifaceted physiological phenomenon. In vitro digestion models have been confirmed as valuable substitute models for in vivo digestion in both animals and humans and can be employed as screening instruments to address issues pertaining to food digestion, including digestibility, bioavailability, release of bioactive compounds, and alterations in food structure. Protein hydrolysis stands out as the most significant change in protein digestion, and the extent of protein hydrolysis is commonly considered an indicator of the extent of protein digestion. Despite the acknowledged limitations, the *o*-phthalaldehyde (OPA) method remains the predominant approach employed by scholars for assessing the extent of hydrolysis. Consequently, the objective of this chapter is to furnish a comprehensive exposition of the aforementioned assay, encompassing a static simulation of the digestive processes and the quantification of protein hydrolysis within a specified sample.

Key words In vitro digestion, Protein hydrolysis, The degree of protein hydrolysis, The *o*-phthalaldehyde (OPA) method, Static simulation

1 Introduction

Dietary digestion is an important source of nutrients for the human body. In recent years, diet-related diseases have been on the rise, and understanding the fate of meals in the gastrointestinal tract has received much attention. Among them, protein, as a key nutrient, can provide a large amount of essential amino acids for the human body [1]. Not only that, as an important part of the diet, protein plays an important role in maintaining the structure and texture of the diet [2]. When protein is ingested, it is first decomposed into large peptides by pepsin in the stomach area under the action of gastric acid; then it is gradually degraded into small peptides and free amino acids in the intestinal tract area under the combination of various pancreatic enzymes [3]. The product of protein digestion

Yonghui Li (ed.), *Plant-Based Proteins: Production, Physicochemical, Functional, and Sensory Properties*, Methods and Protocols in Food Science, https://doi.org/10.1007/978-1-0716-4272-6_12,

contains small peptides that have been shown to have antihypertensive, antithrombotic, antioxidant [4], and allergenic activity or are known to induce satiety [5].

As a gold standard, protein digestion should be studied in vivo animal or human feeding methods, which usually could provide the most accurate results. However, these processes are time-consuming, costly, and laborious and may involve some serious ethical issues, which is why much effort has been put on the development of in vitro procedures for simulated digestion as alternatives to in vivo experiments [6]. Although no in vitro method could completely mimic all these complex in vivo processes in vitro models are recognized as a useful alternative to animal or human models, which may be used as initial screening tools for rapid food ingredients for allowing condition controlling and easy sampling [7, 8]. In vitro digestion protocols could be largely divided into two categories, namely, static and dynamic approaches [9]. The static models have largely focused on reproducing biochemical aspects found in the stomach or small intestine, such as the composition of digestive secretions, activity of enzymes, pH, and temperature [10]. In addition to biochemical aspects, the dynamic models try to reproduce the main dynamic digestion behavior within peristalsis, gastric emptying, absorption of nutrients, and gastrointestinal transit [11, 12]. Despite the simplicity of static in vitro models, convincing evidence exists for an important role in predicting the outcomes of the in vivo digestion [13–15], which explains a much broader use (Fig. 1). They typically involve

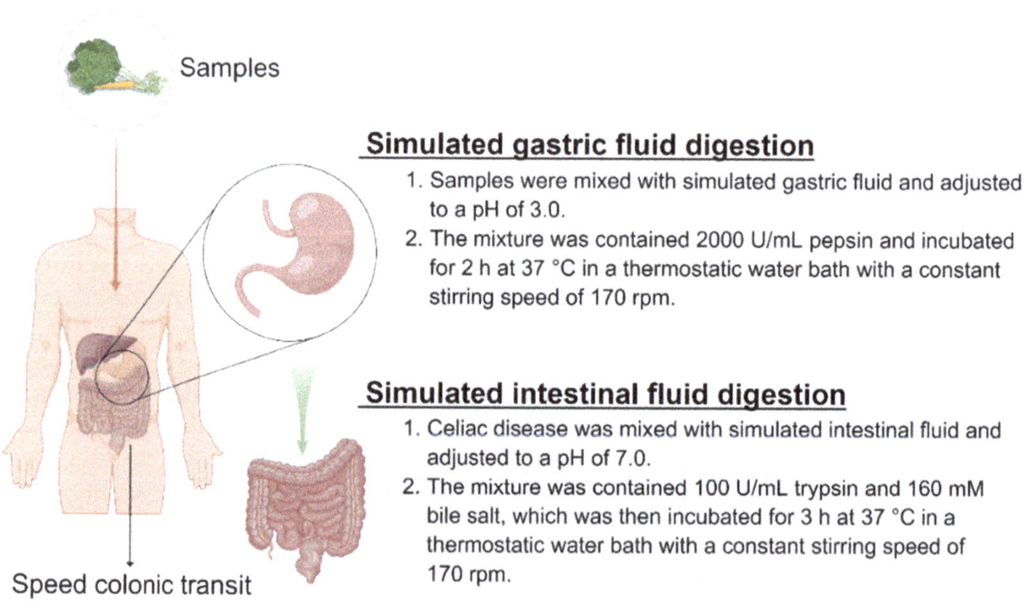

Fig. 1 Static model of in vitro gastrointestinal digestion of protein. (The figure was drawn by FigDraw (www.figdraw.com))

digesting for dietary in simulated gastric fluid (SGF) and simulated intestinal fluid (SIF). Most of models account for the results of previous studies but differ considerably in biochemical aspects of the procedures, such as electrolyte concentration, pH control, and digestion time, causing results from different studies impossible to be compared [16, 17]. To overcome the lack of homogeneity in the protocols, a consensus static protocol was first proposed in 2014 [18] and was most recently updated in 2019 [10], known as INFOGEST methods. It was analyzed theoretically and verified by some experiments in which milk proteins were used in the swine model [19] and human digests for estimation [20].

Enzymatic protein hydrolysis is a complex process. Among these, the extent of it is routinely quantified as the degree of hydrolysis (DH), which stands for the percentage of peptide bonds cleaved [21]. Over the period of protein hydrolysis, the determination methods of DH include pH-stat, trinitrobenzene-sulfonic acid (TNBS), o-phthalaldehyde (OPA), trichloroacetic acid-soluble nitrogen (SN-TCA), ninhydrin, and a variety of other methods. pH-stat is a commonly used method that involves neutralizing the protons released by protein hydrolysis by titrating basic solution in the neutral or alkaline range. The DH could be monitored based on the volume of solution-titrated consumptions in real time [12]. Nevertheless, a technical limitation of pH-stat often precluded direct measurements of DH, which makes the method used remain challenging for enzymatic digestion under acidic conditions. Soluble nitrogen index is a simple and feasible method, which is to measure the soluble nitrogen content of the supernatant after centrifugation of the digested product [22]. However, there are several limitations associated with this method that are not restricted to problems with complete precipitation of all intact proteins in TCA, which impact the accuracy of quantification [23]. Unlike the approaches mentioned above, the ninhydrin, TNBS, and OPA methods are based on the reaction of free amino groups released during hydrolysis with them to form intermediates, which are then quantified by spectroscopic measurements. Ninhydrin reactions are one of the most fundamental and biochemical studies, which as a chromogenic reagent was applied frequently to analyze DH in agriculture and biochemical analysis. However, a narrow physiological range of pH 4.5–5.0 of the diluents used limits this method, in which the poor solubility of proteins has been repeatedly described [24, 25]. It has been documented that the TNBS and OPA methods are the most reliable outcome and the most commonly accepted measure of the DH [23, 26]. Unfortunately, the TNBS method is much maligned and usually discussed as the major bottleneck for various time- and labor-consuming incubation and cooling steps [27]. In addition, this TNBS reagent is unstable in water and toxic, which must be handled with extreme care [28]. One of the key advantages of OPA method over TNBS

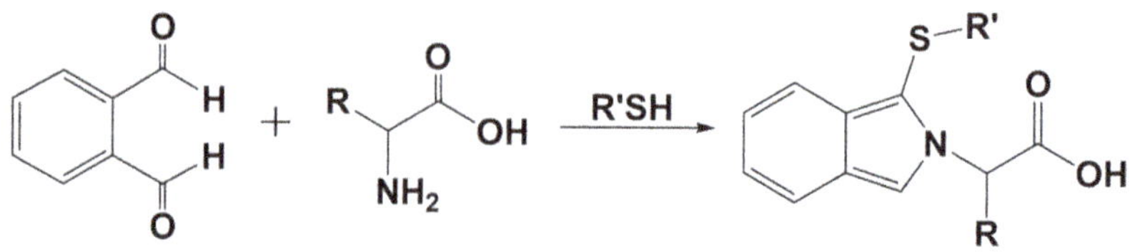

Fig. 2 Specific labeling of free amino acids by *o*-phthalaldehyde

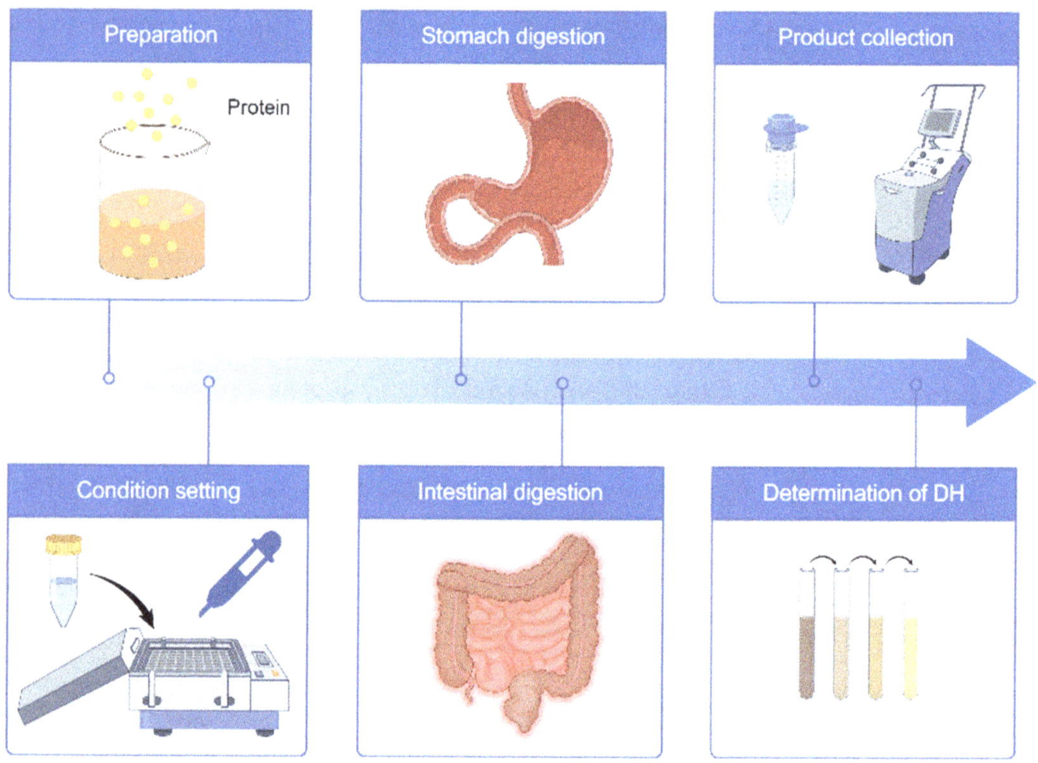

Fig. 3 Flow diagram of static digestion model set up and degree of hydrolysis of protein. (The figure was drawn by FigDraw (www.figdraw.com))

method is that it can provide more accurate data and takes less time for DH analysis (Fig. 2). Moreover, this OPA reagent is even less toxic and more stable [29].

A flow diagram of the measurement set up is shown in Fig. 3. Therefore, the methodology proposed in the present chapter for construction of the static digestion model, reagent configuration, and determination of DH of digested products simultaneously follows the methodology adopted in the literature [18, 29]. Subsequently, we have also obtained similar results by the method described in this chapter [30, 31].

2 Materials

Unless otherwise specified, all solutions should be prepared using DI water. Reagents should be prepared and stored at 4 °C. It is important to diligently adhere to waste disposal regulations when disposing of waste materials.

2.1 Configuration of Acid pH-Adjusting Solution

1. Acid pH-adjusting solution (1 M): To prepare 1 L, 9 mL of HCl (37%) was taken with a pipette and diluted to 1 L with DI water.

2.2 Configuration of Alkali pH-Adjusting Solution

1. Alkali pH-adjusting solution (1 M): To prepare 1 L, 40.01 g of NaOH was dissolved in 1 L of DI water.

2.3 Configuration of the Simulated Gastric Fluid (SGF) Electrolyte Solution

To assure the stability of the digestive solution, it was used immediately or stored at 4 °C until further use.

1. KCl stock solution (69 mM): 5.14 g of KCl was accurately weighed and subsequently dissolved in 100 mL of DI water, thereby serving as the designated stock solution.
2. KH_2PO_4 stock solution (9 mM): 1.24 g of KH_2PO_4 was accurately weighed and subsequently dissolved in 100 mL of deionized DI water, thereby serving as the designated stock solution.
3. $NaHCO_3$ stock solution (250 mM): 21.00 g of $NaHCO_3$ was accurately weighed and subsequently dissolved in 100 mL of DI water, thereby serving as the designated stock solution.
4. NaCl stock solution (472 mM): 27.58 g of NaCl was accurately weighed and subsequently dissolved in 100 mL of deionized DI water, thereby serving as the designated stock solution.
5. $MgCl_2(H_2O)_6$ stock solution (1 mM): 0.20 g of $MgCl_2(H_2O)_6$ was accurately weighed and subsequently dissolved in 100 mL of DI water, thereby serving as the designated stock solution.
6. $(NH_4)_2CO_3$ stock solution (5 mM): 0.59 g of $(NH_4)_2CO_3$ was accurately weighed and subsequently dissolved in 100 mL of DI water, thereby serving as the designated stock solution.
7. $CaCl_2(H_2O)_2$ stock solution (3 M): 0.44 g of $CaCl_2(H_2O)_2$ was accurately weighed and subsequently dissolved in 100 mL of DI water, thereby serving as the designated stock solution.

Subsequently, the electrolyte solutions were mixed, and the final volume was adjusted to 1 L using DI water. Ultimately, the SGF electrolyte solution consists of the following concentrations:

6.9 mM KCl, 0.9 mM KH_2PO_4, 25 mM $NaHCO_3$, 47.2 mM NaCl, 0.1 mM $MgCl_2(H_2O)_6$, 0.5 mM $(NH_4)_2CO_3$, and 0.3 M $CaCl_2(H_2O)_2$ (*see* **Note 1**).

2.4 Configuration of the Simulated Intestinal Fluid (SIF) Electrolyte Solution

To assure the stability of the digestive solution, it was used immediately or stored at 4 °C until further use.

1. KCl stock solution (68 mM): 5.06 g of KCl was accurately weighed and subsequently dissolved in 100 mL of DI water, thereby serving as the designated stock solution.
2. KH_2PO_4 stock solution (8 mM): 1.08 g of KH_2PO_4 was accurately weighed and subsequently dissolved in 100 mL of DI water, thereby serving as the designated stock solution.
3. $NaHCO_3$ stock solution (850 mM): 71.40 g of $NaHCO_3$ was accurately weighed and subsequently dissolved in 100 mL of DI water, thereby serving as the designated stock solution.
4. NaCl stock solution (384 mM): 22.44 g of NaCl was accurately weighed and subsequently dissolved in 100 mL of DI water, thereby serving as the designated stock solution.
5. $MgCl_2(H_2O)_6$ stock solution (3.3 mM): 0.67 g of $MgCl_2(H_2O)_6$ was accurately weighed and subsequently dissolved in 100 mL of DI water, thereby serving as the designated stock solution.
6. $CaCl_2(H_2O)_2$ stock solution (3 M): 0.44 g of $CaCl_2(H_2O)_2$ was accurately weighed and subsequently dissolved in 100 mL of DI water, thereby serving as the designated stock solution.

After that, follow the same steps as in Subheading 2.3. The final SIF electrolyte solution contains 6.8 mM KCl, 0.8 mM KH_2PO_4, 85 mM $NaHCO_3$, 38.4 mM NaCl, 0.33 mM $MgCl_2(H_2O)_6$, and 0.3 M $CaCl_2(H_2O)_2$ (*see* **Note 1**).

2.5 Configuration of OPA Reagent

1. OPA diluent solution: The OPA diluent solution was prepared by weighing 7.62 g of di-Na-tetraborate decahydrate and 200 mg of Na-dodecyl-sulfate (SDS). After that, the components were combined through mechanical stirring and diluted with 150 mL of DI water.
2. OPA test solution: To prepare the OPA test solution, 160 mg of *o*-phthalaldehyde (OPA) was accurately weighed and dissolved in 4 mL of ethanol. Subsequently, the aforementioned solutions were mixed together, followed by the addition of 176 mg of dithiothreitol (DTT). The resulting OPA test solution was vigorously shaken to ensure uniform mixing and then brought to a final volume of 200 mL using DI water (*see* **Note 2**).

3. Serine standard solution: 50 mg of serine was dissolved in 500 mL of DI water to give a concentration of 0.9516 meqv/L as a reference standard (*see* **Note 3**).

3 Methods

All procedures should be maintained at room temperature, except as noted. The digestion experiments should be carried out according to the method proposed by Minekus et al. [18] and improved with reference to the method proposed by Jiang et al. [30]. As aforementioned, since protein digestibility is usually calculated using the degree of hydrolysis (DH), it should be measured using the conventional OPA-based spectrophotometric method described by Nielsen et al. [29].

3.1 Sample Solution Preparation

The sample solution: 5 g of sample was weighted and dissolved in 100 mL of DI water. The sample solution was stirred for 2 h at room temperature and set aside for the next step.

3.2 Simulated in Vitro Gastrodigestion

1. The sample solution and the SGF electrolyte solution were mixed together in equal proportion by volume. The pH of the mixture was adjusted to 3.0 with 1.0 M HCl (*see* **Note 4**).
2. Subsequently, pepsin was added to the mixture to reach a concentration of 2000 U/mL under stirring (*see* **Note 5**), which was used to digest protein at 37 °C in water bath (*see* **Note 6**). The gastric digestion lasted for 2 h with constant stirring at 170 rpm, after which the reaction was stopped by adjusting the pH to 7.0.
3. The supernatant was collected by centrifugation (10,000 × *g*, 20 min, 4 °C) after gastrodigestion period, which was used for further analysis.

3.3 Simulated in Vitro Intestinal Digestion

1. The gastric chyme was mixed in equal proportions with the SIF electrolyte solution. Then, bile salts were added and made its concentration to be 160 mM (*see* **Note 7**). The pH was further adjusted with 1 M NaOH to pH 7.0 (*see* **Note 4**).
2. The quantity of pancreatin incorporated into the mixture was determined by the trypsin activity, which corresponded to a trypsin activity of 100 U/mL (*see* **Note 5**).
3. The mixture was incubated in a water bath at 37 °C for 3 h and continuously stirred at 170 rpm (*see* **Note 6**). We performed quantitative sampling from a digested mixture at a given time. Afterward, the digestion was then terminated by placing the solution in ice water for 30 min.

4. After termination of digestion, the digested mixture was centrifuged (10,000 × *g*, 20 min, 4 °C), and supernatants were collected.

3.4 OPA Method

The reaction between OPA reagent and primary amino groups exclusively takes place in the presence of a thiol and exhibits increased efficiency at basic pH. In such circumstances, the formation of 1-alkylthio-2-alkyl-substituted isoindoles exhibits significant absorption at 340 nm. In this method, the amount of the isoindoles formed from sample by action of gastrointestinal digestive enzymes is measured at 340 nm (*see* **Note 8**). The method is as follows:

1. Procedure: 3 mL OPA reagents were added to all test tubes.
2. Standard measuring: 400 μL serine standard solution was added to test tubes. The mixtures were oscillated for 5 s and incubated at room temperature for 2 min prior to reading at 340 nm in the spectrophotometer with an optical path length of 10 mm.
3. Blank measuring: 400 μL DI water was added to test tubes and was performed as previously described.
4. Sample measuring: Samples were prepared from 400 μL of the supernatant of simulated digestion samples.
5. Expression of results:
   (a) The effective concentration (meqv) of serine-NH_2 groups per gram of protein can be calculated using the following equation:

$$\text{Serine} - NH_2 = \frac{OD_{\text{sample}} - OD_{\text{blank}}}{OD_{\text{standard}} - OD_{\text{blank}}} \times \frac{0.9516\, meqv/L \times 0.1 \times 100}{X \times P}$$

   Where:

   - X represents the weight of the sample (g).
   - 0.1 is the sample volume (L).
   - P signifies the mass fraction of protein concentration in the sample in weight percentage (wt%).
   - OD represents the measured absorbance of the standards, blank, or samples.

   (b) Calculate the number of hydrolyzed peptide bonds, using the following equation:

$$h = \frac{\text{Serine} - NH_2 - \beta}{\alpha}$$

Where:

- The values for α and β, which serve as reference parameters, can be found in Table 1.

(c) The degree of hydrolysis can be calculated as a percentage using the following equation:

$$DH = \frac{h}{h_{tot}} \times 100\%$$

Where:

- The h_{tot} values for the samples can be found in Table 1.

6. In Fig. 4, you can see two examples of DH of soy protein isolate (SPI) after in vitro gastrointestinal digestion.

Table 1
Value of constants of α, β, and h_{tot} for different protein raw materials [32]

Protein	α	β	h_{tot}
Soy	0.970	0.342	7.8
Gluten[a]	1.00	0.40	8.3
Casein	1.039	0.383	8.2
Whey[a]	1.00	0.40	8.8
Gelatin	0.796	0.457	11.1
Meat	1.00	0.40	7.6
Fish	1.00	0.40	8.6

[a]When raw material has not been examined, then α and β are estimated to be 1.00 and 0.40, respectively

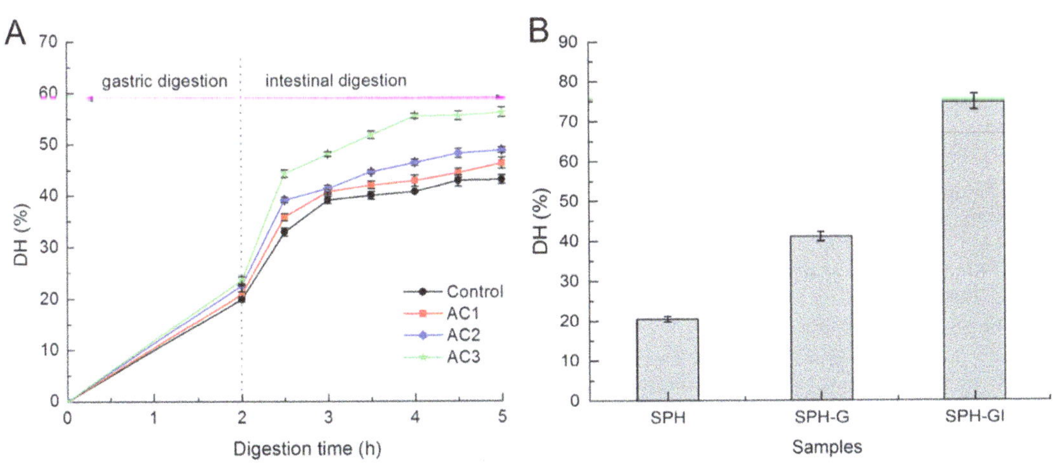

Fig. 4 (**a**) DH of SPI and SPI-anthocyanins conjugates after in vitro gastrointestinal digestion. Control, AC1, AC2, and AC3 represent the ratio of anthocyanins to SPI of 0:1, 1:60, 1:40, and 1:20 (w/w), respectively [30]. (**b**) DH of soy protein hydrolysate (SPH), gastric-digested soy protein hydrolysate (SPH-G), and gastrointestinal-digested soy protein hydrolysate (SPH-GI) [31]

4 Notes

1. The allocation of the electrolyte solution is contingent upon the vessel size, as dictated by the environmental situation. It is important to note that the final concentration of the electrolyte solution, whether SGF or SIF, should include the respective metal ion.
2. The OPA test solution was prepared freshly in a diluent solution and subsequently degassed in an ultrasonic bath for 30 s to prevent oxidation of OPA. If not immediately utilized, the solution was stored at a frozen temperature of −20 °C until needed.
3. A fresh stock of the standard solution was prepared prior to conducting experiments and kept at a temperature of −20 °C until the studies were conducted.
4. pH monitoring during digestion should be conducted utilizing a pH titrator and stabilized at pH 7.0 by employing 0.1 M HCl or NaOH as appropriate.
5. Prior to measurement, activity assays should be conducted for digestive enzymes.
6. To guarantee enzyme activity, relevant digestive enzymes can be incubated in a water bath at 37 °C for 5 min before digestion.
7. If the bile salt purchased exhibits difficulty in dissolving, it can be pre-dissolved into the electrolyte solution for incubation.
8. To prevent sample oxidation, measurements of the samples were expeditiously performed at room temperature following mixing.

References

1. Sui XN, Zhang TY, Jiang LJ (2020) Soy protein: molecular structure revisited and recent advances in processing technologies. Annu Rev Food Sci Technol 12:119–147
2. Zhang TY, Dou W, Zhang X, Zhao Y, Zhang Y, Jiang LJ, Sui XN (2021) The development history and recent updates on soy protein-based meat alternatives. Trends Food Sci Technol:702–710
3. Ding J, Wen JY, Wang JY, Tian R, Yu LL, Jiang LZ, Zhang Y, Sui XN (2020) The physicochemical properties and gastrointestinal fate of oleosomes from non-heated and heated soymilk. Food Hydrocoll 100:105418
4. Zhang QZ, Tong XH, Sui XN, Wang ZJ, Qi BK, Li Y, Jiang LZ (2018) Antioxidant activity and protective effects of Alcalase-hydrolyzed soybean hydrolysate in human intestinal epithelial Caco-2 cells. Food Res Int 111:256–264
5. Zhang S, Vardhanabhuti B (2014) Effect of initial protein concentration and pH on in vitro gastric digestion of heated whey proteins. Food Chem 145:473–480
6. Dupont D, Alric M, Blanquet-Diot S, Bornhorst G, Cueva C, Deglaire A, Denis S, Ferrua M, Havenaar R, Lelieveld J, Mackie AR, Marzorati M, Menard O, Minekus M, Miralles B, Recio I, Van den Abbeele P (2019) Can dynamic in vitro digestion systems mimic the physiological reality? Crit Rev Food Sci Nutr 59(10):1546–1562
7. Sousa R, Portmann R, Dubois S, Recio I, Egger L (2020) Protein digestion of different

protein sources using the INFOGEST static digestion model. Food Res Int 130:108996

8. Hur SJ, Lim BO, Decker EA, McClements DJ (2011) In vitro human digestion models for food applications. Food Chem 125(1):1–12
9. Guerra A, Etienne-Mesmin L, Livrelli V, Denis S, Blanquet-Diot S, Alric M (2012) Relevance and challenges in modeling human gastric and small intestinal digestion. Trends Biotechnol 30(11):591–600
10. Brodkorb A, Egger L, Alminger M, Alvito P, Assunção R, Ballance S, Bohn T, Bourlieu-Lacanal C, Boutrou R, Carrière F, Clemente A, Corredig M, Dupont D, Dufour C, Edwards C, Golding M, Karakaya S, Kirkhus B, Le Feunteun S, Lesmes U, Macierzanka A, Mackie AR, Martins C, Marze S, McClements DJ, Ménard O, Minekus M, Portmann R, Santos CN, Souchon I, Singh RP, Vegarud GE, Wickham MSJ, Weitschies W, Recio I (2019) INFOGEST static in vitro simulation of gastrointestinal food digestion. Nat Protoc 14(4):991–1014
11. Wojtunik-Kulesza K, Oniszczuk A, Oniszczuk T, Combrzyński M, Nowakowska D, Matwijczuk A (2020) Influence of in vitro digestion on composition, bioaccessibility and antioxidant activity of food polyphenols—a non-systematic review. Nutrients 12(5):1401
12. Mat DJL, Cattenoz T, Souchon I, Michon C, Le Feunteun S (2018) Monitoring protein hydrolysis by pepsin using pH-stat: in vitro gastric digestions in static and dynamic pH conditions. Food Chem 239:268–275
13. Cardoso C, Afonso C, Lourenço H, Costa S, Nunes ML (2015) Bioaccessibility assessment methodologies and their consequences for the risk–benefit evaluation of food. Trends Food Sci Technol 41(1):5–23
14. Bohn T, Carriere F, Day L, Deglaire A, Egger L, Freitas D, Golding M, Le Feunteun S, Macierzanka A, Menard O, Miralles B, Moscovici A, Portmann R, Recio I, Rémond D, Santé-Lhoutelier V, Wooster TJ, Lesmes U, Mackie AR, Dupont D (2018) Correlation between in vitro and in vivo data on food digestion. What can we predict with static in vitro digestion models? Crit Rev Food Sci Nutr 58(13):2239–2261
15. Pälchen K, Michels D, Duijsens D, Gwala S, Pallares Pallares A, Hendrickx M, Van Loey A, Grauwet T (2021) In vitro protein and starch digestion kinetics of individual chickpea cells: from static to more complex in vitro digestion approaches. Food Funct 12(17):7787–7804
16. Colombo R, Ferron L, Frosi I, Papetti A (2021) Advances in static in vitro digestion models after the COST action Infogest consensus protocol. Food Funct 12(17):7619–7636
17. Wang R, Mohammadi M, Mahboubi A, Taherzadeh MJ (2021) In-vitro digestion models: a critical review for human and fish and a protocol for in-vitro digestion in fish. Bioengineered 12(1):3040–3064
18. Minekus M, Alminger M, Alvito P, Ballance S, Bohn T, Bourlieu C, Carrière F, Boutrou R, Corredig M, Dupont D, Dufour C, Egger L, Golding M, Karakaya S, Kirkhus B, Le Feunteun S, Lesmes U, Macierzanka A, Mackie A, Marze S, McClements DJ, Ménard O, Recio I, Santos CN, Singh RP, Vegarud GE, Wickham MSJ, Weitschies W, Brodkorb A (2014) A standardised static in vitro digestion method suitable for food—an international consensus. Food Funct 5(6): 1113–1124
19. Egger L, Schlegel P, Baumann C, Stoffers H, Guggisberg D, Brügger C, Dürr D, Stoll P, Vergères G, Portmann R (2017) Physiological comparability of the harmonized INFOGEST in vitro digestion method to in vivo pig digestion. Food Res Int 102:567–574
20. Sanchón J, Fernández-Tomé S, Miralles B, Hernández-Ledesma B, Tomé D, Gaudichon C, Recio I (2018) Protein degradation and peptide release from milk proteins in human jejunum. Comparison with in vitro gastrointestinal simulation. Food Chem 239:486–494
21. Conesa C, Fitz Gerald RJ (2013) Total solids content and degree of hydrolysis influence proteolytic inactivation kinetics following whey protein hydrolysate manufacture. J Agric Food Chem 61(42):10135–10144
22. Zhang Y, Chen S, Qi BK, Sui XN, Jiang LZ (2018) Complexation of thermally-denatured soybean protein isolate with anthocyanins and its effect on the protein structure and in vitro digestibility. Food Res Int 106:619–625
23. Rutherfurd SM (2010) Methodology for determining degree of hydrolysis of proteins in hydrolysates: a review. J AOAC Int 93(5): 1515–1522
24. Cornish-Bowden AJ, Knowles J (1969) The pH-dependence of pepsin-catalysed reactions. Biochem J 113(2):353–362
25. Pelegrine DHG, Gasparetto CA (2005) Whey proteins solubility as function of temperature and pH. LWT Food Sci Technol 38(1):77–80
26. Cheison SC, Zhang S-B, Wang Z, Xu S-Y (2009) Comparison of a modified spectrophotometric and the pH-stat methods for

determination of the degree of hydrolysis of whey proteins hydrolysed in a tangential-flow filter membrane reactor. Food Res Int 42(1): 91–97

27. Spellman D, McEvoy E, O'Cuinn G, FitzGerald RJ (2003) Proteinase and exopeptidase hydrolysis of whey protein: Comparison of the TNBS, OPA and pH-stat methods for quantification of degree of hydrolysis. Int Dairy J 13(6):447–453
28. Beaubier S, Framboisier X, Ioannou I, Galet O, Kapel R (2019) Simultaneous quantification of the degree of hydrolysis, protein conversion rate and mean molar weight of peptides released in the course of enzymatic proteolysis. J Chromatogr B 1105:1–9
29. Nielsen PM, Petersen D, Dambmann C (2001) Improved method for determining food protein degree of hydrolysis. J Food Sci 66(5):642–646
30. Jiang LZ, Liu YJ, Li L, Qi BK, Ju MN, Xu Y, Zhang Y, Sui XN (2019) Covalent conjugates of anthocyanins to soy protein: unravelling their structure features and in vitro gastrointestinal digestion fate. Food Res Int 120:603–609
31. Zhang QZ, Tong XH, Qi BK, Wang ZJ, Li Y, Sui XN, Jiang LZ (2018) Changes in antioxidant activity of Alcalase-hydrolyzed soybean hydrolysate under simulated gastrointestinal digestion and transepithelial transport. J Funct Food 42:298–305
32. Adler-Nissen J (1986) Enzymic hydrolysis of food proteins. Elsevier Applied Science Publishers

Chapter 13

Trypsin and Chymotrypsin Inhibitor Activities

Keshun Liu

Abstract

Trypsin inhibitors and chymotrypsin inhibitors are protease inhibitors of protein nature. Occurring naturally in legumes, cereals, and some other species, they are considered key antinutritional factors, since they can interfere with protein digestions, induce hypersecretion of trypsin and chymotrypsin in animal digestive tracts, and lead to pancreatic enlargement and growth suppression of animals. Paradoxically, during the past few decades, some protease inhibitors have been found to be beneficial in humans and animals by exerting therapeutic effects. Therefore, protease inhibitor activities remaining in food and feed products have been considered important parameters, due to their significant effects on animal nutrition and human health. Historically, trypsin inhibitor activity has been a priority parameter for measurement, but lately determination of chymotrypsin inhibitor activity has also gained attention. This chapter covers the latest method for measuring trypsin inhibitor activity and the latest method for measuring chymotrypsin inhibitory activity, after providing a brief overview of the method development over the past half century for measuring activities of the protease inhibitors. The latest methods feature improved sensitivity and accuracy and can express inhibitory activities in both arbitrary units and absolute amounts of trypsin or chymotrypsin inhibited for easy comparison among studies. A detailed description of the latest method for determining chymotrypsin inhibitor activity is also provided.

Key words Trypsin inhibitor, Chymotrypsin inhibitor, Protease inhibitor, Activity, Soybean

1 Introduction

Protease inhibitors of protein nature, such as inhibitors of trypsin and chymotrypsin (the two main digestive tract proteases) and inhibitors of other proteases, are naturally present in legumes [1, 2] and cereals [3]. There are two types of proteinase inhibitors: the Kunitz trypsin inhibitors (KTI) and Bowman-Birk inhibitors (BBI). The former is characterized by having a molecular mass of about 21 kDa with two disulfide bonds, one reactive site, and inhibition mainly toward trypsin. In contrast, BBI has a molecular mass of 7–9 kDa with seven disulfide bonds, two reactive sites, and strong inhibitions toward both trypsin and chymotrypsin. Therefore, BBI is also called a double-headed inhibitor [1, 4].

Yonghui Li (ed.), *Plant-Based Proteins: Production, Physicochemical, Functional, and Sensory Properties*,
Methods and Protocols in Food Science, https://doi.org/10.1007/978-1-0716-4272-6_13,

The nutritional importance of protease inhibitors is demonstrated by a century-old observation that the nutritional quality of soybeans could be significantly improved upon heat treatment when fed to rats. The improvement was attributed to the thermal inactivation of protease inhibitors, which presumably interfere with protein digestion. Later investigators showed that the growth suppression by protease inhibitors results from not only reduced protein digestibility but also hypersecretion of trypsin and chymotrypsin which leads to pancreatic enlargement [1, 4]. For this reason, protease inhibitors have long been considered key antinutritional factors in food and feed of legume origin [1]. Yet, during the past few decades, some protease inhibitors, such as soybean BBI, have been found beneficial to humans and animals by exerting therapeutic effects (such as antioxidant, anti-inflammatory, and/or anticarcinogenic activities) [5, 6]. This has created a paradox for protease inhibitors with respect to their nutritional significance and health implications. It may be preferable to retain some protease inhibitors in foods because of their anticarcinogenic and anti-inflammatory effects [4].

At the present, the inhibitors of trypsin and chymotrypsin, the two major proteases in digestive tracts, are of major interest. These inhibitors can be inactivated to certain lower levels, but their complete elimination is hard to achieve. Therefore, there is a great need of methods that can measure trypsin inhibitor activity in food and feed with good sensitivity and accuracy. There is also a great need to have methods that can measure chymotrypsin inhibitor activity (CIA) with good sensitivity and accuracy. This chapter deals with the latest development of these methods.

Since almost all protease inhibitors found in legume seeds and cereal grains have inhibitory activities toward trypsin, trypsin inhibitor activity (TIA) has been measured as an important quality parameter for proteinaceous food and feedstuff of plant origin, particularly those from soybeans [1] and other legumes [2]. Most reported assays for measuring TIA are colorimetric, based on measuring the difference in absorbance between the absence and the presence of an inhibitor, resulting from the cleaving action of trypsin on a peptide substrate (natural or synthetic) [7–12]. A few years ago, Liu [13, 14] reported a significantly improved method for TIA assay, based on the American Oil Chemists Society Official Method Ba 12-75 [12]. Major modifications include (1) using the enzyme-last sequence instead of the substrate-last sequence, (2) using a duplicated single level of a dilute sample extract instead of a series of levels, (3) reducing the total assay volume from 10 to 5 mL, and (4) expressing TIA in either an arbitrary unit or absolute amounts of trypsin inhibited per unit sample. At the same time, an international collaborative study was conducted to validate the 5 mL method. It showed that the new method has good repeatability and reproducibility [15]. Subsequently, the AOCS

Uniform Methods Committee approved it as a new AOCS Official Method Ba 12a-2020 [16]. It has superseded the old AOCS Method Ba 12-75 [12]. For detailed description of the new Official Method, readers can refer to Liu [13, 14], Liu et al. [15], and AOCS [16]. For copyright protection reasons, this chapter omits the section on this aspect.

Although TIA has historically been measured as a key quality parameter for legumes and their products, there is also a growing need to measure chymotrypsin inhibitor activity (CIA) in various protein products. The driving forces for this trend include an increasing use of plant proteins for food and feed and increasing knowledge gained about the difference between the two types of inhibitors. Over the years, many methods for measuring CIA have been reported [17–19], but there is a great variation among them, in terms of reagent types, concentrations and volumes of reagents, reaction temperature, reaction time, total reaction volume, wavelength for measuring absorbance, and calculation and expression of the results. Unlike the TIA measurement, no official and standardized method is available for the CIA assay. Therefore, it is impossible to compare CIA results among reports.

Moreover, almost all previously reported methods for measuring CIA follow the substrate-last sequence. Yet, as early as 1990, Liu and Markakis [18] found that, during the CIA assay, the sequence of adding reagents also had a profound effect on the results. This sequence effect of adding reagents is like the sequence effect observed in the TIA assay: The enzyme-last sequence gives a constant inhibition value, but the substrate-last sequence produces a value that varies greatly with pH of the premix and preincubation times. Therefore, the enzyme-last sequence is preferred.

To address several issues described above relating to the CIA assay, Liu [20] conducted a study that led to development of an optimized method for accurately assaying CIA. It features N-benzoyl-L-tyrosine p-nitroanilide (BTpNA or simply BTNA) as a substrate, dimethylformamide (DMF) as an organic solvent, the enzyme-last sequence of adding reagents, and a 5 mL total assay volume. Like most previously reported methods, the new method expresses CIA in an arbitrary unit as chymotrypsin units inhibited (CUI)/mg sample. Just like the arbitrary unit for TIA, although the arbitrary unit for CIA expression is easy to define, it gives values that vary with methods. This makes comparison of results measured by different methods impossible. To address this new issue, Liu [21] conducted a follow-up study by determining a conversion factor between an arbitrary unit and μg chymotrypsin inhibited and standardizing it with a reference chymotrypsin. The standardized conversion factor between the two units was determined as 1.5, whereas chymotrypsin has a specific activity of 50 BTEE (N-benzoyl-L-tyrosine ethyl ester) units/mg protein. Consequently, when the optimized method of Liu [20] is strictly

Sample
↓
Reduce particle size to pass U.S. No. 50 Mesh (0.30 mm openings)
↓
Weigh 0.8 g sample and extract with 40 mL 10 mM NaOH for 3 hr
↓
Dilute with water, 1 mL causes 40-60% trypsin inhibition
↓
Add 0.4, 0.6, 0.8, 1.0 mL diluted extract to each of 4 test tubes (sample readings)
Add 0.6, 0.4, 0.2 mL water to make up to 1.0 mL for each of the first 3 tubes
Add 1.0 mL water to each of 2 test tubes (reference readings)
Add 2.5 BTpNA working solution, mix, and place in 37 °C water bath
Add 1 mL enzyme solution, 10 min later, add 0.5 mL acetic acid
For sample and reference blanks, add acetic acid before adding the substrate
↓
Centrifuge and read absorbances at 400 nm
↓
Define one chymotrypsin unit (CU) as 0.01 A400
Calculate CIA as chymotrypsin units inhibited (CUI)/mg sample
Calculated mg chymotrypsin inhibited (CId)/mg by dividing CUI/mg by 1.5

Fig. 1 Flowchart of the method for measuring chymotrypsin inhibitor activity (CIA) in legume and other plant products. (Adapted from Liu [20, 21])

followed, one can express results in mg chymotrypsin inhibited (CId)/g sample by simply dividing the arbitrary unit (CUI/mg sample) by 1.5 [21].

The following sections provide the detailed description of materials and methods for the CIA measurement in legumes, cereals, and other plant products, based on the two recent reports [20, 21] (Fig. 1).

2 Materials

2.1 Reagents

1. α-Chymotrypsin from bovine pancreas with specific activity of ≥40 N-benzoyl-L-tyrosine ethyl ester (BTEE) units/mg protein.
2. Nα-benzoyl-L-tyrosine-p-nitroanilide (BTpNA).
3. Dimethylformamide (DMF).
4. Tris base (hydroxymethyl) amino-methane.
5. Calcium chloride dihydrate.
6. HCl.
7. NaOH pellets.

All chemicals should be ACS grade except for chymotrypsin preparation.

2.2 Reagent Solutions

1. NaOH solution, 10 mM, for sample extraction.
2. HCl solution, 1 mM, containing 5 mM $CaCl_2$, for making chymotrypsin solutions. Dissolve 0.736 g $CaCl_2{\cdot}2H_2O$ in 900 mL water, add 0.5 mL 2 M HCl solution, and dilute to 1000 mL. The pH of the solution shall be 3.0 ± 0.2.
3. Tris buffer, 50 mM, containing 20 mM $CaCl_2$, pH, 8.2, for making BTpNA working solution. Dissolve 6.05 g Tris and 2.94 g $CaCl_2{\cdot}2H_2O$ in 900 mL water, adjust to pH 8.2 with 2 M HCl, and dilute to 1000 mL.
4. BTpNA stock solution (20 mg/mL): Dissolve 200 mg of BTpNA in 10 mL DMF. When not in use, the solution can be stored in the dark at room temperature (avoid refrigeration) for up to 6 weeks.
5. BTpNA working solution (0.2 mg/mL) containing 15% (v/v) DMF: Add 0.5 mL of the BTpNA stock solution to a prewarmed (37 °C) premix consisting of 7 mL DMF and 42.5 mL Tris buffer (total volume of 50 mL). Prepare fresh daily, and keep in the water bath at 37 °C during use.
6. Chymotrypsin stock solution (0.2 mg/mL): Dissolve 10 mg of bovine chymotrypsin having a measured specific activity of at least 40 BTEE units/mg protein, in 50 mL of the HCl solution. When not in use, keep the stock solution at 5 °C for up to 3 weeks.
7. Chymotrypsin working solution (40 μg/mL): Dilute 5 mL of the cold chymotrypsin stock solution to 25 mL with the HCl solution. Prepare freshly right before each run, and then keep in room temperature.
8. Acetic acid solution (30%v/v) for stopping the colorimetric reaction (also known as reaction stop solution): Mix 30 mL of glacial acetic acid with 70 mL water.

3 Methods

3.1 Sample Preparation and Extraction

1. Grind dry samples repeatedly until the entire sample passes through a US Standard No. 50 mesh screen, equivalent to 0.300 mm diameter sieve openings. Avoid subjecting test samples to temperatures over 60 °C at any time by starting with a small sample size and grinding in intervals for cooling.
2. Weigh 0.8 ± 0.05 g of test sample, and record the exact amount to the nearest 0.1 mg.
3. Transfer the weighed sample into a 50 mL test (or centrifuge) tube containing an octagonal magnetic stirring bar (12.5×8 mm), add 40.0 mL of 10 mM NaOH solution,

cover the tube with parafilm, and mix for 3 h at room temperature at 600 rpm. After 3 h, stop mixing, and let the extract settle for 10 min prior to using the supernatant for dilutions.

4. Dilute an aliquot of the sample extract (which, strictly speaking, is a suspension) with water so that 1.0 mL of the dilute sample extract causes 40–60% chymotrypsin inhibition after correction for blank readings (*see* **Note 1**).

3.2 Procedure

1. In a set of test tubes on a rack, add the reagent solutions exactly in the volumes and sequences specified in Table 1 for preparing the four types of reaction mixtures (*see* **Note 2**):
   (a) Sample readings (with inhibitor) in four increasing levels of a dilute extract.
   (b) Sample blank (with inhibitor).
   (c) Reference reading (without inhibitor, using water instead) in duplicate.
   (d) Reference blank (without inhibitor, using water instead).
2. Add the first reagent (water or dilute sample extract) to all test tubes, and then put the rack in a water bath maintained at 37 °C.
3. Add the second reagent, and vortex each tube well for 10 s immediately.
4. Start timing immediately after adding the third reagent to each tube and before vortexing for 10 s. Keep the interval as exact as possible while adding the same reagent to each test tube and between adding the third reagent and the fourth reagent to the corresponding tubes.
5. Exactly 10 min after adding the third reagent, add the fourth reagent, and vortex for 10 s.
6. Centrifuge the reaction mixtures at 1500× *g* for 5 min. Measure absorbance for each reaction mixture supernatant at 400 nm within 1 h, using water as a spectrometer's reference. Light path should be 1.0 cm.

3.3 Expression of CIA in an Arbitrary Unit and Calculation

CIA is first expressed in an arbitrary unit, as chymotrypsin units inhibited (CUI) per mg sample at 40% chymotrypsin inhibition. One chymotrypsin unit (CU) is hereby defined as an increase of 0.01 absorbance at 400 nm under the 5 mL assay condition specified in the procedure. When CIA is expressed as CUI per mg sample at 40% chymotrypsin inhibition, its value is calculated by the following five steps and equations.

First, based on a sample blank (A400SB4) at the fourth level (1.0 mL) of a dilute extract, calculate sample blanks (A400SB1, A400SB2, and A400SB3) for the three lower levels (0.4, 0.6, and 0.8 mL) of the dilute extract by (1) subtracting reference blank (A400RB) from A400SB4, (2) proportionally calculating values

Table 1
The volume and sequence of adding reagent solutions (in parenthesis) for measuring chymotrypsin inhibitor activity in a single sample

Reagent solution and addition sequence	Sample reading 1 (A_{400S1})	Sample reading 2 (A_{400S2})	Sample reading 3 (A_{400S3})	Sample reading 4 (A_{400S4})	Sample blank (A_{400SB4})	Reference reading 1 (A_{400R1})	Reference reading 2 (A_{400R2})	Reference blank (A_{400RB})
A dilute sample suspension	0.4 (1)	0.6 (1)	0.8 (1)	1.0 (1)	1.0 (1)			
Deionized water	0.6 (1)	0.4 (1)	0.2 (1)			1.0 (1)	1.0 (1)	1.0 (1)
The BTpNA working solution	2.5 (2)	2.5 (2)	2.5 (2)	2.5 (2)	2.5 (3)	2.5 (2)	2.5 (2)	2.5 (3)
The enzyme working solution	1.0 (3)	1.0 (3)	1.0 (3)	1.0 (3)	1.0 (4)	1.0 (3)	1.0 (3)	1.0 (4)
The acetic acid solution	0.5 (4)	0.5 (4)	0.5 (4)	0.5 (4)	0.5 (2)	0.5 (4)	0.5 (4)	0.5 (2)
The total assay volume	5.0	5.0	5.0	5.0	5.0	5.0	5.0	5.0

Adapted from Liu [20]

(A400SB–A400RB) based on the ratio of the corresponding level of the dilute extract to 1.0 mL, and (3) adding A400RB back to the calculated values in (2), respectively, using the following general equation, where X mL represents 0.4, 0.6, or 0.8 mL and x represents 1, 2, or 3, corresponding to the dilute sample level of 0.4, 0.6, or 0.8 mL, respectively.

$$\mathrm{A400SB}x = (\mathrm{A400SB4} - \mathrm{A400RB}) \times (X\ \mathrm{mL}/1.0\ \mathrm{mL}) + \mathrm{A400RB} \tag{1}$$

Second, calculate % chymotrypsin inhibition (% CIn) for each of the four levels of the dilute extract, respectively, using the following general equation. Note that the reference reading (A400R) is the average of the duplicate measurements and that x represents the order of the four levels of the dilute extract (1, 2, 3 and 4 represent 0.4, 0.6, 0.8 and 1.0 mL, respectively:

$$\%\mathrm{CIn} = [(\mathrm{A400R} - \mathrm{A400RB}) - (\mathrm{A400S}x - \mathrm{A400SB}x)]/(\mathrm{A400R} - \mathrm{A400RB}) \tag{2}$$

Third, calculate CUI/mg sample for each of the four levels of the dilute extract, using the following general equation:

$$\begin{aligned}\mathrm{CUI/mg} &= \{[(\mathrm{A400R} - \mathrm{A400RB}) - (\mathrm{A400S}x - \mathrm{A400SB}x)]/0.01\}/\\ &\quad [(X\ \mathrm{mL\ of\ a\ dilute\ sample\ extract\ used\ for\ the\ assay})\\ &\quad \times(\mathrm{concentration\ of\ the\ dilute\ sample\ extract\ in\ mg\ sample/mL})]\\ &= \{[(\mathrm{A400R} - \mathrm{A400RB}) - (\mathrm{A400S}x - 400\mathrm{SB}x)] \times 100\}/\\ &\quad (\mathrm{mg\ sample\ in\ the\ assay})\end{aligned} \tag{3}$$

where A400R = reference reading, A400RB = reference blank, A400S = sample reading, A400SB = sample blank, x = the order of the four levels of the dilute extract, and X = mL dilute sample extract used for the assay = 0.4, 0.6, 0.8, or 1 mL (Table 1).

Fourth, calculate the linear relationship between CUI/mg and % CIn shown below, based on the four levels of the dilute extracts calculated in the second and third steps, by plotting the two sets of data in Excel and adding trend line with equation. Alternatively, the relationship can be calculated directly within Excel using function equations of Slope (CUI/mg cell range, % CIn cell range), Intercept (CUI/mg cell range, % CIn cell range), and RSQ (CUI/mg cell range, % CIn cell range):

$$y = ax + b \tag{4}$$

where y = CUI/mg sample, x = % CIn, a = slope, and b = intercept.

Fifth, using the above equation, calculate CUI/mg at 40% CIn:

$$\mathrm{CUI/mg\ at\ 40\%CIn} = 0.4a + b \tag{5}$$

3.4 Expression of CIA in Amounts of Chymotrypsin Inhibited (see Note 3)

For easier comparison among studies, CIA is preferably expressed as absolute amounts of chymotrypsin inhibited, based on CUI/mg sample at 40% CIn, using the standardized conversion factor of 1.5 determined between CUI and μg CId, where the reference chymotrypsin has a fixed specific activity of 150 BTNA [equivalent to 50 BTEE units/mg protein [21]:

$$\text{mg CId/g sample (same as } \mu\text{g/mg) at } 40\%\text{CIn} \\ = \text{CUI/mg} \times 1/1.5 = \text{CUI/mg} \times 0.667 \quad (6)$$

4 Notes

1. For having an accurate measurement, it is very important to properly dilute a sample extract with water so that 1.0 mL of the dilute sample extract causes 40–60% chymotrypsin inhibition, through preliminary trials.
2. Since in the same run, reference reading and reference blank can be shared, for each additional sample in the same run, only five more test tubes are required.
3. Although expression of CIA in amounts of chymotrypsin inhibited (mg CId/g sample) is preferred over expression of CIA in the arbitrary unit (CUI/mg sample), for easy calculation of mg CId/g sample, the arbitrary unit needs to be calculated first, followed by using the conversion factor of 1 μg CId = 1.5 CUI [21].

References

1. Liener IE (1994) Implications of antinutritional components in soybean foods. Crit Rev Food Sci Nutr 34(1):31–67
2. Guillamón E, Pedrosa MM, Burbano C, Cuadrado C, Sanchez MC, Muzquiz M (2008) The trypsin inhibitors present in seed of different grain legume species and cultivar. Food Chem 107:68–74
3. Boisen S (1983) Protease inhibitors in cereals, occurrence, properties, physiological role, and nutritional influence. Acta Agric Scand 33: 369–381
4. Losso JN (2008) The biochemical and functional food properties of the Bowman-Birk inhibitor. Crit Rev Food Sci Nutr 48:94–118
5. Kennedy AR (1998) The Bowman-Birk inhibitor from soybeans as an anticarcinogenic agent. Am J Clin Nutr 68:1406S–1412S
6. Cid-Gallegos MS, Corzo-Ríos LJ, Jiménez-Martínez C, Sánchez-Chino XM (2022) Protease inhibitors from plants as therapeutic agents- a review. Plant Foods Hum Nutr 77: 20–29. https://doi.org/10.1007/s11130-022-00949-4
7. Kakade MI, Simons N, Liener IE (1969) An evaluation of natural vs. synthetic substances for measuring the antitryptic activity of soybean samples. Cereal Chem 46:518–526
8. Kakade ML, Rackis JJ, McGhee JE, Puski G (1974) Determination of trypsin inhibitor activity of soy products: a collaborative analysis of an improved procedure. Cereal Chem 51: 376–382
9. Smith C, van Megen W, Twaalfhoven L, Hitchcock C (1980) The determination of trypsin inhibitor levels in foodstuffs. J Sci Food Agric 31:341–350
10. Hamerstrand GE, Black LT, Glover JD (1981) Trypsin inhibitors in soy products: modification of the standard analytical procedure. Cereal Chem 58:42–45
11. Liu K, Markakis P (1989) An improved colorimetric method for determining antitryptic

activity in soybean products. Cereal Chem 66: 415–422

12. AOCS (2017) American Oil Chemists' Society. Official methods and recommended practices. In: Method Ba 12-75. Trypsin inhibitor activity. First approval 1975; Reapproved 2017, 7th edn. The Society, Champaign, IL, USA
13. Liu K (2019) Soybean trypsin inhibitor assay: further improvement of the standard method approved and reapproved by American Oil Chemists' Society and American Association of Cereal Chemists International. J Am Oil Chem Soc 96:635–645
14. Liu K (2021) Trypsin inhibitor assay: expressing, calculating, and standardizing inhibitor activity in absolute amounts of trypsin inhibited or trypsin inhibitors. J Am Oil Chem Soc 98:355–373. https://doi.org/10.1002/aocs.12475
15. Liu K, Seegers S, Cao W, Wanasundara J, Chen J, da Silva AE, Ross K, Franco AL, Vrijenhoek T, Bhowmik P, Li Y, Wu X, Bloomer S (2021) An international collaborative study on trypsin inhibitor assay for legumes, cereals and related products. J Am Oil Chem Soc 98:375–390. https://doi.org/10.1002/aocs.12459
16. AOCS (2021) Assaying trypsin inhibitor activity. Method Ba 12a-2020. First Approval, 2020. In: Official methods and recommended practices of the AOCS. American Oil Chemists' Society, Urbana, Illinois, USA.
17. Tan NH, Rahim ZH, Khor HT, Wong KC (1984) Chymotrypsin inhibitor activity in winged beans (Psophocarpus tetragonolobus). J Agric Food Chem 32:163–166.
18. Liu K, Markakis P (1990) Effect of reactant mixing sequence on the chymotrypsin inhibition assay. Analyst 115:1143–1144.
19. Awosika T, Aluko RE (2019) Enzymatic pea protein hydrolysates are active trypsin and chymotrypsin inhibitors. Food Secur 8:200. https://doi.org/10.3390/foods8060200
20. Liu K (2022) Method development and optimization for measuring chymotrypsin and chymotrypsin inhibitor activities. J Food Sci 87: 2018–2033. https://doi.org/10.1111/1750-3841.16141
21. Liu K (2023) Chymotrypsin inhibitor assay: expressing, calculating, and standardizing inhibitory activity in absolute amounts of chymotrypsin inhibited. Sustain Food Prod 1(1): 30–44.

Chapter 14

Antinutrients in Pulses

Bhupinder Kaur and Kaustav Majumder

Abstract

Pulses stand out in the dietary realm due to their rich reservoir of both macro- and micronutrients. Beyond their nutritional benefits, these foods also harbor a complex array of compounds known as antinutrients (ANs). This group of ANs, which includes proteins, glycosides, tannins, saponins, and alkaloids, has the potential to hinder the gastrointestinal absorption of essential nutrients, thereby introducing possible health concerns. Particularly salient antinutrients found in pulses are phytic acid (PA), lectins, saponins, and various enzyme inhibitors, notably trypsin, chymotrypsin, and α-amylase inhibitors. This chapter meticulously explores the historical progression of methodologies designed to evaluate the levels of phytic acid, lectins, and saponins in pulses. Highlighting the advancements achieved over the last 50 years, the contemporary techniques highlighted herein not only boast improved sensitivity and accuracy but also facilitate meticulous quantification. This precision paves the way for comprehensive comparative analyses, especially relevant when examining plant-based proteins derived from pulses.

Key words Antinutrients, Phytic acid, Lectins, Saponins

1 Introduction

Amid growing consumer and research interest in plant-based protein sources, pulses have emerged as a focal point. Derived from legume plants, pulses offer a cost-effective reservoir of plant-based proteins, containing about two to three times more protein per serving compared to staple cereal grains like rice, barley, and wheat [1–3]. Moreover, a single cup serving of pulses provides nearly half of the daily recommended intake of fiber, encompassing both soluble and insoluble fibers, alongside resistant starches [4]. Consequently, pulses have garnered significant attention in dietary discussions, owing to their robust nutritional profile. Nevertheless, pulses contain a diverse array of compounds, encompassing proteins, glycosides, tannins, saponins, and alkaloids, which are collectively referred to as antinutrients (ANs) [5–7]. These ANs have the potential to hinder the gastrointestinal absorption of vital nutrients, leading to detrimental physiological outcomes [8, 9]. In severe

Yonghui Li (ed.), *Plant-Based Proteins: Production, Physicochemical, Functional, and Sensory Properties*,
Methods and Protocols in Food Science, https://doi.org/10.1007/978-1-0716-4272-6_14,

cases, such interference may contribute to developmental delays, stunting, and other adverse health effects. Notable ANs prevalent in pulses include phytic acid (PA), lectins, saponins, and various enzyme inhibitors such as trypsin, chymotrypsin, and α-amylase inhibitors.

Phytic acid (PA), a naturally occurring component, serves as the principal reservoir of phosphorus in plant seeds, constituting approximately 70% of the total phosphorus content [7, 10, 11]. Structurally, PA is characterized by six phosphate groups anchored to an inositol ring which enables it to form stable complexes with two or more positively charged minerals at the same time [12]. Intriguingly, some research has associated PA consumption with a plethora of health benefits, including diminished risks of cancer, heart disease, diabetes, and kidney stones, benefits previously predominantly attributed to dietary fiber [13–16]. Yet, the inherent negative charge of PA's phosphate groups tends to form complexes with positively charged mineral ions, thereby inhibiting their intestinal absorption.

Specifically, PA and phytate have the potential to bind with certain divalent minerals, notably iron, zinc, and calcium of nutritional importance [12, 17, 18]. This binding process forms insoluble complexes, hindering the absorption of these minerals in the gastrointestinal tract. Consequently, prolonged consumption of foods rich in PA or phytate may contribute to mineral deficiencies, particularly iron, zinc, and calcium, even if these minerals are present in adequate quantities in the diet [19]. PA's mineral molar ratio dictates its ability to elicit mineral deficiency and mineral bioavailability [20]. Mineral molar ratios less than critical levels (PA/Ca, 0.2; PA/Fe, 0.4; PA/Zn, 10; PA × Ca/Zn, 0.5) suggest no effect on mineral absorption and bioavailability, while levels above the critical levels suggest PA may elicit mineral deficiencies [12, 21].

The spatial configuration of the phosphate groups on the inositol ring plays a pivotal role in modulating their physiological interactions, with InsP6 emerging as the primary mineral-chelating entity, accounting for about 83% of the total inositol phosphates [7]. Five out of six phosphate groups in PA are present in thermodynamically stabilized equatorial position and can be transformed to axial orientation from equatorial orientation between pH 9.0 and 9.5 [22]. The hydrolysis of InsP6 within dietary sources carries significant nutritional ramifications; as phosphate groups dissociate from the inositol ring, the mineral-binding capacity of InsPn diminishes. This process augments the bioavailability of dietary minerals, facilitating their enhanced absorption within the body. From an analytical perspective, chromatographic methodologies predominantly serve to quantify PA levels, highlighting its indispensable role and complex interplay in the broader spectrum of dietary and physiological contexts. The most success has been obtained by chromatographic techniques, which allow the effective separation of isomers of inositol phosphates.

Lectins, also known as hemagglutinins, represent a significant class of storage and defense proteins, renowned for their capacity to agglutinate red blood cells. While their structural characteristics can be diverse, they typically feature at least one carbohydrate-binding domain. This domain enables lectins to engage in reversible and non-covalent interactions with specific dietary sugars or the carbohydrate components of certain enzymes present on the gastrointestinal tract's surface, thereby impeding digestion and nutrient absorption [23]. Interestingly, emerging research suggests potential therapeutic applications for lectins in various medical conditions, encompassing roles in immunomodulation for cancer [24–26], diabetes management [27, 28], HIV treatment [25, 26], and obesity mitigation [28, 29].

However, the multifaceted impact of lectins also encompasses potential toxicity, as evidenced by their capacity to hinder physiological development. Such interference extends to hormonal functionality, normal nutrient absorption, and the enzymatic activity integral to digestion. These adverse effects may compromise the integrity of the small intestine's epithelial layer, induce pancreas hypertrophy, elevate inhibitor levels, and influence lymphocyte myogenesis [28, 30, 31]. Consequently, it becomes imperative to ascertain the lectin concentration in any food product or derivative intended for human consumption, particularly for novel food items lacking an established consumption history. The hemagglutination assay stands as the predominant technique employed to assess the hemagglutination activity of foods [32].

Saponins, characterized by their unique structure comprising a hydrophobic steroid or triterpene group linked to a hydrophilic sugar moiety, possess a bitter taste and can exhibit toxicity at elevated concentrations [33]. This distinctive configuration enables saponins to interact with cell membranes, thereby impeding normal digestive system functions and forming complexes with minerals that further disrupt digestive processes and impede nutrient absorption [34, 35]. Additionally, saponins can induce hemolysis by interacting with cholesterol in erythrocyte membranes. Given their pronounced surface activity, saponins have garnered attention for their potential beneficial biological effects.

Enhancing the nutritional profile of pulses and mitigating their underutilization rely significantly on the reduction of ANs. A spectrum of conventional techniques, including dehulling, germination, cooking, soaking, boiling, membrane filtration, irradiation, fermentation, and enzyme treatments, has demonstrated effectiveness in diminishing ANs such as PA, lectins, tannins, saponins, and enzyme inhibitors. The selection of these methods, whether applied individually or in combination, is contingent upon achieving a specific target level of AN removal. However, it's imperative to recognize that none of these methods guarantees the complete elimination of all ANs, emphasizing the need to ascertain the quantity of these substances in diverse food formulations.

While specific regulatory rules regarding ANs levels in food are absent, a general recommendation persists to maintain a low threshold of these antinutrients to optimize nutritional benefits. As we pursue optimal nutritional quality, continuous research and comprehension of diverse methods for AN elimination will remain pivotal in advancing the incorporation of pulses into our diets.

Numerous essential compounds derived from pulses play a pivotal role in shaping nutrition and potentially influencing the absorption of vital dietary components. Although reducing the prevalence of certain compounds to a certain degree may be achievable, completely eliminating them poses a formidable challenge. Consequently, their presence within the food system highlights the urgent need for methodologies adept at precisely and sensitively quantifying their existence in both isolated pulse proteins and food products derived from pulse proteins.

Accurate and reliable methods for isolating and quantifying ANs in food products or derivatives intended for human consumption are imperative, particularly for innovative food items lacking established consumption histories. Such meticulous assessments facilitate valuable evaluations of their physiological implications. Various techniques have been employed for the quantitative assessment of antinutritional factors in foods, as referenced in studies by diverse authors. Notable methodologies include the determination of hemagglutination activities as per Jaffe [36], phytate quantification following Maga [37] protocols, tannin analysis as described by Dawra et al. [38], saponin assessment based on Brunner [39], and trypsin inhibitor activities are determined according to Liener [40]. Subsequent to these reports, numerous scientific articles have delved into the modifications, optimization, and standardization of methods for detecting ANs. The popular traditional techniques for the detection of ANs were precipitation, spectrophotometry, electrophoresis, potentiometric titrations, and colorimetry. These methods limited their application due to the tedious procedures, higher experimental errors, environmental factors and (thermal) pretreatment, and requirement of large quantities of samples (in titration). In response to these limitations, novel techniques emerged [7, 22, 41, 42], particularly chromatographic methods, offering improved accuracy, reduced complexity, and a more specific approach with robustness. This book chapter aims to summarize the most employed and adapted techniques for quantifying antinutritional factors.

2 Materials

Prepare all reagents/solutions using distilled water or analytical grade solvents, and store at the recommended storage conditions. Carefully follow all waste disposal regulations when disposing of waste materials.

2.1 Sample Preparation for Analysis

Materials needed for the sample preparation are dryer (preferably tray dryer or freeze dryer), cutting board, knife, professional mill, and sieve analyzer.

3 Methods

3.1 General Sample Preparation for Analysis

(a) Clean the sample properly.

(b) Cut the sample into small uniform pieces and mill it properly.

3.2 Phytic Acid Determination

3.2.1 Working Principle

Liquid chromatography separates components in a mixture based on their interactions with a stationary phase inside a column. The components are not retained but are washed along with the eluent at different flow rates. In a packed chromatography column, slight differences in the solute-solid affinities result in varied retention times, allowing the separation of complex compounds with high purities. The separated components are detected and quantified, providing valuable information about the composition of the sample. It provides relative abundance information (peak areas); however, precise quantification often requires calibration with known standards and consideration of response factors.

3.2.2 Extraction

1. Mix the 0.5 g sample with 10 mL HCl (0.5 M), and stir for 1 h (Fig. 1).
2. Centrifuge the extracted solution at 18000*g* for 10 min, and collect the supernatant.
3. Filter an aliquot of 1 mL of supernatant containing phytate with a 1-mL tuberculin syringe and a 13 mm/0.45 μm syringe filter.

3.2.3 Assay Procedure

1. Inject 1 μL of extract, and perform the elution of phytate using a binary liquid chromatography system with 50 × 4.6 mm PL-SAX strong anion-exchange column (Polymer Laboratories, Amherst, MA) equipped with a 7.5 × 4.6 mm guard column.
2. Achieve the elution of phytate with a 30-min linear gradient of 0.01 M 1-methylpiperazine, pH 4.0, added to 0.5 M NaNO3 in 0.01 M 1-methylpiperazine, pH 4.0, at a flow rate of 1 mL/min.

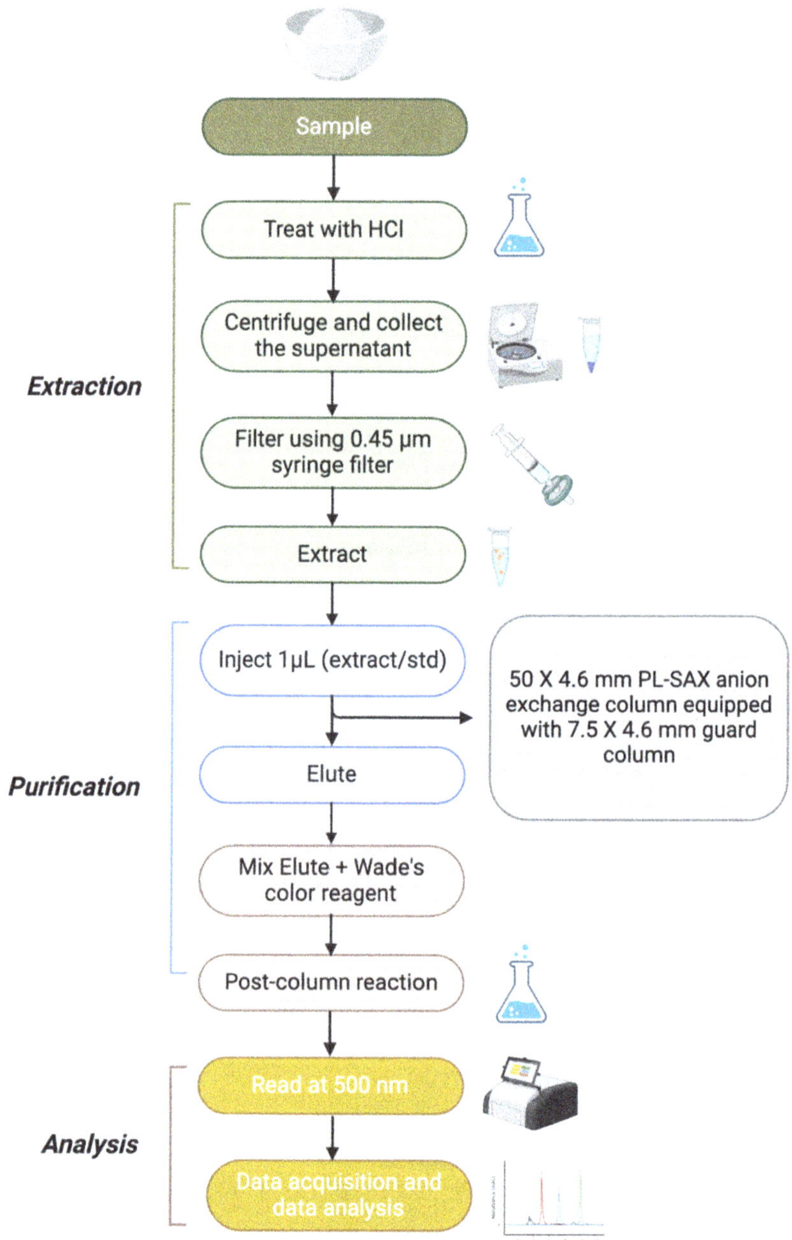

Fig. 1 Chromatographic approach for phytic acid analysis

3. Mix Wade's color reagent consisting of 0.015% (w/v) FeCl3 and 0.15% (w/v) 5-sulfosalicylic acid (also at a flow rate of 1 mL/min) and phytate eluted from the column in a mixing tee with inline check valves for both eluents installed prior to the mixing tee to prevent backflow.
4. Allow the post-column reaction to take place in a 0.05 × 210 cm polyetheretherketon tubing at the combined flow rate of 2 mL/min.

5. Record the absorbance at 500 nm using a UV detector, and process/integrate the detector signals and/or phytate peaks using the chromatographic data acquisition system by using sodium phytate as internal standard.

3.3 Lectin or Hemagglutination Determination

3.3.1 Working Principle

The hemagglutination assay is based on the ability of lectins to agglutinate (clump together) red blood cells. The degree of agglutination indicates lectin activity, providing qualitative information about lectin presence and activity in a sample.

3.3.2 Preparation of Erythrocyte Cell Suspension

1. Collect A, B, and O human blood types from the local blood transfusion center or blood bank (Fig. 2).
2. Take 2 mL of blood suspension, and centrifuge at 140*g* for 5 min at room temperature.
3. Remove the supernatant (approximately 1 mL) very carefully in order to obtain erythrocytes.
4. Wash the erythrocytes three times with 2 mL of cold PBS buffer, followed by a second centrifugation at 140*g* for 5 min at room temperature.
5. After washing, add 9 mL of PBS buffer to obtain 10% erythrocyte suspension.
6. Add 2 mL of a 10% erythrocyte suspension to 6 mL of PBS buffer to produce a 2.5% erythrocyte suspension, and store on ice for same day use.

3.3.3 Lectin Extraction

1. Add 30 g PBS buffer to a 2.5 g sample material, 5 g positive control, and 5 g negative control followed by incubation with strong agitation for 15 min and a sonication step for 5 min.
2. Centrifuge the slurry at 4300*g* for 20 min at 4 °C.
3. Transfer 5 mL of the supernatant to new 10 mL tubes, and centrifuge using same conditions as above, prior to the preparation of the dilution series.

3.3.4 Preparation of Dilution Series

1. Mix 500 μL of PBS buffer and 500 μL of undiluted extract or lectin stock solution.
2. After a 5 s whirl mix at 500 g, add 500 μL of above mix to 500 μL of PBS buffer to produce the next dilution.
3. Perform subsequent dilutions in the same manner.
4. Prepare dilutions for the samples and the negative control to obtain 1/2, 1/4, 1/8, 1/16, 1/32, 1/64, 1/128, 1/256, 1/512, 1/1024, and 1/2048 and for the positive control to obtain 1/4096, 1/8192, and 1/16,384 times dilution.
5. The purified lectin stock solutions (1 mg/mL) were diluted 23 times to reach a 1/8,388,608 dilution.

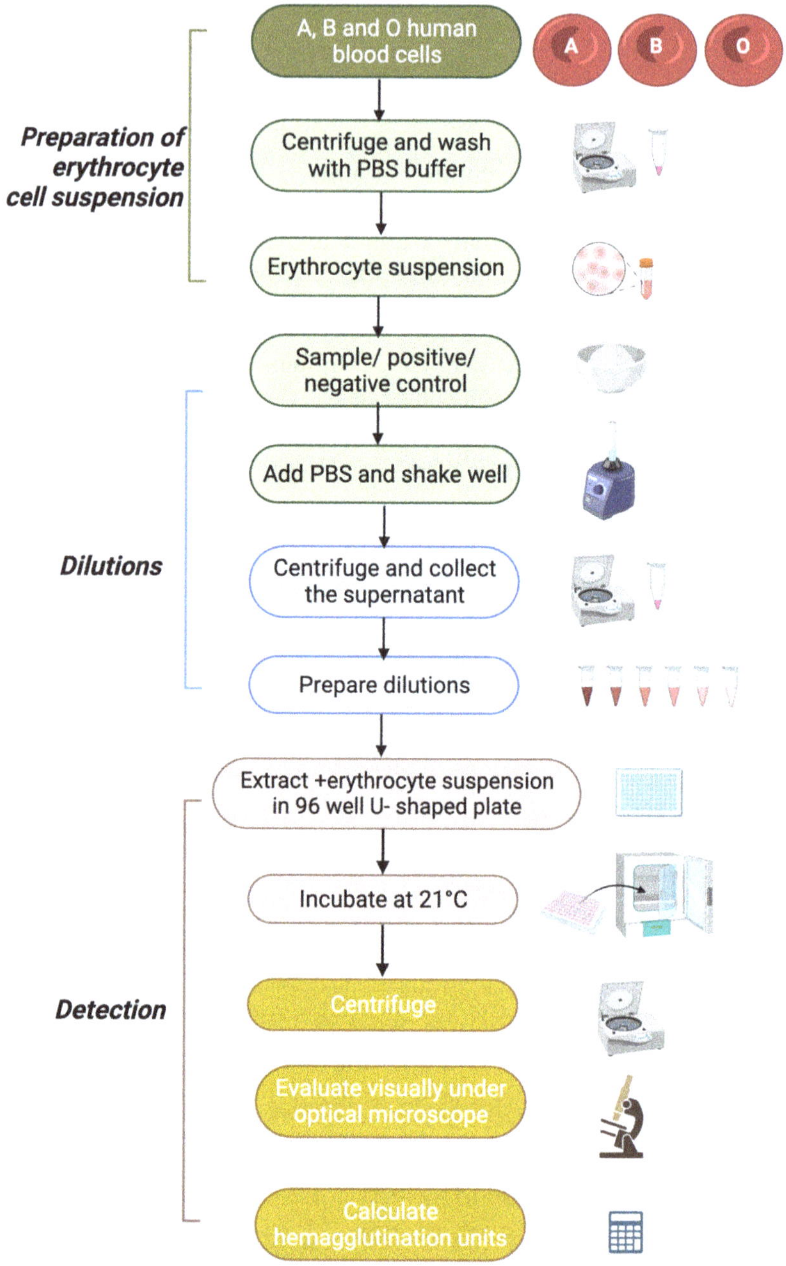

Fig. 2 An outline of a systematic protocol to analyze the ability of lectins to agglutinate red blood cells via hemagglutination assay

3.3.5 Assay Procedure

1. Perform hemagglutination assay in 96-well U-shaped microtiter plates as was described earlier.
2. Put 50 μL of the undiluted/diluted extracts or lectin stock dilutions or positive/negative control/blank control to wells in a descending dilution order.

3. Then, add 50 μL of 2.5% erythrocyte suspension to each well and shake gently.
4. Incubate the plates at 21 °C in a climate chamber for 1–20 h prior to centrifugation at 140*g* for 3 min at room temperature.
5. Immediately after centrifugation, evaluate the plates visually evaluated under an optical microscope.
6. As a prerequisite for plate quality approval, the blank and dilution series of the negative control must show no hemagglutination in all wells, and the dilution series of the positive control has to show hemagglutination between 1/512 and 1/1024 dilutions.
7. A pellet appearing as a dot on the bottom of the well confirms the absence of lectins or their inactive form.
8. Erythrocytes spread out covering the surface of the bottom of the well with no pellet formation, indicating the confirmation of the presence of active lectins.
9. Express the hemagglutination activity of the lectin as HAU/g dry material and calculate as the reciprocal of the highest dilution that caused hemagglutination multiplied by the initial dilution factor of the sample.
10. For example, if a sample causes hemagglutination at a maximum dilution of 1:128 and the initial dilution factor of the sample is 1, the titer of the stock solution is 128 [43].

3.4 Saponin Determination

The detection of saponins can be accomplished through two primary analytical methods: UV spectrophotometry and gas chromatography-mass spectrometry (GC-MS).

3.4.1 Working Principle of UV-Vis Spectrophotometry

UV-vis spectrophotometry allows for quantitative analysis of saponins and relies on the specific reaction between saponins and the reagent, ensuring that only saponins contribute to the color formation. It allows the quantitative analysis of saponins based on the measured absorbance, providing information about the saponin concentration in the sample.

3.4.2 Working Principle of GC-MS

Gas chromatography (GC) separates saponins based on their volatility, while mass spectrometer (MS) identifies and quantifies these compounds by their mass-to-charge ratios and fragmentation patterns. The precise quantification often requires comparison of relative abundance information of peak areas with known standard. It provides highly specific information about the molecular structure of saponins, aiding in accurate identification.

3.4.3 Extraction Step

1. Defat the sample using petroleum ether according to the AACCI method (30-25.01).
2. Extract 5 g of defatted sample with 80% methanol (MeOH (4): H2O (1)) for 3 h in Soxhlet reflux extractor (Fig. 3).

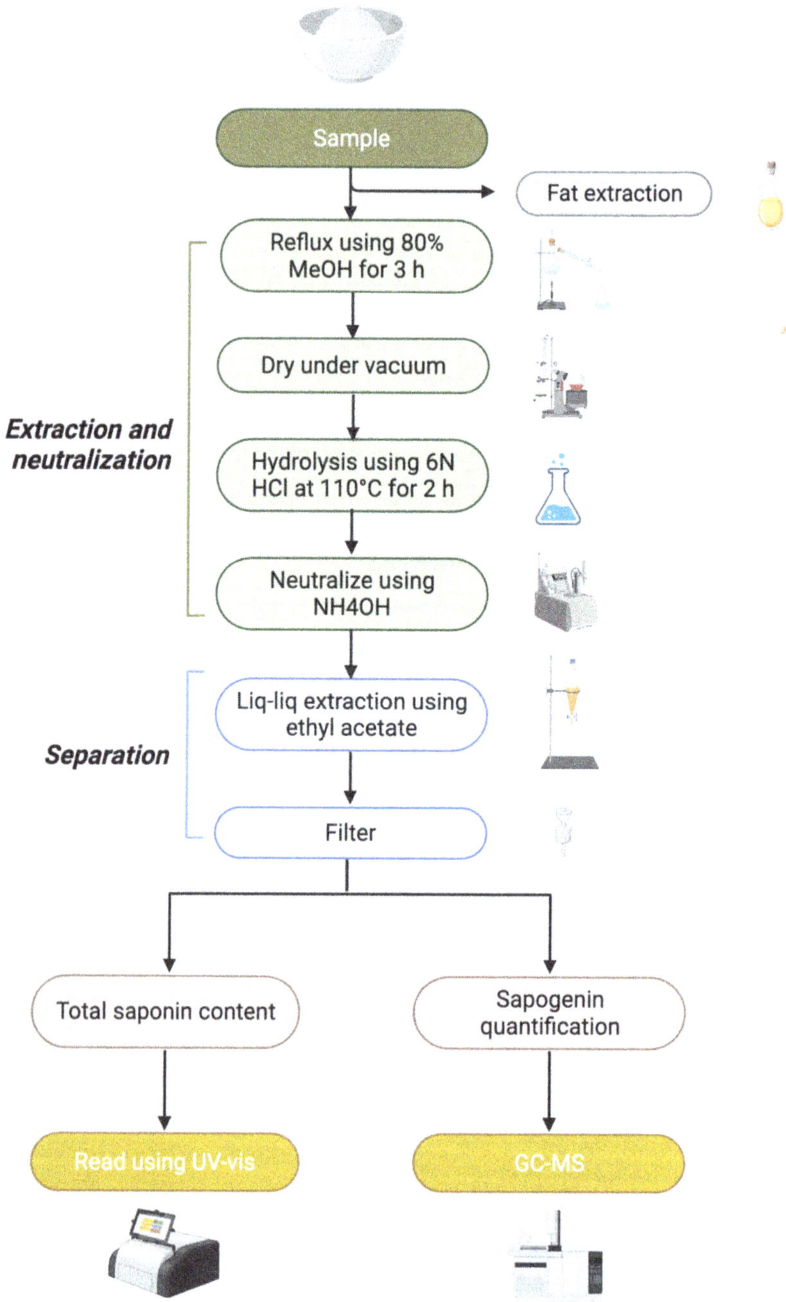

Fig. 3 An overview of analytical technique for the determination of saponins

3. Concentrate the crude extract under reduced pressure.
4. Hydrolyze 5 mL aliquot of the crude concentrate with 6 N HCl at 110 °C for 2 h.
5. Cool the hydrolysate, and neutralize with ammonia water (NH_4OH) followed by centrifugation at 3000*g* for 5 min.

6. Dissolve the concentrate in 4 mL of distilled water followed by liquid-liquid partitioning with 3 mL of ethyl acetate.
7. Repeat this step three times, combine the ethyl acetate fractions, and filter with sodium sulfate anhydrous (Na2SO4) to obtain Saponin extract.

3.4.4 Total Saponin Content by UV-Vis Spectrophotometry

1. Add 0.25 mL of saponin extract of sample solution to 1 mL of reagent mixture (glacial acetic acid/sulfuric acid 1:1 v/v) for color development.
2. Vortex the mixture and heat at 60 °C in a water bath for 30 min and then cool.
3. Record the absorbance of the sample at 527 nm in a spectrophotometer using glacial acetic acid as blank.
4. Calculate total saponin content, expressed as mg/mL of oleanolic acid equivalents by comparing with standard curve of oleanolic acid (0–1000 μg/mL).

3.4.5 Triterpenoid Sapogenin Quantification by GC-MS

1. Dry an aliquot of the final extract under a N_2 stream to analyze quinoa triterpenoid profile.
2. To derivatize the extracted saponins, add 100 μL of anhydrous pyridine, 100 μL of bis(trimethylsilyl)trifluoroacetamide (BSTFA), and 10 μL of cholesteryl decanoate (internal standard, IS) to the dried extract, and heat at 70 °C for 1 h.
3. Inject the derivatized extracts (2 μL) into GC-MS fitted with a Zebron ZB-5HT (Phenomenex, Torrance, CA) capillary column (30 × 0.25 mm × 0.25 μm).
4. Set the injector and detector temperatures as 350 °C.
5. Program the oven temperature from 160 to 220 °C at 15 °C/min, from 220 to 290 °C at 10 °C/min, for 7 min, from 290 to 330 °C at 8 °C/min hold for 15 min.
6. Use spitless mode and carrier pressure of 100 kPa.

3.4.6 Mass Spectrometry Conditions

1. Identify and confirm the structure of saponins by an ISQ single quadrupole mass spectrometer (Thermo Scientific, Waltham, MA) in an electron impact mode (70 eV).
2. Record the MS data in a full scan mode (mass range of m/z 50–900 amu).
3. Set MS transfer line and ion source temperature at 350 °C.
4. Perform the analyses in triplicates, and quantify sapogenin content quantified as peak area under the curve relative to that of the internal standard.

Acknowledgments

The figures used in this chapter were created using BioRender.com under a paid subscription plan.

References

1. Rafique H, Dong R, Wang X, Alim A, Aadil RM, Li L, Zou L, Hu X (2022) Dietary-nutraceutical properties of oat protein and peptides. Front Nutr 9:950400. https://doi.org/10.3389/fnut.2022.950400
2. Lignicka I, Graci (Balgalve) A, Zīdere-Laizāne AM (2022) Nutritious lentil and rice meal for sustainable vegan and pescatarian diet. Agron Res 20(1):229–234. https://doi.org/10.15159/AR.22.021
3. Samaranayaka A (2017) Lentil: revival of poor man's meat. In: Sustainable protein sources. Academic Press, Cambridge, pp 185–196. https://doi.org/10.1016/B978-0-12-802778-3.00011-1
4. Gurusamy S, Vidhya CS, Khasherao BY, Shanmugam A (2022) Pulses for health and their varied ways of processing and consumption in India – a review. Appl Food Res 2(2):1–9. https://doi.org/10.1016/j.afres.2022.100171
5. Purohit P, Rawat H, Verma N, Mishra S, Nautiyal A, Anshul, Bhatt S, Bisht N, Aggarwal K, Bora A, Kumar H, Rawal P, Kumar A, Kapoor R, Sehrawat J, Rather M, Naik B, Kumar V, Rustagi S, Preet MS, Jha AK, Gupta A (2023) Analytical approach to assess anti-nutritional factors of grains and oilseeds: a comprehensive review. J Agric Food Res 14:100877. https://doi.org/10.1016/j.jafr.2023.100877
6. Salim R, Nehvi IB, Mir RA, Tyagi A, Ali S, Bhat OM (2023) A review on anti-nutritional factors: unraveling the natural gateways to human health. Front Nutr 10:1215873. https://doi.org/10.3389/fnut.2023.1215873
7. Kumar Y, Basu S, Goswami D, Devi M, Shivhare US, Vishwakarma RK (2021) Anti-nutritional compounds in pulses: implications and alleviation methods. Legum Sci 4(2):e111. https://doi.org/10.1002/leg3.111
8. Nath H, Samtiya M, Dhewa T (2022) Beneficial attributes and adverse effects of major plant-based foods anti-nutrients on health: a review. Hum Nutr Metab 28:200147. https://doi.org/10.1016/j.hnm.2022.200147
9. López-Moreno M, Garcés-Rimón M, Miguel M (2022) Antinutrients: lectins, goitrogens, phytates and oxalates, friends or foe? J Funct Foods 89:104938. https://doi.org/10.1016/j.jff.2022.104938
10. Vashishth A, Tehri N, Tehri P, Sharma A, Sharma AK, Kumar V (2023) Unraveling the potential of bacterial phytases for sustainable management of phosphorous. Biotechnol Appl Biochem 70(5):1690–1706. https://doi.org/10.1002/bab.2466
11. Wieczorek D, Żyszka-Haberecht B, Kafka A, Lipok J (2022) Determination of phosphorus compounds in plant tissues: from colourimetry to advanced instrumental analytical chemistry. Plant Methods 18:22. https://doi.org/10.1186/s13007-022-00854-6
12. Dashti B, AL-Waalan T, AL-Fili B, Khashawi R, AL-Azmi B, Ejaz M (2023) Establishment of a phytate database in Kuwait for frequently consumed traditional and composite dishes in Kuwait: a study on the role of phytate in the bioavailability of iron and zinc using phytate-mineral molar ratios. J Food Compos Anal 121:105387. https://doi.org/10.1016/j.jfca.2023.105387
13. Pujol A, Sanchis P, Grases F, Masmiquel L (2023) Phytate intake, health and disease: "let thy food be thy medicine and medicine be thy food". Antioxidants (Basel) 12(1):146. https://doi.org/10.3390/antiox12010146
14. Brouns F (2021) Phytic acid and whole grains for health controversy. Nutrients 14(1):25. https://doi.org/10.3390/nu14010025
15. Raboy V (2020) Low phytic acid crops: observations based on four decades of research. Plan Theory 9(2):140. https://doi.org/10.3390/plants9020140
16. Nissar J, Ahad T, Naik HR, Hussain SZ (2017) A review phytic acid: as antinutrient or nutraceutical. J Pharmacogn Phytochem 6(6):1554–1560
17. Zhang YY, Stockmann R, Ng K, Ajlouni S (2022) Revisiting phytate-element interactions: implications for iron, zinc and calcium bioavailability, with emphasis on legumes. Crit Rev Food Sci Nutr 62(6):1696–1712. https://doi.org/10.1080/10408398.2020.1846014
18. Gupta RK, Gangoliya SS, Singh NK (2015) Reduction of phytic acid and enhancement of bioavailable micronutrients in food grains. J

Food Sci Technol 52(2):676–684. https://doi.org/10.1007/s13197-013-0978-y

19. Abdulwaliyu I, Arekemase SO, Adudu JA, Batari ML, Egbule MN, Okoduwa SIR (2019) Investigation of the medicinal significance of phytic acid as an indispensable antinutrient in diseases. Clin Nutr 28:42. https://doi.org/10.1016/j.yclnex.2019.10.002
20. Hassan LG, Abdulmumin U, Umar KJ, Ikeh OP, Aliero AA (2014) Nutritional and antinutritional composition of Strychnos innocua Del. (Monkey Orange) fruit pulp grown in Zuru, Nigeria. Nig J Basic Appl Sci 22(1–2): 33. https://doi.org/10.4314/njbas.v22i1.6
21. Karkle ENL, Beleia A (2010) Effect of soaking and cooking on phytate concentration, minerals, and texture of food-type soybeans. Food Sci Technol 30(4):1056–1060. https://doi.org/10.1590/S0101-20612010000400034
22. Marolt G, Kolar M (2020) Analytical methods for determination of phytic acid and other inositol phosphates: a review. Molecules 26(1): 174. https://doi.org/10.3390/molecules26010174
23. Muzquiz M, Varela A, Burbano C, Cuadrado C, Guillam'on E, Pedrosa MM (2012) Bioactive compounds in legumes: pronutritive and antinutritive actions. Implications for nutrition and health. Phytochem Rev 11: 227–244. https://doi.org/10.1007/s11101-012-9233-9
24. Kitaguchi D, Oda T, Enomoto T, Ohara Y, Owada Y, Akashi Y, Furuta T, Yu Y, Kimura S, Kuroda Y, Kurimori K, Miyazaki Y, Furuya K, Shimomura O, Tateno H (2020) Lectin drug conjugate therapy for colorectal cancer. Cancer Sci 111(12):4548–4557. https://doi.org/10.1111/cas.14687
25. Tsaneva M, van Damme EJM (2020) 130 years of plant lectin research. Glycoconj J 37(5): 533–551. https://doi.org/10.1007/s10719-020-09942-y
26. Akkouh O, Ng TB, Singh SS, Yin C, Dan X, Chan YS, Pan W, Cheung RCF (2015) Lectins with anti-HIV activity: a review. Molecules 20(1):648–668. https://doi.org/10.3390/molecules20010648
27. Govindappa M, Sadananda T, Ramachandra YL, Chandrappa CP, Rai PS, Prasad SK (2015) In vitro and in vivo antidiabetic activity of lectin (N- acetyl-galactosamine, 64 kDa) isolated from endophytic fungi, Alternaria species from Viscum album on alloxan induced diabetic rats. Integr Diabetes Obes 1(1): 11–19. https://doi.org/10.15761/IOD.1000104
28. van Buul VJ, Brouns FJPH (2014) Health effects of wheat lectins: a review. J Cereal Sci 59(2):112–117. https://doi.org/10.1016/j.jcs.2014.01.010
29. Stienstra R, Dijk W, van Beek L, Jansen H, Heemskerk M, Houtkooper RH, Denis S, van Harmelen V, Willems van Dijk K, Tack CJ, Kersten S (2014) Mannose-binding lectin is required for the effective clearance of apoptotic cells by adipose tissue macrophages during obesity. Diabetes 63(12):4143–4153. https://doi.org/10.2337/db14-0256
30. Santos AFS, da Silva MDC, Napoleão TH, Paiva PMG, Correia MTS, Coelho LCBB (2014) Lectins: function, structure, biological properties and potential applications. Curr Top Pept Protein Res 15:41–62
31. Gemede HF, Ratta N (2014) Antinutritional factors in plant foods: potential health benefits and adverse effects. Int J Food Sci Nutr 3(4): 284–289. https://doi.org/10.11648/j.ijnfs.20140304.1
32. Trombetta CM, Remarque EJ, Mortier D, Montomoli E (2018) Comparison of hemagglutination inhibition, single radial hemolysis, virus neutralization assays, and ELISA to detect antibody levels against seasonal influenza viruses. Influenza Other Respir Viruses 12(6): 675–686. https://doi.org/10.1111/irv.12591
33. Timilsena YP, Phosanam A, Stockmann R (2023) Perspectives on Saponins: food functionality and applications. Int J Mol Sci 24(17):13538. https://doi.org/10.3390/ijms241713538
34. Zhang Z, Liu C, Wu S, Ma T (2023) The non-nutritional factor types, mechanisms of action and passivation methods in food processing of kidney bean (*Phaseolus vulgaris* L.): a systematic review. Foods 12(19):3697. https://doi.org/10.3390/foods12193697
35. Johnson IT, Gee JM, Price K, Curl C, Fenwick GR (1986) Influence of saponins on gut permeability and active nutrient transport in vitro. J Nutr 116(11):2270–2277. https://doi.org/10.1093/jn/116.11.2270
36. Jaffe WG (1979) Haemagglutinins. In: Toxic constituents of plant foodstuffs. Academic Press, New York, pp 161–182
37. Maga JA (1983) Phytate: its chemistry, occurrence, food interaction, nutritional significance and methods of analysis. J Agric Food Chem 30(1):1–9. https://doi.org/10.1021/jf00109a001
38. Dawra RK, Makkar HP, Singh B (1988) Protein-binding capacity of microquantities of

tannins. Anal Biochem 170(1):50–53. https://doi.org/10.1016/0003-2697(88)90088-7

39. Brunner JH (1984) Direct spectrophotometer determination of Saponin. Anal Chem 34: 1314–1326
40. Liener IE (1980) Heat-labile anti-nutritional factors. In: Advances in legume science. Royal Botanic Gardens, Kew, pp 157–170
41. Menéndez-Rey A, González-Martos R, Ye P, Quiroz-Troncoso J, Alegría-Aravena N, Sánchez-Díez M, Maestu-Unturbe C, Bensadon-Naeder L, Ramírez-Castillejo C (2021) Quantification of lectins in Synsepalum dulcificum and comparison with reference foods. Food Chem 352:129341. https://doi.org/10.1016/j.foodchem.2021.129341
42. Uematsu Y, Hirata K, Saito K, Kudo I (2000) Spectrophotometric determination of saponin in Yucca extract used as food additive. J AOAC Int 83(6):1451–1454
43. Kawsar SMA, Matsumoto R, Fujii Y, Yasumitsu H, Dogasaki C, Hosono M, Nitta K, Hamako J, Matsui T, Kojima N, Ozeki Y (2009) Purification and biochemical characterization of a D-galactose binding lectin from Japanese sea hare (Aplysia kurodai) eggs. Biochemistry (Mosc) 74(7):709–716. https://doi.org/10.1134/s0006297909070025

Part IV

Physicochemical Characterization of Plant-Based Proteins

Chapter 15

Particle Size and Distribution Analysis of Protein Powders

Eric Nkurikiye, Bipin Rajpurohit, and Yonghui Li

Abstract

The usage of protein isolates and concentrates is increasing globally in the food industry. Protein isolates and concentrates are typically produced in a powder form to facilitate their incorporation into commonly consumed food products. Similar to other powders and flours, determining particle size is crucial for product developers. Proper particle size ensures that protein isolates and concentrates can be incorporated into food product without significantly altering system stability, as particle size influences water absorption, product flowability, and distribution. Particle size also influences the handling process of the powders and flours. In this chapter, three main methodologies for determining particle size and distribution are described. The first method is the traditional sieve shaker technique, which uses mechanical shaking and rotation movement to separate particles by their size. The second method is the jet sieve technique, which uses the power of pressurized air and vacuum to break agglomerated particles and sort them by size. The third method described here is the particle size and distribution analysis using imaging systems. This approach uses high-speed, high-definition cameras to capture the images of the particles as they are purged with pressurized air. This chapter details step-by-step guide to determining the particle size and distribution of protein -rich powders.

Key words Particle size, Particle distribution, Ro-tap, Jet sieve, Imaging analysis

1 Introduction

The demand for sustainable protein products is increasing, with plant-based options like pea, soybean, and wheat protein isolates and concentrates dominating the market. Other protein sources like lentils, chickpeas, and mung beans are also becoming popular [1].

To make transportation and usage easier, most protein isolates and concentrates are sold as powders. Just like any other powdered substance, it's crucial to understand the size and distribution of particles in these powders. This knowledge is essential for determining how to process the powder effectively and figuring out its possible uses [2]. Particle analysis is a discipline that provides essential insights into the proper handling of materials, particularly

Yonghui Li (ed.), *Plant-Based Proteins: Production, Physicochemical, Functional, and Sensory Properties*, Methods and Protocols in Food Science, https://doi.org/10.1007/978-1-0716-4272-6_15,

flour and powders. The complex properties of particles carry a significant influence over their behavior, directing the trajectory of industrial processes, pharmaceutical formulations, and food application [3, 4].

The historical importance of particle size in material grading highlights its significance. Particle size has long been utilized as a fundamental criterion for assessing the quality and appropriateness of materials for various applications, spanning fields as diverse as agriculture, construction, and pharmaceuticals. The ramifications of particle analysis stretch even further, contributing to the optimization of powder utilization. In food applications, particle size influences processes ranging from ingredients blending to texture modification. It is well-established that the interaction dynamics of large particles differ markedly from those of their smaller counterparts. This distinction, gleaned through meticulous particle analysis, empowers researchers and engineers to fine-tune their processes, kindling innovation across multiple sectors. Understanding how different particle sizes interact within a mixture is essential for achieving the desired sensory attributes, such as mouthfeel and flavor release, as well as for enabling precise ingredient dispersion [4–7].

Within the production landscape, particle size analysis assumes a pivotal role in the pursuit of standardization. By quantifying particle dimensions, manufacturers can establish robust quality control protocols, ensuring a consistent and dependable output. This level of harmonization is particularly critical in industries where even the slightest deviations can lead to significant variations in performance [3, 8]. In this chapter we will discuss three main particle analysis techniques: the sieve shaker, jet sieve, and imaging method.

1.1 The Sieve Shaker

The sieve shaker is a simple old technology used to determine the particle distribution in powder and other particles, yet it is widely used in industries globally. The sieve shaker uses the rotation and tapping actions coupled with gravity to separate varying size materials. Generally, particles percolate through different mesh until they are too big to pass through. The sieve shaker applications range from construction to food and pharmaceuticals industries. The sieve shaker principle was used to develop a standard method for analyzing the fineness of the feed materials [9]. In this chapter, that standard method was based to develop the protein isolate or concentrate measuring method.

1.2 Alpine Air Jet Sieve

Particle analysis plays a crucial role in a wide range of industries, from pharmaceuticals and food processing to construction and environmental monitoring. Accurate characterization of particle size is essential for understanding material properties, optimizing processes, and ensuring product quality. However, this task can be challenging due to various factors that influence particle behavior, such as aggregation and caking [10].

Agglomerated particles tend to cluster together, making them appear larger than they actually are, thus complicating traditional particle analysis methods. Inaccurate results from such techniques can have significant implications for research, development, and production processes. To address this issue and improve particle analysis, scientists and engineers have developed the innovative technique of air jet sieve particle analysis. The jet sieve particle analysis method leverages the power of vacuum air to efficiently break down aggregated particles, ensuring better separation and more precise size determination. This sophisticated approach has shown a tremendous promise in overcoming the limitations of conventional particle analysis methods. By applying vacuum extraction, jet sieve particle analysis effectively disperses agglomerated particles, providing a clearer and more accurate representation of particle size distribution.

1.3 Imaging System

Traditionally, powder analysis relied on manual methods such as the sieve shaker, which proved time-consuming, labor-intensive, and susceptible to human and material errors. However, propelled by rapid advances in imaging technology, a new era of powder analysis has dawned. Imaging equipment, such as QICPIC, has revolutionized powder analysis techniques. QICPIC, based on cutting-edge image analysis principles, offers a rapid, precise, and automated approach to characterizing powder particles. The imaging system employs high-speed cameras in conjunction with pressurized air to disperse clustered materials, propelling individual particles across the camera lens for comprehensive particle analysis. This innovative approach has demonstrated remarkable efficiency and speed in particle analysis. Moreover, it introduces the groundbreaking capability to measure particle size and distribution within emulsions, thereby revolutionizing the field of particle analysis [11, 12].

The imaging system technology gives the researcher clear observation of the morphology of particles being analyzed, and with more sophisticated equipment, the exact number of particles is reported [13]. The step-by-step guide of how to properly use the imaging system for particle analysis is detailed below.

2 Material

2.1 The Sieve Shaker Materials

1. Protein isolate, concentrate.
2. Ro-tap RX-29 (W.S. Tyler Inc., Mentor, OH, USA) (Fig. 1).
3. Sound Enclosure R-30050 (W.S. Tyler Inc., Mentor, OH, USA) (Fig. 2), which can minimize the noise generated.
4. Sieve agitator rubber balls (Fig. 3, right).

Fig. 1 Picture representing the Tyler Ro-tap with different sieves

Fig. 2 Tyler Ro-tap with closed sound enclosure (left) and with open sound enclosure (right)

Fig. 3 Image of sieve arrangement (left) and rubber bounce ball on the sieve (right)

5. Weighting balance.
6. Sieves with different opening (at least six different, must include the sieve with 106, 180, and 210 μm opening) (Fig. 3, left) as most protein particles are less than 200 μm.

2.2 Alpine Air Jet Sieve Materials

1. Protein isolates or concentrate.
2. Alpine air jet sieve (Hosokawa, E200 LS, Northborough, MA, USA) (Fig. 4).
3. Alpine air jet sieve screens (Hosokawa, Northborough, MA, USA).
4. Weighting balance.

2.3 Imaging System Materials

1. Protein isolates, or concentrate.
2. QICPIC particle analyzer (Sympatec GmbH System-partikel-technik, Clausthal-Zellerfeld, Germany) (Fig. 5).
3. RODOS dispenser (Sympatec GmbH System-partikel-technik, Clausthal-Zellerfeld, Germany) shown in Fig. 6.
4. Weighting balance.

3 Methods

3.1 The Sieve Shaker Method

1. Mix thoroughly at least 100 g of the representative protein powder sample.
2. Arrange the sieves in an ascending order with the pan on the bottom and the largest opening sieve on the top (preferably in

Fig. 4 Alpine air jet sieve instrument

Fig. 5 QICPIC particle analyzer equipped with RODOS

Fig. 6 Image (**a**) feeding the sample into the hopper and (**b**) sample automatically flowing and falling into the funnel of RODOS dry dispenser

this order: pan, 44, 106, 149, 180, 210, 297, and 425 μm mesh opening size sieves) as shown in Fig. 3.

3. Add the bouncing ball on the top of each sieve with the opening below 300 μm sieve to separate agglomerated powders of different sizes.
4. Put the powder sample on the top sieve (425 μm in this case).
5. Put the cover lid on the top sieve, and place the tapping baton over the cover lid.
6. Close the sound-insulating platform if available.
7. Set the run time to 10 min, and start the tapping and rotating mechanism.
8. Measure the amount of powder retained on the top of each sieve, and record the values using the format in Table 1. Precaution are detailed in **Note 1**.
9. With the use of computing software, plot the graph of particle distribution.
10. Use the following formula (Eq. 1) to measure the geometrical particle size as indicated by the ASABE method ANSI/ASAE S319.4 [9]:

Table 1
Powder retained by each sieve using the sieve shaker method

Sieve opening (μm)								
Sample	**25**	**44**	**106**	**149**	**180**	**210**	**425**	**GMD (μm)**
Pea protein isolate	0	4.1	5	1.5	4.4	84.1	0	261.6
Soy protein isolate	0	3.4	4.9	0.4	7.2	83.2	0	263.1
Wheat gluten	0	12	3.8	0.8	0.8	82.2	0.1	240.1

Powder mass in g
GMD: geometric mean diameter

$$d_{gw} = \log^{-1}\left[\frac{\sum_{i=1}^{n}(W_i \log \overline{d}_i)}{\sum_{i=1}^{n} W_i}\right] \tag{1}$$

d_{gw}: geometric mean diameter (GMD), μm
W_i: mass of flour retained on the *i*th sieve, *g*
d_i: aperture size of the *i*th sieve, μm
n: total number of sieves +1 (pan)

The Sieve Shaker Method Sample Result Data

Table 1 shows in detail how powders were retained on each sieve using the sieve shaker method. The results show that most powder was retained on the sieve with the opening of 210 μm. This means that most particles had particle size greater than 210 μm. Even with the use of sieve ball, protein isolate powders were very light and cohesive so they couldn't percolate properly to bottom sieves; instead they agglomerated together and got stuck on the 210 μm sieve. Despite the challenges posed by exceptionally lightweight and cohesive materials, the sieve shaker method remains a valuable tool for comparative analysis. It allows for a broad understanding of how different protein isolates interact with the sieving process, even if it may not provide a granular insight into the precise particle sizes. The sieve shaker method indicated that among the protein powder tested (pea, soy, and wheat), wheat gluten had relatively smaller particles (240.1 μm), while soy (263.1 μm) and pea (261.6 μm) were very comparable. The particle size distribution is shown in Fig. 7.

3.2 Alpine Air Jet Sieving Method

1. Power on the Alpine air jet sieve device.
2. Mix thoroughly a maximum of 100 g of the representative powder sample.
3. Place the sieve starting from the smallest opening on the jet sieve (25 μm sieve in this case).

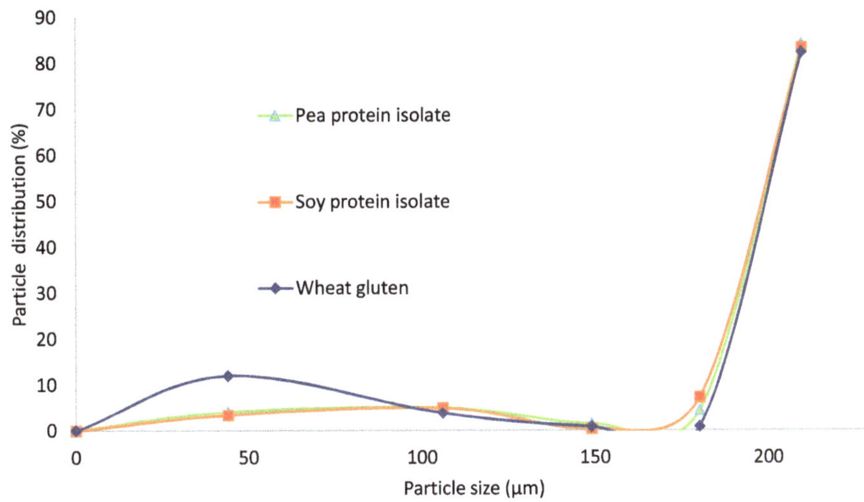

Fig. 7 Particle distribution of different protein isolates using the sieve shaker method

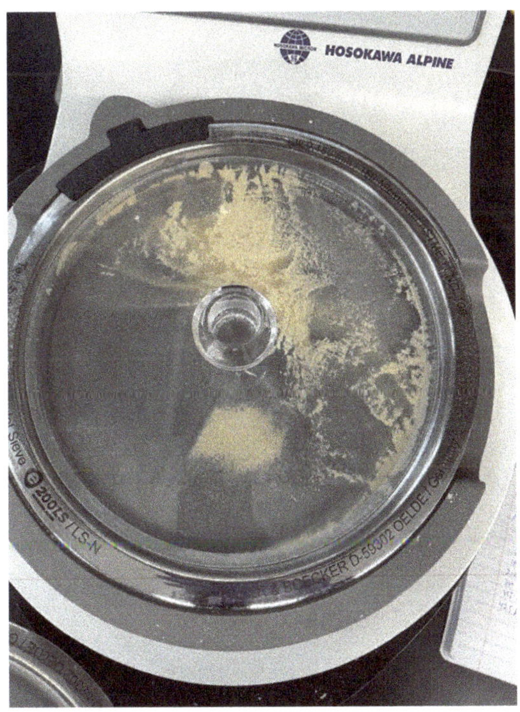

Fig. 8 The lid on the top of the sieve

4. Place your powder on the sieve top, distribute it evenly across the surface, and place on the lid (Fig. 8). Make sure there is no open space between the lid and the sieve as that will minimize the vacuum and lower the machine extraction efficiency.

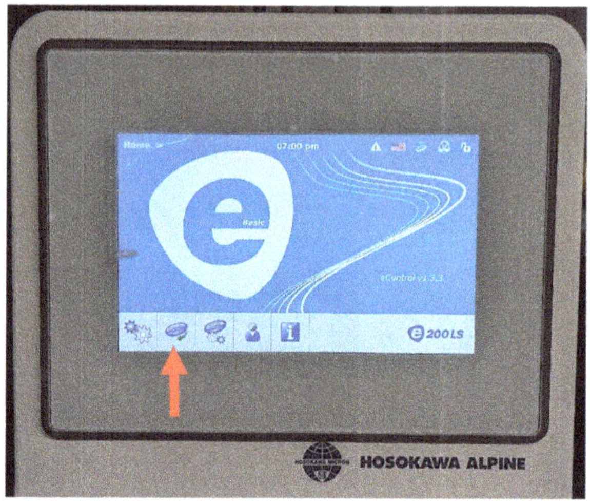

Fig. 9 Manual sieving button indicated by an arrow (second from bottom left) on the home screen

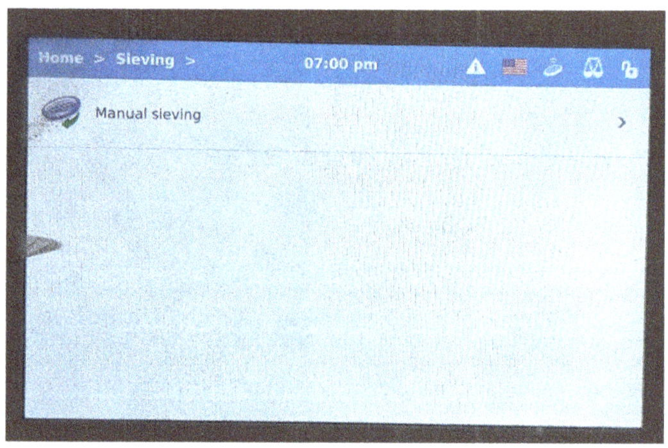

Fig. 10 Screen appears after hitting the manual sieving button on the home screen

5. Press the manual sieving button (second from bottom left, Fig. 9) on the device home screen. Press manual sieving on the next screen (Fig. 10).
6. The device will detect the sieve placed on the top and then input the weight of the sample placed on the sieve.
7. Click the blue start button to start the extraction (Fig. 11).
8. After the extraction, weigh the sample in the glass jar, and record it (as the pass through) using the format shown in Table 2.

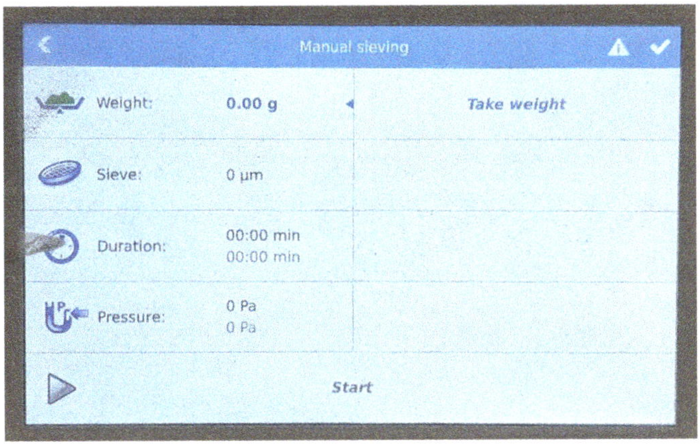

Fig. 11 Final screen where weight needs to be input and the start button

Table 2
Powder passing through each sieve using the jet sieve method

Sieve opening (µm)							
Sample	**25**	**106**	**149**	**180**	**210**	**425**	**GMD (µm)**
Pea protein isolate	0.1	1.8	77.7	19	1.0	0.0	130.4
Soy protein isolate	0.1	0.8	78	19.1	1.2	0.0	131.8
Wheat gluten	0.3	32.8	61.4	5.3	0.0	0.0	94.8

Powder mass in g
GMD: geometric mean diameter

9. Carefully take out the sieve, and replace it with the next in line on the device top.
10. Transfer the retained sample from the first sieve to the next sieve (now on the top of the device).
11. Repeat **steps 7–10** until finishing the final sieve according to the plan. Precaution are detailed in **Note 2**.
12. With the use of computing software, plot the graph of particle distribution.
13. Use Eq. 1 to measure the geometrical particle size of the powder (the passed-through powder mass of the certain sieve (x) is used as the retained powder mass of the previous sieve ($x-1$)).

Alpine Air Jet Sieving Sample Results

Table 2 and Fig. 12 detail the passed-through powder from each sieve and show the GMD for the protein isolates tested. With the application of pressurized air and vacuum, the jet sieve method was

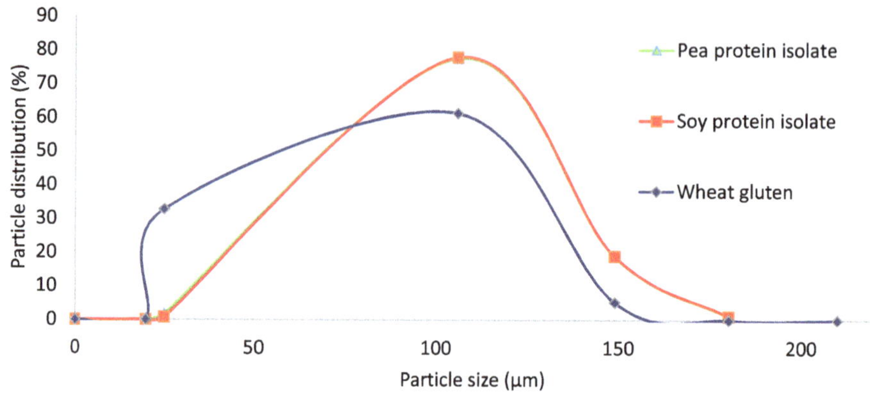

Fig. 12 Particle size distribution of protein isolate using the jet sieve method

able to better portray the particle distribution compared to the sieve shaker method. Wheat gluten materials seeped deeper compared to pea and soy protein isolate materials. The GMD data also indicated that wheat gluten (94.8 μm) materials were smaller in size compared to pea (131.4 μm) and soy (131.8 μm) protein isolates. This method is a big improvement in characterizing the powder's particle size and distribution as it can disintegrate agglomerated materials compared to the sieve shaker method, and the geometric mean diameter was reduced to a half compared to the sieve shaker method which shows how agglomerated particles can misrepresent the materials in general. Like the sieve shaker method, jet sieve showed that pea and soy protein isolates were very comparable, and they were larger in size than wheat gluten.

3.2.1 Imaging System Method

1. Power on the instrument and open the QICPIC software. Open the gas regulator of the compressed air to a pressure of between 50 and 70 psi.
2. The measuring range should be set to M6 (5–1705 μm) (Fig. 13).
3. The feeding rate should be set to 30% (Fig. 13).
4. The purging air pressure is set to three bars.
5. Measure 5 g of the powder, and transfer that to the hopper on the top of the equipment (Fig. 6).
6. Start the measurement by pressing the green button at the top left.
7. The measurement will be done automatically, and the live image can be observed on the small window on the right if enabled.
8. After measurement, the report will be generated by the software (Fig. 14). Precaution tips are detailed in **Note 3**.

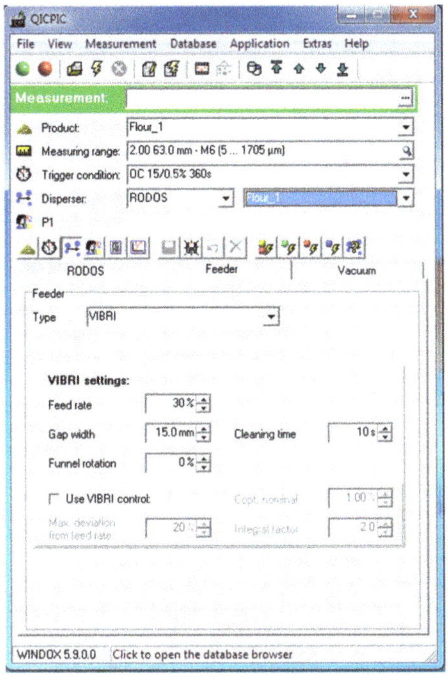

Fig. 13 Image showing equipment configuration before starting the particle analysis

9. Within the report different particle measurements are shown:

   X10 represents 10% of the total sample size.

   X50 represents 50% of the total sample size.

   X90 represents 90% of the total sample size.

   VMD (volumetric mean diameter) represents the average particle diameter based on the volume of the sample.

   SMD (Sauter's mean diameter) represents the average particle diameter basing on the surface area of the particles.

Imaging System Sample Results

The imaging system method for particle analysis showed again that wheat gluten had smaller particle size materials compared to pea and soy protein isolates (Table 3 and Fig. 15). The jet sieve system showed relatively comparable analysis to the imaging system implying that breaking down the agglomerated particles is crucial before doing detailed particle analysis. The particle distribution observed with the imaging system detailed how wheat gluten particles were composed by numerous fine particles in comparison to pea and soy protein isolates. The quickness and accuracy of the imaging systems put them as the best choice for particle size and distribution analysis.

Sympatec GmbH
System-Partikel-Technik

QICPIC Particle Size Analysis
WINDOX 5

QICPIC (QP0191) & RODOS/L, 2.00 63.0 mm - M6 (5 ... 1705 µm)
Flour_1 **2023-04-19, 10:13:40**.693

x_{10} = 23.12 µm **x_{50} = 66.63 µm** **x_{90} = 115.27 µm** **SMD = 51.11 µm** **VMD = 69.54 µm**
x_{16} = 33.76 µm x_{84} = 103.28 µm x_{99} = 168.40 µm

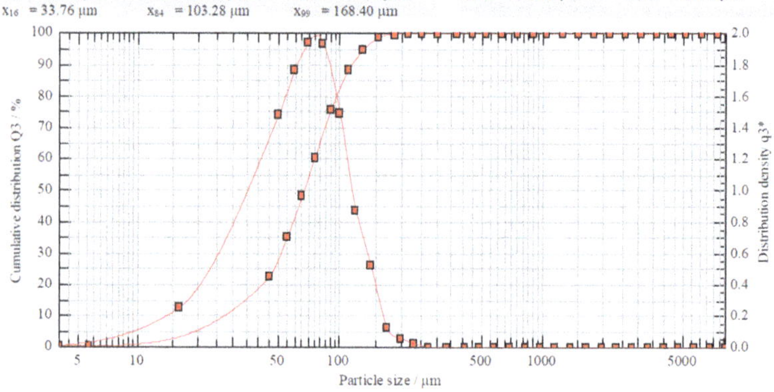

cumulative distribution

x_0/µm	Q_3/%	x_0/µm	Q_3/%	x_0/µm	Q_3/%	x_0/µm	Q_3/%
5.64	0.14	155.00	98.57	625.00	100.00	2550.00	100.00
45.00	22.34	185.00	99.54	750.00	100.00	3050.00	100.00
55.00	35.21	215.00	99.88	900.00	100.00	3650.00	100.00
65.00	48.03	250.00	100.00	1050.00	100.00	4350.00	100.00
75.00	60.11	300.00	100.00	1250.00	100.00	5150.00	100.00
90.00	75.41	375.00	100.00	1500.00	100.00	6150.00	100.00
110.00	88.34	450.00	100.00	1800.00	100.00	7350.00	100.00
130.00	94.62	525.00	100.00	2150.00	100.00	8750.00	100.00

distribution density (log.)

x_m/µm	q_3lg	x_m/µm	q_3lg	x_m/µm	q_3lg	x_m/µm	q_3lg
3.99	0.00	141.95	0.52	572.82	0.00	2341.47	0.00
15.93	0.25	169.34	0.13	684.65	0.00	2788.82	0.00
49.75	1.48	199.44	0.05	821.58	0.00	3336.54	0.00
59.79	1.77	231.84	0.02	972.11	0.00	3984.66	0.00
69.82	1.94	273.86	0.00	1145.64	0.00	4733.13	0.00
82.16	1.93	335.41	0.00	1369.31	0.00	5627.83	0.00
99.50	1.48	410.79	0.00	1643.17	0.00	6723.28	0.00
119.58	0.87	486.06	0.00	1967.23	0.00	8019.51	0.00

Fig. 14 Sample report from the QICPIC system

Table 3
Particle size of different protein isolates measured using the imaging method

Sample	X_{10} (µm)	X_{50} (µm)	X_{90} (µm)	SMD (µm)	VMD (µm)
Pea protein isolate	37.98	87.13	148.74	67.03	91.25
Soy protein isolate	33.94	89.41	185.72	67.19	101.74
Wheat gluten	23.12	66.63	115.27	51.11	69.54

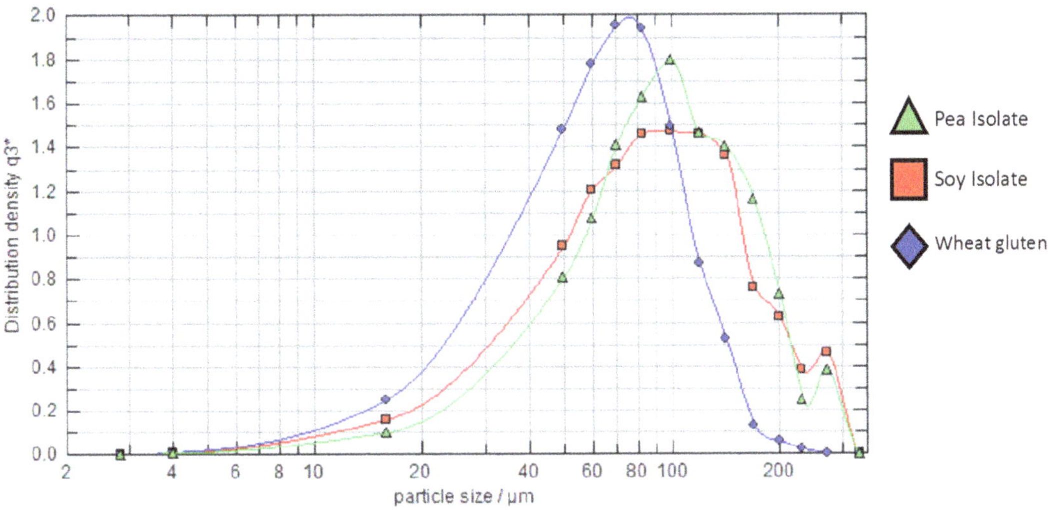

Fig. 15 Particle distributions of different protein isolates using the imaging method

4 Notes

1. The sieve shaker:
   - Some materials may get stuck on larger sieve even if they are fine. Using different cleaning brushes coupled with rubber sieve ball is recommended for such materials.
   - It is efficient to determine the largest approximate sieve to use to avoid using sieves that are too large for the sample being tested.
   - The use of the sieve shaker is not highly recommended for determining particle size of fine materials such as protein powders.
2. Alpine air jet sieve:
   - Always ensure that the collecting jar is attached to the system before running the analysis.
3. QICPIC Imaging system:
   - The imaging systems are relatively complex, and incorrect system configuration will halt the analysis.
   - The report format can be adjusted by the user.
   - Different report templates serve different purposes; always use the appropriate template for the needed data.

Acknowledgments

This is contribution no. 24-261-B from the Kansas Agricultural Experimental Station. This work was supported in part by the USDA National Institute of Food and Agriculture Hatch project (Grant Accession No. 7003330). The authors thank Dr. Kaliramesh Siliveru for providing the Alpine air jet sieve instrument used in this study.

References

1. Fernando S (2022) Pulse protein ingredient modification. J Sci Food Agric 102(3): 892–897
2. Sozer N, Holopainen-Mantila U, Poutanen K (2017) Traditional and new food uses of pulses. Cereal Chem 94(1):66–73
3. Bhandari B, Bansal N, Zhang M, Schuck P (2013) Handbook of food powders: processes and properties. Elsevier
4. Dodds J (2013) Techniques to analyse particle size of food powders. In: Handbook of food powders. Elsevier, pp 309–338
5. Chen J, Khandelwal N, Liu Z, Funami T (2013) Influences of food hardness on the particle size distribution of food boluses. Arch Oral Biol 58(3):293–298
6. Robins MM (2006) Particle size analysis in food. In: Encyclopedia of analytical chemistry: applications, theory and instrumentation. https://doi.org/10.1002/9780470027318.a1021
7. Schubert H (1987) Food particle technology. Part I: properties of particles and particulate food systems. J Food Eng 6(1):1–32
8. Fitzpatrick JJ, Barringer SA, Iqbal T (2004) Flow property measurement of food powders and sensitivity of Jenike's hopper design methodology to the measured values. J Food Eng 61(3):399–405
9. ASABE method ANSI/ASAE S319.4 (2008) Method of determining and expressing fineness of feed materials by sieving. American Society of Agricultural and Biological Engineers. https://www.asabe.org/
10. Siliveru K, Kwek JW, Lau GM, Ambrose RK (2016) Image analysis approach to understand the differences in flour particle surface and shape characteristics. Cereal Chem 93(3): 234–241
11. Brown L (2009) Imaging particle analysis: resolution and sampling considerations. Fluid Imaging Technologies, Inc, Scarborough
12. Khan A, Munir MT, Yu W, Young BR (2020) A review towards hyperspectral imaging for real-time quality control of food products with an illustrative case study of milk powder production. Food Bioprocess Technol 13:739–752
13. Yu W, Hancock BC (2008) Evaluation of dynamic image analysis for characterizing pharmaceutical excipient particles. Int J Pharm 361(1–2):150–157

Chapter 16

Bulk Properties of Protein Powders Using Hosokawa Powder Tester PT-R

Rania Marie Buenavista and Kaliramesh Siliveru

Abstract

Plant-based protein powders exhibit varying flow characteristics during processing, handling, and storage. Various factors, such as physicochemical characteristics, processing conditions, and environmental conditions, determine the ease in powder flowability. Thus, it is critical to consider the flow behavior of protein powders in the design of equipment and optimization of processing conditions. This chapter provides an overview on characterization of the flow behavior of protein powders in terms flow properties accounted in Carr's flowability and floodability indices using the Hosokawa Micron Powder Tester. Specifically, this chapter encompasses standard protocols for measuring aerated and packed bulk density, compressibility, angle of repose, angle of spatula, cohesion, angle of fall, angle of difference, dispersibility, and uniformity. This chapter also evaluates and interprets the flowability and floodability indices of select plant-based protein powders, including pea, soy, and wheat gluten.

Key words Hosokawa powder tester, Plant protein powder, Flowability, Floodability, Bulk density, Angle of repose

1 Introduction

Plant-based protein ingredients are commercially produced in a powder form. Due to their reduced water content, plant protein powders offer advantages over their liquid counterparts, such as increased shelf-life, which translates to longer preservation of their nutritional, physicochemical, and organoleptic properties [1]. Plant protein powders can be classified depending on its purity: concentrates (50–80% protein) and isolates (greater than 80%). The difference in chemical composition of protein concentrates and isolates from varying sources can influence difference in cohesiveness and particle-to-particle interactions [2]. Due to the increasing variety of plant protein sources, plant protein powders are insufficiently characterized.

Yonghui Li (ed.), *Plant-Based Proteins: Production, Physicochemical, Functional, and Sensory Properties*, Methods and Protocols in Food Science, https://doi.org/10.1007/978-1-0716-4272-6_16,

At an industrial level, the flow behavior of plant protein powders, regardless of its classification, plays a crucial role in the design of process flow and equipment and use of flow aids to ensure its efficient bulk handling and storage [3]. In both processing and handling operations, protein powders are subjected from its stationary state to its moving state. As the operation becomes more complex, plant protein powders undergo more transition stages within an individual or a series of equipment. Hence, it is important to determine the flow characteristics of plant protein powders during its stationary condition and transition stages.

Powder flowability is a measure of the powder's tendency to transition from its stationary condition to its moving condition. The physical conditions of the powder influence its flow characteristics, which include particle size distribution, particle shape, cohesiveness, adhesiveness, electrostatic charges, compressibility, void fraction, moisture content, environmental temperature and humidity, and friction coefficient of particle surface [4, 5]. Carr's flowability index is a method of evaluation of powder flowability that accounts for various flow properties such as angle of repose, compressibility, angle of spatula, and cohesiveness. On the other hand, Carr's floodability index is a method of evaluation of powder floodability—the tendency to exhibit an unstable liquid-like behavior that happens when powder is aerated [4]. Carr's indices were empirically attained after the analysis of nearly 3000 types of powder systems. Hosokawa Micron Powder Tester has been developed to measure the flow properties needed for determining Carr's flowability and floodability indices. The use of this instrument allows standardization and mechanization of the measurement methods of powder flow characteristics, thus minimizing human errors. Good flowability is exhibited by powders characterized by obtuse angles of repose and spatula, low compressibility, and low cohesiveness index. On the other hand, higher tendency for powder to flood is observed for powders with high angle of difference, high dispersibility index, and high angle of fall [5]. Tables 1 and 2 show the point scores for powder flowability and floodability evaluation based on Carr's measurement method, respectively. Aside from flowability and floodability indices, Hosokawa Micron Powder Tester is also developed to measure both aerated and packed bulk density of powders, which are important in determining appropriate size of hopper, conveying, or bin capacity [6].

This chapter provides a step-by-step guide in using the Hosokawa Microns Powder Tester PT-R for measuring flowability and floodability properties of protein powders. Moreover, flowability and floodability properties specific to pea, soy, and wheat gluten powder are also presented in this chapter.

Table 1
Carr's point scores for powder flowability evaluation [6]

Degree of flowability	Flowability index	Necessity of bridge-breaking measures	Angle of repose		Compressibility		Angle of spatula		Uniformity[a]		Cohesion[b]	
			Degree	Index	%	Index	Degree	Index	No.	Index	%	Index
Very good	90–100	Not required	>25	25	<5	25	<25	25	1	25		
			26–29	24	6–9	23	26–30	24	2–4	23		
			30	22.5	10	22.5	31	22.5	5	22.5		
Fairly good	80–89	Not required	31	22	11	22	32	22	6	22		
			32–34	21	12–14	21	33–37	21	7	21		
			35	20	15	20	28	20	8	20		
Good	70–79	Sometimes vibration is required	36	19.5	16	19.5	39	19.5	9	19		
			37–39	18	17–19	18	40–44	18	10–11	18		
			40	17.5	20	17.5	45	17.5	12	17.5		
Normal	60–69	Bridging will take place at the marginal point	41	17	21	17	46	17	13	17		
			42–44	16	22–24	16	47–59	16	14–16	16		
			45	15	25	15	60	15	17	15	<6	15
Not good	40–59	Required	46	14.5	26	14.5	61	14.5	18	14.5	-9	14.5
			47–54	12	27–30	12	62–74	12	19–21	12	10–29	12
			55	10	31	10	75	10	22	10	30	10
Bad	20–39	Powerful measures should be provided	56	9.5	32	9.5	76	9.5	23	9.5	31	9.5
			57–64	7	33–36	6	77–89	7	24–26	7	32–54	7
			65	5	37	5	90	5	27	5	55	5
Very bad	0–9	Special apparatus and techniques are required	66	4.5	38	4.5	91	4.5	28	4.5	56	4.5
			67–89	2	39–45	2	92–99	2	29–35	2	57–59	2
			90	0	>45	0	>99	0	>35	0	>79	0

[a] Use these values for granules or granular powder wherein uniformity can be measured
[b] Use these values for fine and cohesive powders wherein cohesion can be measured

Table 2
Carr's point scores for powder floodability evaluation [6]

Degree of floodability	Floodability index	Measure for flushing prevention	Flowability		Angle of fall		Angle of difference		Dispersibility	
			Degree	Index	%	Index	Degree	Index	No.	Index
Very high	80–100	Rotary seal must be used	>60	25	<10	25	>30	25	<50	25
			59–56	24	11–19	23	29–28	24	49–44	24
			55	22.5	20	22.5	27	22.5	43	22.5
			54	22	21	22	26	22	42	22
			53–50	21	22–24	21	25	21	41–36	21
			49	20	25	20	24	20	35	20
Fairly high	60–79	Rotary seal is required	48	19.5	26	19.5	23	19.5	34	19.5
			47–45	19.5	27–29	18	22–20	18	33–29	18
			44	19.5	30	17.5	19	17.5	28	17.5
			43	19.5	31	17	18	17	27	17
			42–40	19.5	32–39	16	17–16	16	26–21	16
			39	19.5	40	15	15	15	20	15
Tends to flush	40–59	Sometimes rotary seal is required	38	14.5	41	14.5	14	14.5	19	14.5
			37–34	12	42–49	12	12	12	18–11	12
			33	10	50	10	10	10	10	10
May flush	23–39	Rotary seal is necessary depending on flow speed and feed conditions	32	9.5	51	9	9.5	9.5	9	9.5
			31–29	8	52–56	8	8	8	8	8
			28	6.25	57	7	6.25	6.25	7	6.25
Won't flush	0–24	Not required	27	4	58	6	6	6	6	6
			26–23	3	59–64	5–1	3	3	5–1	3
			<23	0	>64	0	0	0	0	0

2 Materials

- Protein powder (≥100 g for each parameter test).
- Fine paint brush or duster.
- Hosokawa Micron Powder Tester PT-R.
- PT-R accessory kit: sieve, spacer, sieve extension, sieve holding bar, funnels for angle of repose, table, pan, scoop, cup extension, cup for bulk density, scraper, vibrating chute, stationery chute, spatula pan holder, spatula angle measurement unit, seal cap, spatula pan, watch glass, impacting device, tap lifting bar, and electric balance.

3 Methods

Below are the step-by-step procedures in operating specific flowability and floodability tests using the Hosokawa Micron Powder Tester PT-R. These methodologies were conducted for pea, soy,

Table 3
Flowability and floodability characteristics of pea, soy, and wheat gluten protein powders

Flowability properties	Pea	Soy	Wheat gluten
Angle of repose (°)	84.70 ± 1.75	84.83 ± 0.21	85.00 ± 0.10
Aerated bulk density (kg/m^3)	374.80 ± 2.82	306.80 ± 2.71	393.63 ± 5.60
Packed bulk density (kg/m^3)	563.00 ± 0.10	536.37 ± 5.37	726.7 ± 2.42
Compressibility (%)	33.42 ± 0.49	42.79 ± 1.01	45.83 ± 0.90
Angle of spatula (°)	71.27 ± 3.11	72.73 ± 3.81	69.40 ± 0.75
Uniformity (%)	2.70	3.34	3.41
Floodability properties			
Angle of fall (°)	80.97 ± 1.75	79.23 ± 4.08	77.80 ± 3.01
Angle of difference (°)	3.73 ± 2.28	5.60 ± 4.20	7.20 ± 2.91
Dispersibility (%)	14.05 ± 5.79	26.24 ± 6.42	24.96 ± 4.19

and wheat gluten proteins in determining the flowability and floodability parameters and are summarized in Table 3. Follow **Notes 1** and **2** for each bulk property test to be performed.

3.1 Starting and Shutting Down the Hosokawa Micron Powder Tester PT-R

1. Connect the powder tester to a powder source.
2. Press the power button at the lower right side of the powder tester. The screen of the powder tester will show its home page with various flowability and floodability parameters.
3. To shut down the powder tester, return to the home page and then press the power button. Make sure to clean and return the powder tester and its accessories to the storage drawer.

3.2 Aerated and Packed Bulk Density Tests

Bulk density is the mass of powder that can occupy a specific volume. It is the ratio of the powder mass and the volume it occupies. Two measurement methods can be used to evaluate bulk density of a powder: aerated and packed bulk densities [7].

Aerated and packed bulk density tests are usually performed sequentially. After measuring aerated bulk density, the same container holding the sample can be tapped for a certain number of times to measure the packed bulk density.

3.2.1 Aerated Bulk Density

Aerated bulk density is the density of a loose powder that is free of any consolidation force that is directly applied to the sample. The powder is allowed to disperse and settle in a sample cup by means of gravity. Lower bulk density is usually observed in powders with a strong structural strength as these particles resist the occurrence of collapse when dispersed in the cup. Meanwhile, lower structural strength of a powder causes collapse, thus resulting in high bulk density. Moreover, lower particle-to-particle and particle-to-surface frictions lead to an increase in bulk density [7]. The setup for this test includes a cylindrical steel container with known volume, a 750 μm sieve with extension, and a vibrating chute. The powder will be vibrated at a frequency of 1 m/m until complete filling of cylinder and formation of powder heap are achieved. Excess powder will be removed from the top of sample cup. Aerated bulk density is computed by dividing the mass of the powder by the volume of the container [4]. Below is the step-by-step procedure to perform this test using the Hosokawa powder tester:

1. Click "Aerated Bulk Density" icon in the home screen.
2. Assemble the parts as shown in Fig. 1. Then, press "Next" after the assembly.
3. Tare the weighing balance by pressing the "Balance" button on the screen. Wait until the balance shows "0 g."
4. Place the cup on the weighing balance to measure its weight. The screen will show the cup weight once measurement is done. Press the "Next" button to proceed. The screen proceeds to the "Vibration time" tab.
5. Set the vibration time. The standard time is set to 30 s.
6. Press "Next" to proceed to "Sample feed" tab.
7. Pour the powder sample until it fills the sieve.

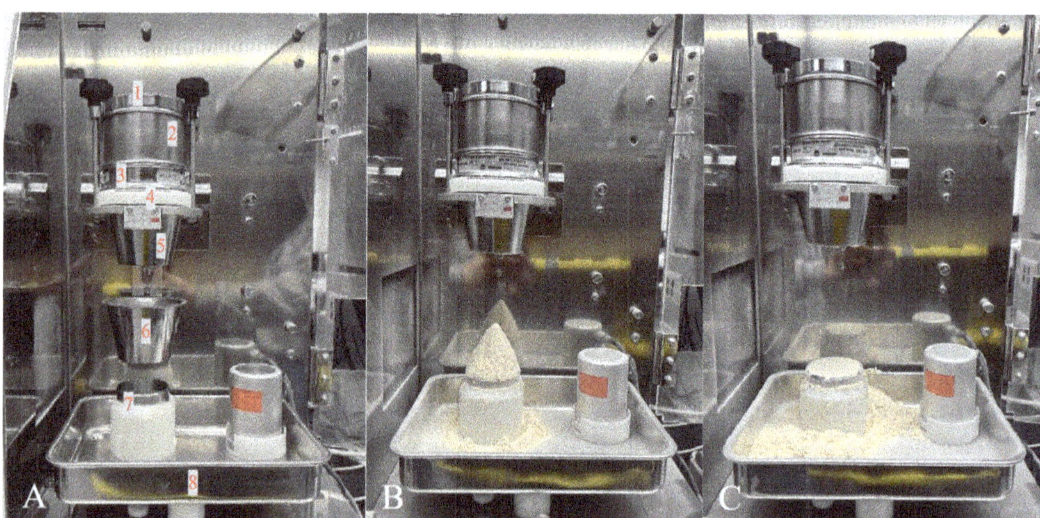

Fig. 1 Aerated bulk density test: (**a**) Assembly of parts (1, sieve holding bar; 2, sieve extension; 3, sieve (710 μm); 4, spacer; 5, vibration chute; 6, fixed chute; 7, cup; and 8, pan). (**b**) B Complete feeding of powder sample. (**c**) Leveled top surface of the sample cup prior to measurement

8. Press the "Start" button to start the vibration. During vibration, set the vibration intensity by turning the knob (located at the lower right of the front side of the powder tester) clockwise until the right triangle image screwed at the sieve holder becomes two triangles intersecting at point 1 as shown in Fig. 2. *See* **Notes 3** and **4** to make sure that the vibration is set correctly.
9. Wait until the sample cup is filled with powder forming a mound on top. Sufficient sample feeding is shown in Fig. 1. In case the sample feeding is insufficient, stop the test, load more sample into the sieve, and then press "Vibration" again.
10. After sufficient sample feeding, use the scraper to remove excess powder, and level the sample to the rim of the sample cup. Hold the scraper at its shortest side vertically, and scrape the powder by pressing the scraper against the rim of the sample cup. Follow **Note 6** to make sure leveling the sample is done correctly.
11. Follow screen prompt for the necessary buttons to click prior to putting the sample cup into the weighing balance.
12. Carefully lift the sample cup to the weighing balance to measure the sample weight. Press "Next" to proceed.
13. After measuring the sample weight, the powder tester will automatically compute for the aerated bulk density of the sample, which is the ratio of the weight of the powder and cup volume.

3.2.2 Packed Bulk Density Test

Packed bulk density is the density of a loose powder measured after compression caused by tapping the container, which displaces the entrained air within the particles across its volume. This measurement represents the bulk density of a material following handling, storage, and transportation [7]. In addition to the setup used for the aerated bulk density test, an open-ended plastic cylinder extension is used for the packed bulk density test. After measuring aerated bulk density, this cylinder extension is attached to the

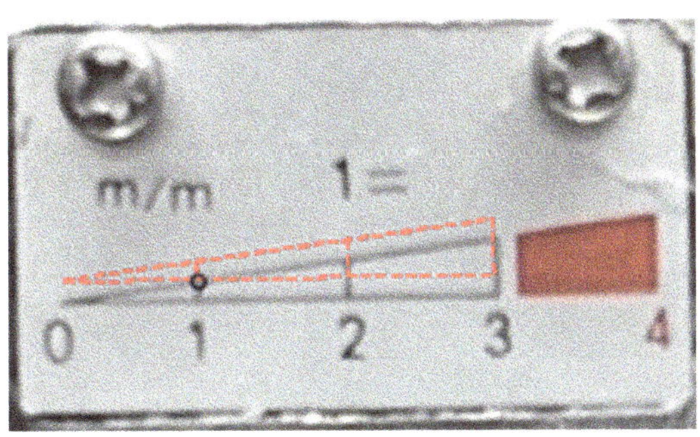

Fig. 2 Vibration frequency for all tests (black circle indicates the target intersection point of the two triangles)

cylindrical steel container. Additional sample is filled into the sieve and vibrated at 1 m/m frequency to fill the cylinder extension. The sample is tapped 180 times, and the resulting mass will be determined. It is computed as the mass of the compacted powder sample over the volume of container [4]. Below is the step-by-step guide in performing the packed bulk density test using the Hosokawa powder tester:

1. Following the aerated bulk density test, do not throw the sample as it will be used for the succeeding packed bulk density test. Click "Packed bulk density" icon in the home screen.
2. Assemble parts as shown in Fig. 4. Press "Next" after assembly.
3. Place the seal cap on the stainless-steel cup (as shown in Fig. 3), and then follow screen prompts in feeding more sample into the seal cap. Press "Next" to proceed to the "Tap count" tab. *See* **Note** 7 to make sure feeding sample is done correctly.
4. Set the number of tapping. The standard number for tapping powders is 180 times. Press "Next" to proceed to tapping.
5. After 180 taps, press the "Prepare" button to tare the weighing balance to 0 g. In this step, do not place anything on the balance.
6. Follow screen prompt for the necessary buttons to click prior to putting the sample cup into the weighing balance.
7. Carefully lift the sample cup to the weighing balance to measure the sample weight. Press "Next" to get the computed packed bulk density.

3.3 Flowability Parameter Test

3.3.1 Compressibility Test

Compressibility is a measure of stability and strength of a powder. It is directly proportional to the fine fraction of powder particles as finer particle size distribution results in more compact powder. Higher compressibility is also observed for powders with higher moisture content. Another powder characteristic influencing

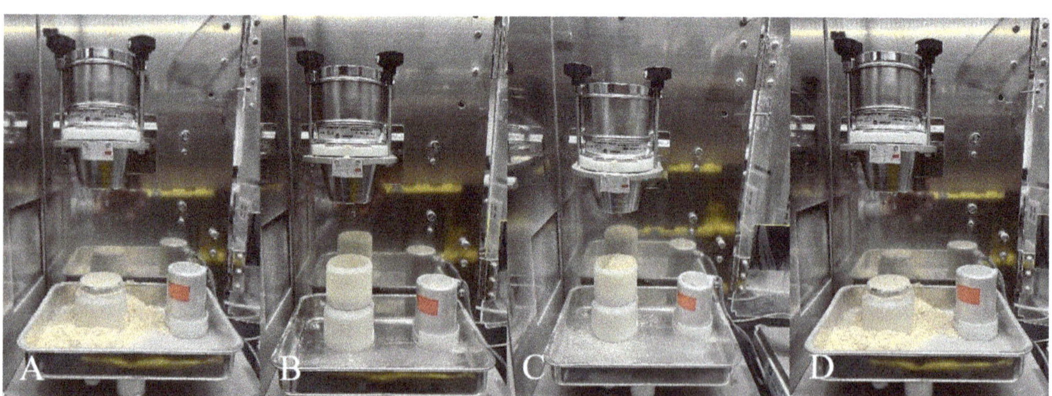

Fig. 3 Packed bulk density test: (**a**) Post-aerated bulk density test. (**b**) Attaching sleeve onto the sample cup. (**c**) Loading more sample onto the sleeve. (**d**) Leveled powder sample after tapping 180 times

compressibility is particle shape, as it influences the arrangement of particles in a container and tendency to collapse through shear as powder undergoes compaction [8]. Below are the steps in performing compressibility test using the Hosokawa powder tester:

1. Press "Compressibility" test in the home screen.
2. Encode aerated and packed bulk densities by clicking the black text box across each parameter. Make sure to enter the values expressed in g per cm^3. Click the "Enter" button after typing the value in each text box.
3. After encoding the aerated and packed bulk densities, the screen will automatically update the value of compressibility (expressed in %) and index. Compressibility is calculated by the powder tester using the following formula:

$$\text{Compressibility } (\%) = \frac{\text{Packed bulk density} - \text{Aerated bulk density}}{\text{Packed bulk density}} \times 100$$

4. At the end of compressibility test, an option will be given to proceed with either "Cohesion" or "Uniformity" test when you press the "Save" button.
5. If you will not perform these tests, press "Main Menu." Otherwise, the "Cohesion" or "Uniformity" button will allow you to proceed to their respective test panes.

3.3.2 Angle of Repose

The angle of repose is the steepest angle of descent at which stability of the sloping surface of the powder material is achieved. It is determined by measuring the angle between the sloping surface and the base of the mound formed by pouring the powder sample onto a flat surface [4]. Based on Carr's classification of powder flowability in terms of angle of repose, powders are very free flowing if less than 30°, free flowing between 30° and 38°, fair to passable flow between 38° and 45°, cohesive between 45° and 55°, and very cohesive or non-flowing if greater than 30° [9]. The setup for the angle of repose measurement method comprises of a stainless-steel table, a 750 μm sieve with extension, and a small-mouthed glass funnel [10]. Figure 4 shows the angle of repose test performed on the three protein powder samples using the powder tester. Below is the step-by-step procedure in performing this test:

1. Click the "Angle of Repose" icon in the home screen.
2. Assemble the parts as shown in Fig. 4.
3. Set vibration time. The standard vibration time is 180 s. Press the "Next" button to proceed.
4. Fill powder sample into the sieve.
5. Press "Start vibration" to start sample feeding until powder forms into a heap. Similar to aerated bulk density, turn the

Fig. 4 Angle of repose test: (**a**) Assembly of parts (1, sieve holding bar; 2, sieve extension; 3, sieve (710 μm); 4, spacer; 5, funnel; 6, impacting device; 7, table; and 8, pan). (**b**) Powder heap formation after complete feeding of powder sample

knob clockwise until the vibrating triangle appears similar to Fig. 2. Complete sample feeding is shown in Fig. 4. Press "Vibration" to continue sample feeding in case the standard vibration time is insufficient for the sample. If powder feeding results in the formation of small tunnel along its conical side (incline), discard the poured sample, and repeat sample feeding. To stop the feeding, press "Vibration" again. *See* **Note 5** to make sure that sample feeding is done correctly.

6. Press "Start" to measure the angle of repose using the built-in laser technology of the powder tester. Wait until the laser completes scanning the powder from top to bottom (*see* **Note 9**).
7. After laser scanning, angle of repose will appear on the screen.

3.3.3 Angle of Spatula

The angle of spatula is a measure of a relative angle to internal friction of a powder. In this test, a spatula lies underneath the powder heap parallel to its base and is raised out of the powder heap to isolate a section [11]. This test involves the use of a pan assembly and spatula from the Hosokawa powder tester accessory kit. Below are the steps in measuring the angle of spatula:

1. Assemble parts as shown in Fig. 5. Then, press "Ready" to initiate raising of the pan until it touches the spatula.

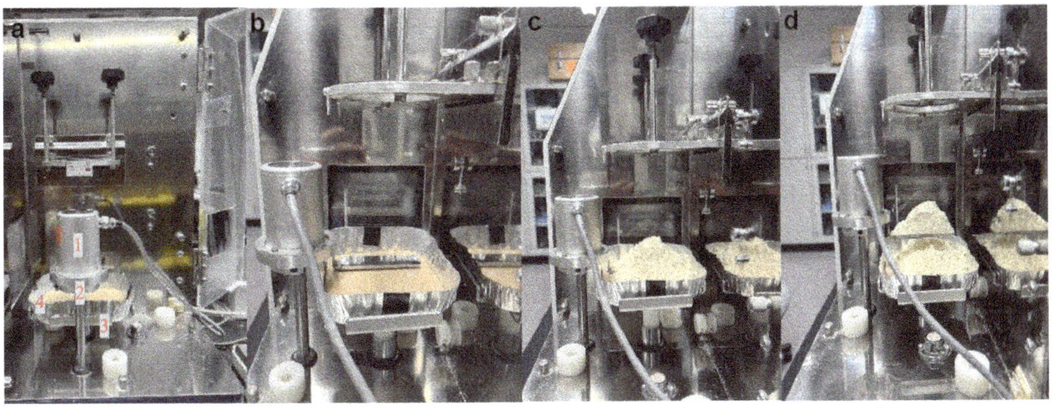

Fig. 5 Angle of spatula test: (**a**) Assembly of parts (1, impacting device; 2, spatula angle measurement unit; 3, pan holder; and 4, pan). (**b**) Raised pan after assembly. (**c**) Poured powder sample until it forms a mound. (**d**) Descended pan, leaving the section of powder sample occupying the spatula

2. As the pan is in the upper position, pour powder sample onto the spatula until it forms a mound. Powder sample would be about 100–300 cm^3. Make sure to pour a similar sample on this step for each measurement. *See* **Note 8** to make sure that pouring of the sample is done correctly.
3. Press "Next" for the spatula to descend automatically, creating a cut along the sides of the powder and leaving only the powder sitting directly on the spatula.
4. If the spatula condition is full in its entire length and shows a heap, press "Measure" to measure the angle of spatula before applying impact. Otherwise, choose "Assembly" to repeat powder formation on spatula if necessary.
5. Once measurement is completed, the angle of spatula before impacting will appear on the screen. Press "Save" button.
6. Apply impact by pressing "Start." The impacting device will produce a one-time impact force that will cause some of the powder to be removed from the spatula, thus reducing the angle of spatula after impact. Press "Measure" to allow the built-in laser technology to measure the angle of spatula (*see* **Note 9**).
7. Once measurement is done, the average angle of spatula before and after impact will be computed by the powder tester. An index will also be shown.

3.3.4 Cohesion and Uniformity Test

Cohesion and uniformity are two flow properties of a powder that indicate general flowability. Cohesion is usually measured in powders with very fine particle size distribution. It can also be used to describe flowability of powders that allow measurement of an effective cohesive force. On the other hand, uniformity coefficient is only applicable to describe flowability of granular materials and powders with coarser particle size distribution. It can also be used for powders where an effective cohesive force cannot be measured [4]. Hosokawa Micron Powder Tester PT-R manual includes a selection guide on whether cohesion or uniformity test is the applicable measurement method for the powder sample. Figure 6 shows a decision tree diagram that can be used in determining the appropriate test between cohesion or uniformity test. It starts with the computation of the mean bulk density (average of bulk and packed bulk densities), followed by determining the degree of fineness of the powder. A combination of sieves is recommended based on the mean bulk density and particle size of the powder. In

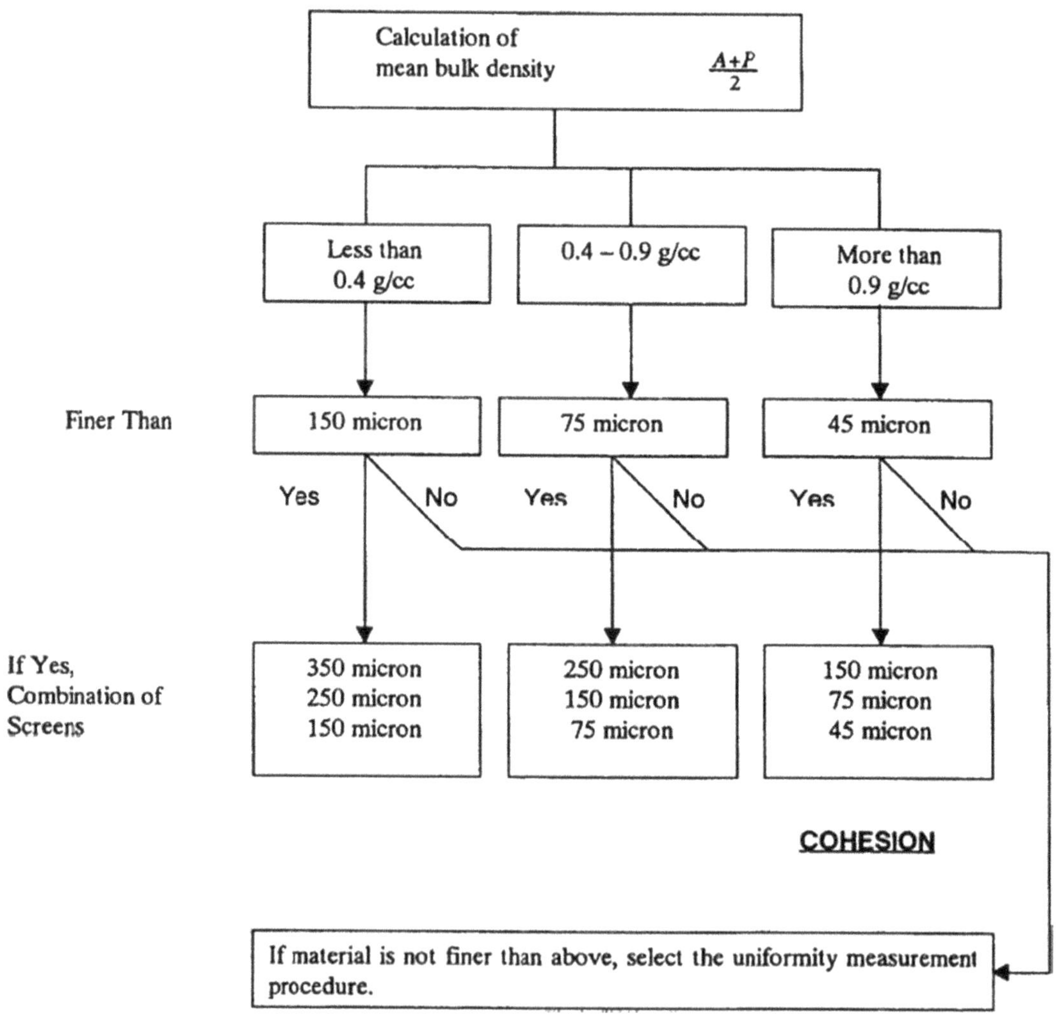

Fig. 6 Decision tree diagram for determining the appropriate test between cohesion and uniformity test for the powder sample [12]

the case of the selected plant protein powders, uniformity test was conducted.

Below are the necessary steps in performing the cohesion test using the Hosokawa powder tester:

1. Press “Cohesion” on the home screen.
2. Press the “Balance” key, and follow the remaining screen prompts in setting up the weighing balance.
3. Encode the aerated and packed bulk density of the powder sample in their respective text boxes. Press “Average” to let the powder tester compute for the average of these two bulk densities.
4. Arrange the assembly of cohesion test as shown in Fig. 7. The sieve arrangement is 355 μm sieve at the top, 250 μm sieve in the middle, and 150 μm sieve at the bottom. Then, press “Start” to vibrate the sample, and set the amplitude to 1 m/m by turning the adjustment knob clockwise.
5. Pour the weighed sample (2 g) into the sieve.

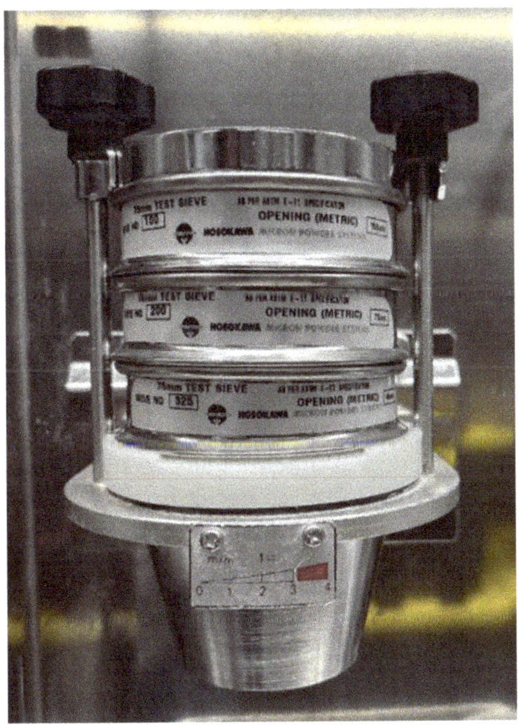

Fig. 7 Cohesion test setup: a stack of three sieves (size depends on the mean bulk density and particle size of the powder sample). Note: Sieve sizes are based on the results from the decision tree diagram in Fig. 7

6. Press the "Time" key to set the vibration time after sample is fed.
7. Press the "Balance" key to turn on balance for adjusting it to 0 g. Follow the remaining screen prompts in weighing the samples for each sieve.
8. Sample weights will be shown on the screen. The powder tester will automatically show the cohesiveness expressed in % on the screen. Below is the formula used to compute cohesiveness manually:

$$\text{Cohesion } (\%) = \left(\frac{U}{N} + \frac{3M}{5N} + \frac{L}{5N} \right) \times 100$$

where U is the mass of the retained sample from the top sieve, M is the mass of the retained sample from the middle sieve, L is the mass of the retained sample from the bottom sieve, and N is the initial sample mass.

In case that uniformity test is applicable for the powder sample, below are the steps that need to be followed:

1. Click "Uniformity" on the home screen.
2. Encode the particle size (μm) of which 10% passes the sieve (D10). Press "Save" (*see* **Note 10**).
3. Encode the particle size (μm) of which 10% passes the sieve (D60). Press "Save" (*see* **Note 10**).
4. The next screen prompt will show the computed uniformity which is the ratio of D60 and D10.

3.4 Floodability Parameter Test

3.4.1 Angle of Fall

The angle of fall is the angle formed between the sloping surface and base of the powder after the application of impact to a carefully built powder heap upon initial formation of an angle of repose. Higher flowability of a powder results in a lower angle of fall, as particles flow down the slow more easily, forming a reduced angle. A floodable powder exhibits a lower angle of fall due to the entrained air inside the formed powder heap, causing it to easily collapse once impact is applied [6]. The angle of fall setup is similar to the angle of repose setup with the addition of the impacting device. Below is the step-by-step guide in measuring angle of fall using the Hosokawa powder tester:

1. Click the "Angle of Fall" icon in the home screen.
2. Assemble parts as shown in Fig. 8. Press "Next" to proceed.
3. Set the vibration time. The standard vibration time is 180 s. Press the "Next" button.
4. Feed the sample into the sieve. Then, press "Start" to proceed with vibration.

Fig. 8 Angle of fall test: (**a**) Assembly of parts (similar to angle of repose test). (**b**) Powder heap formation after complete feeding of sample. (**c**) Powder formation showing angle of fall after application of impact to the powder

5. During vibration test, set the vibration frequency by turning the vibration adjustment knob clockwise. Start with 1/1 (m/m) by following the triangle image in Fig. 2. If feed does not flow continuously, increase vibration frequency, or change the funnel to a bigger size.
6. Apply impact to the powder heap on the table by pressing "Start."
7. The powder tester will prompt "Impacting completed!." In this case, the angle of fall is formed.
8. Press "Start" to measure angle of fall using the built-in laser technology of the powder tester (*see* **Note 9**).
9. The angle of fall measurement is completed when the screen shows the final measurement in degrees.

3.4.2 Angle of Difference

The angle of difference is calculated as the difference between the angle of repose and angle of fall. This property indicates the tendency of the powder to flush during handling [4]:

1. Click the "Angle of Difference" icon in the home screen.
2. Encode the values of angle of repose and angle of fall in their respective black text boxes. Press "Enter" to save the value for each parameter.
3. After encoding the values, angle of difference (expressed in degrees) will be automatically computed by the powder tester as the difference between the angle of repose and angle of fall.

4. If you have not performed the measurements of the angles of repose and fall, press "Assembly (R)" or "Assembly (F)" to proceed with their respective tests. Make sure to save the values for these two parameters to be able to determine the angle of difference.

3.4.3 Dispersibility

Dispersibility is a property that indicates the tendency of the powder to flood and behave like a fluid. A separate dispersing chamber located at the rear part of the powder tester is used for this test [4]:

1. Click the "Dispersibility" icon in the home screen.
2. Press "Balance" to tare the weighing balance to 0 g. Follow the remaining screen prompts to setup the weighing balance.
3. Assemble parts as shown in Fig. 9. Pull the dispersibility unit out front, and pull its shutter upward. Place the watch glass in the receiving container. Press "Next" to proceed.
4. Pour the pre-weighed sample (10 g) into the hopper, while the shutter is closed.
5. After feeding the sample, open the shutter (by clicking the button on top of the setup), and press "Next" to proceed.
6. Put the watch glass containing the sample on the weighing balance and press "Next."
7. Once weight measurement is done, the screen will show the dispersibility (%) result.

Fig. 9 Dispersibility test: (**a**) Assembly of parts. (**b**) Loading of powder sample. (**c**) Retained powder sample over a watch glass after free falling from the sample holder

Table 4
Flowability indices of pea, soy, and wheat gluten protein powders based on Carr [6] indices chart of flowability

Type of protein powder	Flowability index	Degree of flowability	Necessity of bridge-breaking measures
Pea	45	Not good	Required (must agitate or vibrate)
Soy	40	Not good	Required (must agitate or vibrate)
Wheat gluten	38	Bad	Powerful measures should be provided (agitate more aggressively)

Table 5
Floodability indices of pea, soy, and wheat gluten protein powders based on Carr [6] indices chart of floodability

Type of protein powder	Floodability index	Degree of floodability	Measures for flushing prevention
Pea	33	May flush	Rotary seal is necessary (depending on flow speed and feed conditions)
Soy	38	May flush	Rotary seal is necessary (depending on flow speed and feed conditions)
Wheat gluten	36.75	May flush	Rotary seal is necessary (depending on flow speed and feed conditions)

3.5 Flowability and Floodability Indices

After conducting all flowability and floodability tests, respective indices for each parameter will be added. These indices are based on the point scores for powder flowability and floodability evaluation based on Carr's measurement method, as shown in Tables 1 and 2. The computed flowability and floodability indices are shown in Tables 4 and 5 with their corresponding interpretation of degree of flowability and the suggestion for bridge-breaking and flushing prevention measures.

4 Notes

1. Always clean the powder tester prior to starting the test for a different sample.
2. After clicking each test from the home screen, the assembly of each test will be shown in the on-screen prompt. Follow specific sieve sizes and other parts specified in this prompt to properly assemble the setup for each test.

3. For setting the vibration frequency, make sure that your eyes are at the same level as the triangle image before you aim for the intersection at point 1.
4. Make sure that the safety door is closed prior to sample vibration.
5. In sample feeding during angle of repose, make sure the receiving table is in the middle of the funnel spout. This step will minimize formation of small tunnels along the conical side of the powder heap.
6. For aerated bulk density, leveling the sample after sample feeding may be difficult for very loose powder as it may collapse, and the top surface may recede. Hence, carefully run the stainless-steel scraper through the lid of the sample cup. Any presence of holes scraped off the surface, or irregularities indicate that sample feeding must be repeated.
7. For the tapped bulk density, loose powder sample may need feeding more sample than the capacity of the cup sleeve. As tapping continues, monitor the top surface of the powder sample. If it reaches below the stainless-steel cup lid, then further sample feeding is needed. Stop tapping operation and go back to the vibration section of the method. Load more samples and repeat the tapping operation. Do this procedure until the powder sample does not recede below the lid of stainless-steel cup after complete tapping.
8. For the angle of spatula, sample mass should be equal for each sample. Pour the sample approximately in the middle portion of the spatula to form a powder heap. The location of sample feeding may affect the angle of spatula.
9. For angles of repose, fall, and spatula, wipe the area where laser beam passes through and the black mat across it after application of impact. This step will ensure that the laser beam will measure the angle accurately.
10. For uniformity computation, please refer to Chapter 10 for the method in obtaining D10 and D60.

References

1. Fizpatrick JJ, Ahrne L (2005) Food powder handling and processing: industry problems, knowledge barriers and research opportunities. Chem Eng Proc: Proc Intensif 44(2):209–214. https://doi.org/10.1016/j.cep.2004.03.014
2. Jakobson K, Kaleda A, Adra K, Tammik M, Vaikma H, Kriščiunaite T, Vilu R (2023) Techno-functional and sensory characterization of commercial plant protein powder. Food Secur 12(14):2805. https://doi.org/10.3390/foods12142805
3. Fathollahi S, Sacher S, Escotet-Espinoza MS, DiNunzio J, Khinast JG (2020) Performance evaluation of a high-precision low dose powder feeder. AAPS PharmSciTech 21:1–13. https://doi.org/10.1208/s12249-020-01835-5
4. Barretto R, Siliveru K, Casada M (2022) Shape characteristics and particle size distribution: effects on flowability and floodability of select rain dust types. Biosyst Eng 225:13–24. https://doi.org/10.1016/j.biosystemseng.2022.11.005

5. Stoklosa AM, Lipasek RA, Taylor LS, Mauer LJ (2012) Effects of storage conditions, formulation, and particle size on moisture sorption and flowability of powders: a study of deliquescent ingredient blends. Food Res Int 49(2): 783–791. https://doi.org/10.1016/S0260-8774(98)00140-X
6. Carr RL Jr (1965) Evaluating flow properties of solids. Chem Eng 72:163–168
7. Abdullah EC, Geldart D (1999) The use of bulk density measuements as flowability indicators. Powder Technol 102(2):151–165. https://doi.org/10.1016/S0032-5910(98)00208-3
8. Moondra S, Maheshwari R, Taneja N, Tekade M, Tekadle RK (2018) Chapter 6 – bulk level properties and its role in formulation development and processing. In: Tekade RK (ed) Dosage form design parameters. Elsevier, London
9. Beakawi Al-Hashemi HM, Baghabra Al-Amoudi OS (2018) A review on the angle of repose of granular materials. Powder Technol 330:397–427. https://doi.org/10.1016/j.powtec.2018.02.003
10. Barretto R, Buenavista RM, Pandiselvam R, Siliveru K (2021) Influence of milling methods on the flow properties of ivory teff flour. J Texture Stud 53(6):820–833. https://doi.org/10.1111/jtxs.12630
11. Hosokawa Micron Powder Systems News (2021) Understanding the Hosokawa Micron Powder Tester characteristics tester PT-X terminology. Summit, New Jersey. https://www.hmicronpowder.com/news-and-events/news/understanding-the-hosokawa-micron-powder-characteristics-tester-pt-x-terminology/#:~:text=Angle%20of%20Spatula%20is%20an,out%20of%20the%20powder%20sample. Accessed 28 Aug 2023
12. Hosokawa Micron Corporation (n.d.) Instruction manual for Hosokawa Micron Powder Tester PT-R

Chapter 17

Flow Properties of Protein Powders Using FT4™ Powder Rheometer

Jared Rivera and Kaliramesh Siliveru

Abstract

The production of plant-based protein powders from diverse sources leads to products with different handling and flow characteristics. Powders can exhibit agglomeration, cohesion, and segregation behaviors, which inhibit powder flow during processing. Hence, it is important to evaluate the flow characteristics of protein powders to help design more efficient process steps. This book chapter aims to describe the common protocols for measuring the bulk, dynamic, and shear flow properties of protein powder ingredients using the FT4™ Powder Rheometer. The bulk, dynamic, and shear flow properties reflect powder characteristics under stationary, powder flow, and consolidation and low flow conditions. This chapter covers the standard protocols for measuring compressibility, permeability, aeration, stability and variable flow rate, consolidation, wall friction, and shear flow properties of protein powders. This chapter also provides sample test results and analysis on the flow properties on pea, soy, and wheat gluten protein powders obtained using the protocols included in this chapter.

Key words Aeration, Consolidation, Flow properties, Protein powders, Shear flow, Stability

1 Introduction

Food powders comprise a large portion of the total processed food produced worldwide. These are defined as dry, bulk solids composed of discrete granular particles that flow when an external force (e.g., shaking) is applied. Using food powders provides several benefits such as low weight, convenience, diverse applications, and high production rate [1]. Plant protein powders are produced by grinding the raw grain or seed followed by protein extraction via wet (e.g., alkali or acid extraction) and/or dry (e.g., air classification) methods. The protein extracted is then further processed to produce plant-based protein powder. These have been increasingly used as part of protein powder mixes with its market projected to

Yonghui Li (ed.), *Plant-Based Proteins: Production, Physicochemical, Functional, and Sensory Properties*,
Methods and Protocols in Food Science, https://doi.org/10.1007/978-1-0716-4272-6_17,

expand at a compound annual growth rate (CAGR) of 8.8% from 2022 to 2030 [2]. This growth has mainly been fueled by increased consumer awareness to the benefits of plant protein powders in terms of nutrition, environmental concerns, food safety, and personal beliefs (i.e., animal cruelty, climate change, etc.).

Powder flow is defined as the relative motion of bulk quantities of powder particles among neighboring particles or along container wall surfaces. Evaluating the flow properties of plant protein powders is essential to maintaining efficiency during production and handling. Understanding powder flow properties helps in improving equipment design and process parameters which help maintain efficiency (i.e., preventing clogging, agglomeration), product quality, and output consistency. The diversity of raw materials and processes used in producing plant protein powders result in different particle characteristics (e.g., size, shape, structure) that can alter the flow and handling properties of powder. For example, it was reported that larger particles of lactose powders flowed more freely (i.e., better flowability) than smaller-sized particles of the same lactose powder [3]. In another study, it was observed that high protein content milk powders resulted in poorer dynamic and shear flow properties compared to lower protein content milk powders [4]. Elevated temperatures (40 °C) during milk storage also improved the flowability of the powders.

Plant protein powders are often mixed (to ensure homogeneity) with other powdered ingredients to increase palatability and nutritional benefits. Post-mixing steps also present a problem in maintaining homogeneity in powders as segregation could occur because of conveying, transport, and storage processes. Evaluating the flow properties of plant protein powders is needed in designing mixing processes that maximize powder homogeneity. This chapter covers a "hands-on" guide on the basic protocols for measuring the bulk, dynamic, and shear flow properties of protein powders using the FT4™ Powder Rheometer (Freeman Technology; Tewkesbury, UK). Bulk flow properties reflect powder characteristics when it is stationary, which is important in processes such as powder storage. Dynamic flow properties reflect powder characteristics when it is in motion; this is crucial for processes such as powder transport (e.g., screw conveyors, pneumatic transfer, filling). Shear flow properties reflect powder characteristics under consolidated and low flow rate conditions. Shear properties have been used for understanding powder behavior during hopper storage, which is essential in determining suitable hopper dimensions. The results of these flow testing procedures on pea, soy, and wheat gluten powders were also included in this chapter.

2 Materials

- Protein powder (≥500 g needed) (*see* **Note 1**).
- Plastic or aluminum scooper (*see* **Note 2**).
- Fine paint brush or duster.
- Computer with MS Excel, "Powder Rheometer," and "Data Analysis v4" softwares installed and connected to the FT4™ Powder Rheometer (*see* **Note 3**).
- FT4™ Powder Rheometer.
- FT4™ Powder Rheometer Accessory Kit (48 mm blade, vented piston head).
- Shear Cell Head Accessory Kit.
- Aeration control unit (ACU).
- Air compressor (for permeability and aeration test)—any air compressor would be okay as long as it can fit the tubes connected to the ACU.
- Autotapper or jolting volumeter (for consolidation testing).

3 Methods

3.1 Starting and Shutting Down the FT4 Powder Rheometer

1. Check the power connection of the computer and powder rheometer, and make sure that they are connected to a suitable powder source.
2. Press the "Power" button on the upper right corner of the rheometer, and wait for the computer to boot up.
3. Open the "Powder Rheometer" software, and enter your username and password if needed.
4. Lift the safety cover of the Powder Rheometer for testing.
5. Click the "Standard Tests Program" button, and select the required test procedure for the material to be tested.
6. To shut down the rheometer after testing, make sure that all attachments are removed from the rheometer. Return the safety cover to its original position, and shut down the rheometer using the connected PC.

3.2 Managing, Saving, and Reviewing Analysis Results and Exporting Data

1. Create a folder for each sample powder for saving rheometer test results.
2. Make sure that the results are saved in the folder created by selecting it as the file path after testing.
3. Open the required test results with the "Data Analysis v4" software.

4. Export results in the tables and graphs shown to MS Excel.
5. Exporting can be done by right clicking the graph and selecting the "Export data to Excel" option. Do the same for the "Series values table."
6. Save both Excel data from the graph and the series value table in one Excel file to minimize the number of files generated. The graphs can be redrawn in MS Excel.

3.3 Bulk Property Testing

Bulk properties relate to the properties of the powder while stationary and under confined conditions. This section will talk about compressibility and permeability testing, which are the bulk properties of powders commonly evaluated using the FT4 Powder Rheometer.

3.3.1 Compressibility

Compressibility testing using the Powder Rheometer is done by applying increasing levels of compressive force to the conditioned powder bed using the vented piston accessory. The powder compressibility (% change in volume per applied stress) is then measured as the change in powder bed volume as a function of the applied load. Figure 1 shows the proper assembly and sequence of compressibility testing. The following steps are needed to measure the protein powder compressibility, and please refer to **Notes 1–19** for tips in properly conducting the test:

1. From the "Standard Tests Programs" menu, select compressibility, and follow on-screen prompts.
2. Select "50 mm" as the vessel for the analysis.
3. Screw the 48.00 mm blade spindle by hand into the FT4™ rheometer (*see* **Note 6**).

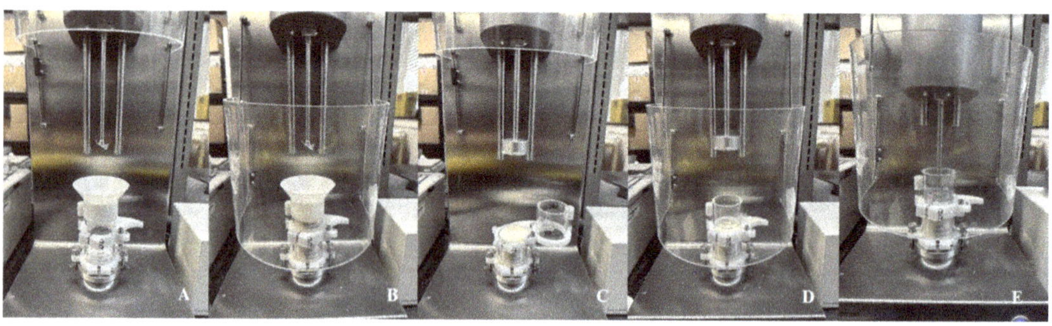

Fig. 1 Summary of steps involved in compressibility testing using FT4™ rheometer. (**a**) Assembled empty vessel. (**b**) Vessel with sample included. (**c**) Splitting of vessel and insertion of vented piston attachment. (**d**) Closed vessel after splitting. (**e**) Compressibility testing. Compressibility testing uses the normal vessel base (white) included in the accessory kit

4. Assemble the 50 mm × 85 mL vessel (uses the white vessel base), and screw the base of the vessel onto the platform of the rheometer. Place the funnel on top of the vessel (*see* **Notes 4**, **7** and **8**).
5. Attach the vented piston attachment into the extra spindle.
6. Tick the check boxes shown on-screen for the 48.00 mm blade and 50 × 85 mL vessel requirements for testing.
7. In the on-screen controls, tare the weighing scale, and fill the vessel with the sample powder until the volume added is between ½ and ¾ of the vessel. Lower the safety cover of the rheometer (*see* **Notes 5** and **9**).
8. On-screen, select "Record sample mass" once the sample weight has stabilized. Enter sample details, and select appropriate file path for the results of the analysis. Press "Start Test" once details entered are correct.
9. The rheometer will then proceed with the pre-conditioning step. Wait for the pre-conditioning step to finish and for the on-screen prompt to appear.
10. Once the on-screen prompt appears, lift the safety cover of the rheometer, remove the funnel, and split the vessel. Split the vessel by pressing down on the spring of vessel connector and releasing the upper half of the vessel. Use a scooper to catch the excess powder from the split vessel.
11. Carefully return the vessel to its original position by moving the upper half of the vessel toward its lower half. Press on the spring, push the vessel all the way in, and release the spring to connect the vessel.
12. Unscrew the blade from the rheometer hub, screw the vented piston spindle by hand, and lower the safety cover.
13. On-screen, click "OK" to proceed with the test analysis. The rheometer will then start the compressibility test.
14. Once the test is done, save the results (refer to Subheading 3.2), and disassemble and clean the accessories attached to the rheometer to prepare for the next test.

3.3.2 Permeability

The permeability test measures the ease with which air can flow through the compressed powder column. This test measures permeability as the resistance (expressed as pressure drop at the bottom and top of the powder column) of the powder to airflow. Figure 2 shows the proper assembly and sequence of the permeability testing procedure. The following steps are needed to measure the protein powder permeability, and please refer to **Notes 1–19** for tips in properly conducting the test:

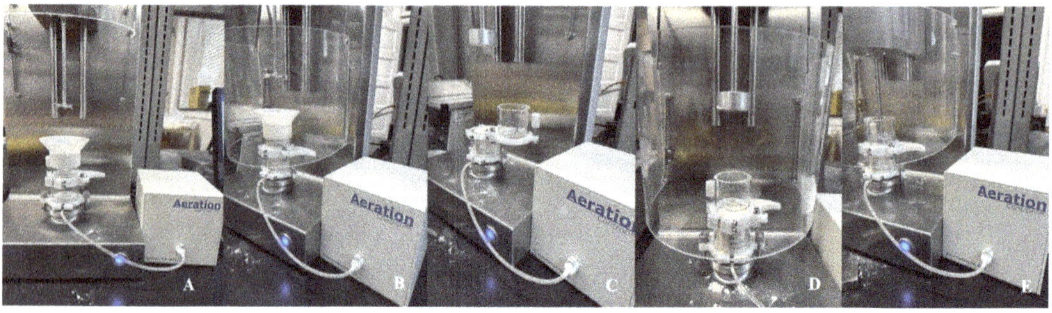

Fig. 2 Summary of steps involved in permeability testing using FT4™ rheometer. (**a**) Assembled empty vessel. (**b**) Vessel with sample included. (**c**) Splitting of vessel and insertion of vented piston attachment. (**d**) Closed vessel after splitting. (**e**) Permeability testing. Permeability testing uses the mesh base (metal) included in the accessory kit

1. From the "Standard Tests Programs" menu, select permeability, and follow the on-screen prompts for the testing procedure.
2. Select the "50 mm" option as the vessel size for the analysis.
3. Screw the 48.00 mm blade spindle by hand into the FT4™ rheometer (*see* **Note 6**).
4. Assemble the 50 mm × 85 mL vessel. This test uses the mesh metal base included in the accessory kit. Place the base at the bottom of the assembled vessel, and place it on the platform of the rheometer (*see* **Notes 4**, **7** and **8**).
5. Assemble the vented piston set up by screwing the piston to the extra spindle.
6. Make sure to connect the rheometer aeration control unit (ACU) tube to the aerated base by fully inserting the ACU tube into the protrusion of the metal base.
7. Turn on the air compressor attached to the ACU.
8. Tick the checkboxes shown on-screen for the 48.00 mm blade and 50 × 85 mL vessel requirements for testing.
9. Place the funnel on the vessel, and tare the weighing scale. Fill the vessel with the powder sample until volume added is within ½ to ¾ of the volume of the vessel. Lower the safety cover of the rheometer (*see* **Notes 5** and **9**).
10. On-screen, select "Record sample mass" once the sample weight has stabilized, enter sample details, and select file path for the results of the analysis. Press "Start Test."
11. The rheometer will then proceed with the pre-conditioning step. Wait for the pre-conditioning step to finish and for the on-screen prompt to appear.
12. Once the on-screen prompt appears, lift the safety cover of the rheometer, remove funnel, and split the vessel. Split the vessel

by pressing down on the spring of vessel connector and moving the upper half of the vessel. Use the scooper to catch the excess powder from the split vessel.

13. Carefully return the vessel to its original position by moving the upper half of the vessel toward its lower half. Press on the spring, push the vessel all the way in, and release the spring to connect the vessel.
14. Unscrew the blade from the rheometer hub, and screw the vented piston spindle by hand. Lower the safety cover.
15. On-screen, click "OK" to proceed with the test analysis. The rheometer will then start the permeability test.
16. Once the test is done, save the results (refer to Subheading 3.2), and disassemble and clean the accessories attached to the rheometer to prepare for the next test.

3.3.3 Sample Data Analysis for Bulk Property Testing

Figure 3 shows the compressibility characteristics of the protein powders at increasing levels of applied normal stress from 0.5 to 15 kPa. The pea protein powder had higher compressibility values at all applied normal stresses compared to the gluten and soy protein powders. Compressibility reflects the changes in powder density as a function of applied normal stress [5]. These values are calculated as the ratio of the density of the powder after compression and conditioned bulk density, which are shown in Eqs. 1 and 2:

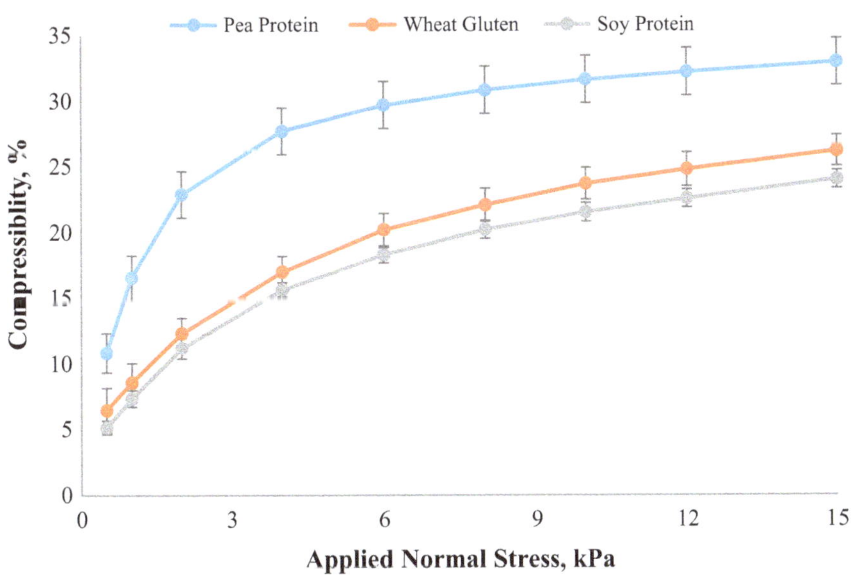

Fig. 3 Compressibility of protein powders measured using the standard compressibility test of the FT4™ Powder Rheometer

$$\text{Conditioned bulk density}\left(\frac{\text{g}}{\text{ml}}\right) = \frac{\text{powder mass after splitting vessel (g)}}{\text{volume after conditoning (ml)}} \quad (1)$$

$$\text{Compressibility Index} = \frac{\text{Density after compression}\left(\frac{\text{g}}{\text{ml}}\right)}{\text{Conditioned bulk density}} \quad (2)$$

This property is affected by many powder characteristics such as particle cohesion, composition, shape, size, and texture. In general, cohesive powders would have higher compressibility values as they are not able to pack efficiently when poured resulting in the formation of more pores throughout the powder bed. These pores are then compressed when normal stress is applied leading to higher compressibility values (i.e., higher volume change). Hence, it can be inferred from the results (shown in Fig. 3) that the pea protein powder packs less efficiently than the gluten and soy protein powders, which explains its higher compressibility values. Powder compressibility characteristics are relevant in many processes (e.g., hopper storage and powder compaction), wherein the powder is in a stationary or bulk condition.

Figure 4 shows the permeability, which is measured as the pressure drop of the protein powder bed under increasing levels of consolidation stresses. Pressure drop is estimated from the difference of the air flow between the bottom and top of the powder column. Permeability reflects the ease of transmitting fluid through

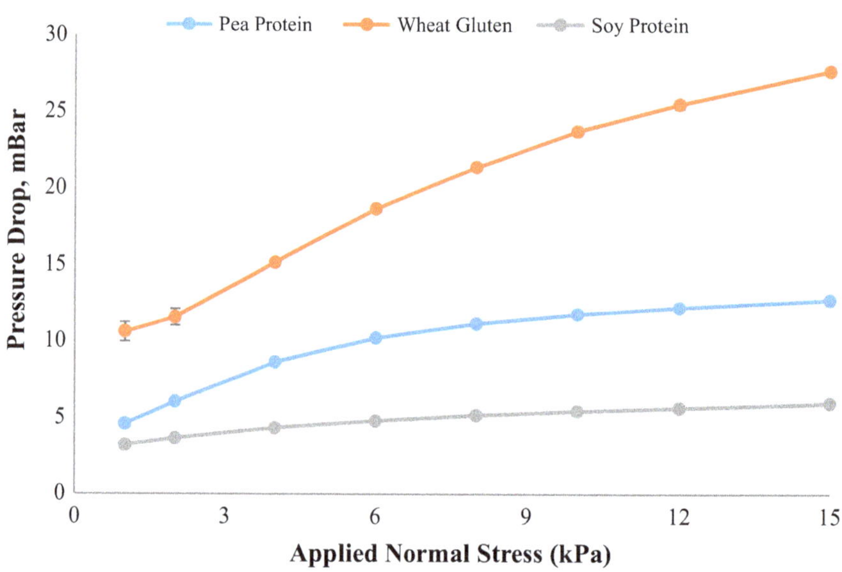

Fig. 4 Permeability of protein powders measured using the standard permeability test (air speed at 2 mm/s) of the FT4™ Powder Rheometer

the powder bed. This is done by simultaneously applying consolidation stress and transmitting air throughout the powder bed. Based on Fig. 4, the wheat gluten powder was the least permeable powder among the protein powders tested. This trend is related to compressibility observations such that the more efficient compression of the powder (wheat gluten) prevents the formation of pores in the powder column. This would inhibit the passage of air through the powder column resulting in higher-pressure drops across the powder bed indicating lower permeability. Permeability characteristics are also influenced by powder properties such as composition, shape, size, and texture [5]. Measuring permeability is important as it has implications in powder handling steps such as pneumatic transfer, hopper discharge, vacuum transfer, and filling steps.

3.4 Dynamic Flow Property Testing

Dynamic flow properties refer to the flow characteristics of the powder during motion. The FT4 Powder Rheometer uses a rotating precision blade that is moved downward and upward through the powder column to measure the resistance of the powder to flow during the dynamic flow property evaluation. Rotational and vertical resistances are quantified by measuring the torque and force from the blade, respectively. This section will cover the standard stability and variable flow rate, aeration, and consolidation tests for the FT4 Powder Rheometer.

3.4.1 Stability and Variable Flow

This test uses a rotating precision blade (at different speeds) that moves through the powder bed to measure the energy required to initiate consistent flow. The basic flowability energy (BFE), stability index (SI), flow rate index (FRI), and specific energy (SE) values of the protein powder are measured in this test. The following steps are needed to measure the stability and variable flow characteristics of protein powders (a summary of these steps is shown in Fig. 5), and please refer to **Notes 1–19** for tips in properly conducting the test:

1. From the “Standard Tests Program” menu, select the “Stability and Variable Flow Rate” test, and follow the on-screen prompts for the testing procedure.
2. In the next screen, select the “50 mm” option as the vessel size.
3. Screw the 48.0 mm blade into the rheometer hub. Assemble the 50 mm × 160 mL vessel needed for testing, and screw the vessel into the rheometer platform. This test uses the white vessel base included in the accessory kit (*see* **Notes 4**, **6**, **7** and **8**).
4. In the screen controls, tick the check boxes shown on-screen for the 48.00 mm blade and 50 × 160 mL vessel requirements for testing.

Fig. 5 Summary of steps involved in stability and variable flow rate testing using FT4™ rheometer. (**a**) Assembled empty vessel. (**b**) Vessel with sample included. (**c**) Splitting of vessel. (**d**) Closed vessel after splitting and stability and variable flow rate testing. Stability and variable flow rate testing uses the normal vessel base (white) included in the accessory kit

5. Place the funnel at the top of the vessel, and tare the weighing scale from the on-screen controls. Fill the vessel with the powder sample until volume added is within ½ to ¾ of the volume of the vessel. Lower the safety cover of the rheometer (*see* **Notes 5** and **9**).
6. On-screen, select "Record sample mass" once sample weight has stabilized, enter sample details, and select file path for the results of the analysis. Press "Start Test."
7. The rheometer will proceed with the pre-conditioning step. Wait for the pre-conditioning step to finish, and follow the on-screen prompt for the next steps.
8. Lift the safety cover of the rheometer, remove the funnel, and split the vessel. Split the vessel by pressing down on the spring of vessel connector, and move the upper half away from the lower half of the vessel. Use the scooper to catch the excess powder from the splitting the vessel.
9. Carefully return the vessel to its original position by moving the upper half of the vessel toward its lower half. Press on the spring, push the vessel all the way in, and release the spring to connect the vessel.
10. On-screen, click "OK" to proceed with the test analysis.
11. Once the test is done, save the results (refer to Subheading 3.2), and disassemble or clean the accessories attached to the rheometer to prepare for the next test needed.

3.4.2 Aeration

Aeration test involves passing controlled air through the mesh base of the test vessel. The use of a controlled air supply allows the simulation of pneumatic powder processing steps such as conveying and drying. This test also allows the evaluation of the cohesiveness

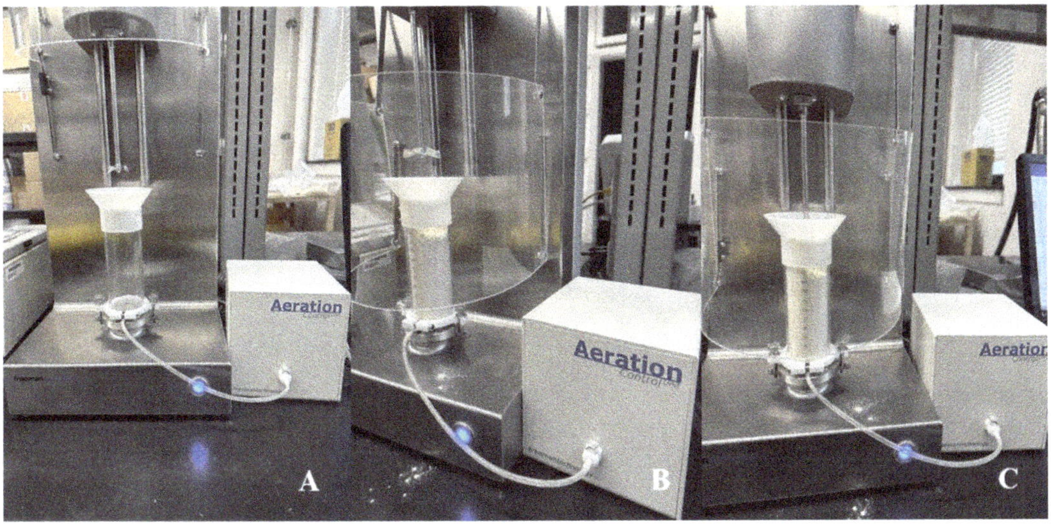

Fig. 6 Summary of steps involved in aeration testing using FT4™ rheometer. (**a**) Assembled empty vessel. (**b**) Vessel with sample included. (**c**) Aeration testing. Aeration testing uses the mesh base (metal) included in the accessory kit

of powders as aeration separates cohesive powder particles allowing them to be fluidized. Cohesiveness is measured through aerated energy (AE). The ratio of the BFE and the AE value of a powder gives the effectiveness of aeration in powder samples. The following steps are needed to measure the aeration properties of powders (a summary of these steps is shown in Fig. 6), please refer to **Notes 1–19** for tips on properly conducting the test:

1. From the "Standard Tests Programs" menu, select aeration, and follow the on-screen prompts.
2. Select the "50 mm" option as the vessel size for the analysis.
3. Screw the 48.00 mm blade spindle by hand into the FT4™ rheometer (*see* **Note 6**).
4. Assemble the 50 mm × 260 mL vessel. The aeration test uses an aerated metal mesh base. Place the mesh base at the bottom of the assembled vessel, and place it in the rheometer platform (*see* **Notes 4**, 7 and **8**).
5. Make sure to connect the rheometer aeration control unit (ACU) tube to the aerated base by fully inserting the ACU tube into the protrusion in the aerated mesh base.
6. Turn on the air compressor attached to the ACU.
7. Tick the check boxes shown on-screen for the 48.00 mm blade and 50 × 260 mL vessel requirements.
8. Place the funnel on the vessel, and tare the weighing scale. Fill the vessel with the powder sample until volume added is within ½ to ¾ of the vessel volume. Lower the safety cover of the rheometer (*see* **Notes 5** and **9**).

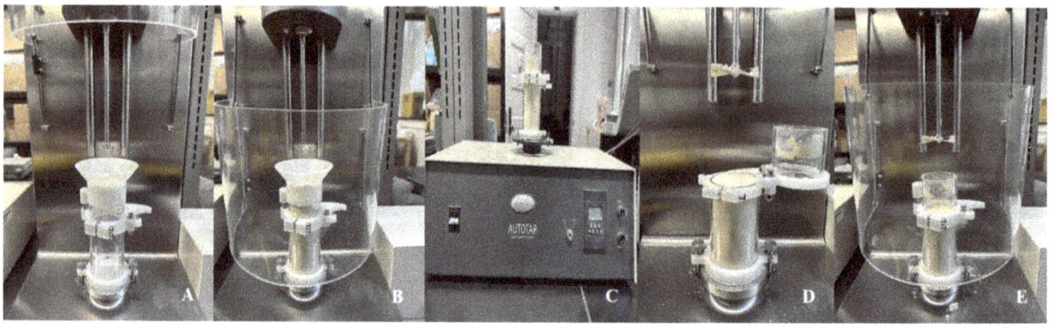

Fig. 7 Summary of steps in consolidation testing using FT4™ rheometer. (**a**) Assembled empty vessel. (**b**) Vessel with sample included. (**c**) Tapping ($n = 50$ taps) of conditioned sample using an autotapper. (**d**) Splitting of vessel after tapping. (**e**) Closed vessel after splitting and consolidation testing. Consolidation testing uses the normal vessel base (white) included in the accessory kit

9. On-screen, select "Record sample mass" once the weight has stabilized. Enter sample details, and select file path for the results of the analysis. Press "Start Test."
10. The rheometer will then proceed with the pre-conditioning step.
11. No splitting of the vessel is needed for this test. The aeration test will start immediately after the pre-conditioning step.
12. Once the test is done, save the results (refer to Subheading 3.2), and disassemble or clean the accessories attached to the rheometer to prepare for the next test.

3.4.3 Consolidation

Consolidation involves testing the powder flowability under consolidated conditions. Powder consolidation can occur in various ways including vibration of the powder bed and the presence of normal stress due to the powder weight [6]. In this test, an autotapper or jolting volumeter is used to consolidate the powder bed through vibration. The following steps are needed to conduct consolidation testing using the FT4™ rheometer (a summary is shown in Fig. 7), and please refer to **Notes 1–19** for tips on properly conducting test:

1. From the "Standard tests program" menu, select the "Consolidation" test, and follow the on-screen prompts.
2. In the next screen, select the "50 mm" option as the vessel size.
3. Screw the 48.00 mm blade into the rheometer hub (*see* **Note 6**).
4. Assemble the 50 mm × 160 mL vessel needed, and screw the vessel into the rheometer platform. This test uses the white base for the vessel (*see* **Notes 4**, **7** and **8**).
5. Place the funnel at the top of the vessel, and tare the weighing scale from the on-screen controls.

6. Fill the vessel with the powder sample until the volume added is within ½ to ¾ of the volume of the vessel. Lower the safety cover of the rheometer (*see* **Notes 5** and **9**).
7. On-screen, select "Record sample mass" once sample weight has stabilized, enter sample details, and select file path for the results of the analysis. Press "Start Test."
8. Lift the safety cover of the rheometer, remove the funnel and the vessel from the rheometer platform. Place the vessel onto the platform of an autotapper or jolting volumeter.
9. Using an autotapper or a jolting volumeter, tap the vessel 50 times. Screw the tapped vessel back into the rheometer platform.
10. After the vessel is tapped, split the vessel by pressing down on the spring of vessel connector, and move the upper half away from the lower half of the vessel. If the powder volume is below the splitting point of the vessel after tapping, restart test and increase powder quantity added.
11. Carefully return the vessel to its original position by moving the upper half of the vessel toward its lower half. Press on the spring, push the vessel all the way in, and release the spring to connect the vessel.
12. On-screen, click "OK" to proceed with the test analysis.
13. Once the test is done, save the test results (refer to Subheading 3.2), and disassemble or clean the accessories attached to the rheometer to prepare for the next test needed.

3.4.4 Sample Data Analysis for Dynamic Flow Property Testing

The stability and variable flow rate tests measure the energy required to make the powder flow. Figure 8 shows the total energy required for the protein powders to flow under different flow rates (100, 70, 40, and 10 mm/s). From these values, different dynamic flow properties for the protein powders are measured (shown in Table 1). The basic flowability energy (BFE) values refer to the energy required to establish a specific flow pattern in a powder when it is forced to flow under confined conditions at a similar flow rate (test 1 to 7). These values are measured during the downward motion of the blade during testing with the closed base of the vessel confining the powder. The stability index (SI), which is calculated as the ratio of total energy at test 7 and test 1, refers to the resistance of the powder to changes in its form (i.e., agglomeration, segregation, or break) during flow [7]. Hence, stable powders would have SI values close to 1. The flow rate index (FRI) value, calculated as the ratio of the total energy at test 8 and at test 11, is an indicator of the stability of the powder to changing flow rates (100 and 10 mm/s). Specific energy (SE) values relate to the energy required for the powder to flow under

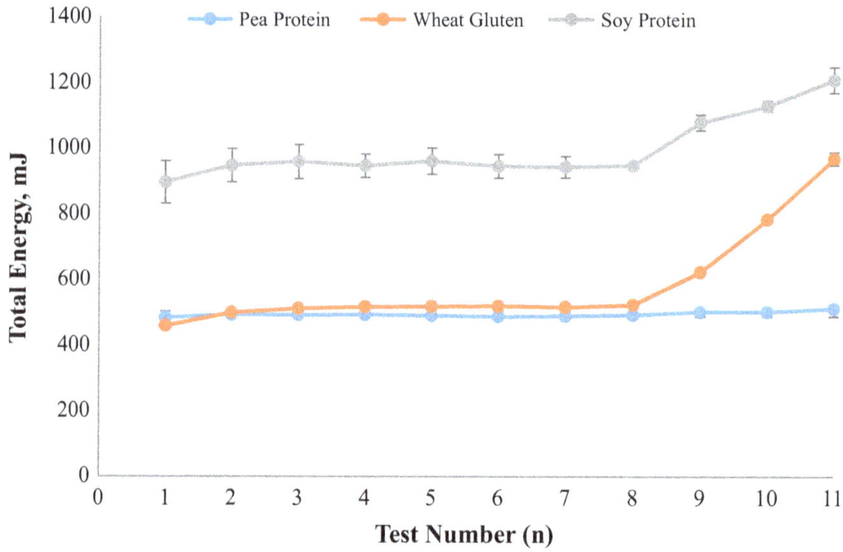

Fig. 8 Total energy of protein powders during flow measured using standard stability and variable flow rate test of the FT4™ Powder Rheometer; blade tip speed for test numbers 1–8 were −100 mm/s, while the subsequent tests were conducted at tip speeds of −70, −40, and −10 mm/s

Table 1
Stability and variable flow rate and consolidation properties of protein powders from standard FT4™ Powder Rheometer test

Protein powder	BFE, mJ	SI	FRI	SE, mJ	CE_{50}, mJ
Pea	489.22 ± 8.14	1.01 ± 0.02	1.04 ± 0.03	9.16 ± 0.11	733.5 ± 2.12
Wheat gluten	516.09 ± 6.54	1.12 ± 0.04	1.85 ± 0.08	10.50 ± 0.33	1173.5 ± 75.7
Soy	942.35 ± 32.42	1.05 ± 0.04	1.28 ± 0.03	21.26 ± 1.04	1044.5 ± 50.2

BFE basic flowability energy, *SI* stability index, *FRI* flow rate index, *SE* specific energy, CE_{50} consolidated energy after tapping (50 taps) of powder bed

unconfined conditions. The lack of a solid surface enclosing the top of the powder column during the upward motion of the blade provides unconfined conditions for testing.

Based on these definitions, it can be inferred from Table 1 that the soy protein powder requires the highest energy to establish a flow pattern as it had the highest BFE and SE values among the protein powders. All the protein powders tested were stable during flow at 100 mm/s as their SI values are close to 1, which indicate that minimal powder consolidation, aggregation, or interlocking occurred during flow. In contrast, the wheat gluten powder was more sensitive to slower flow rates as it had the highest FRI value among the protein powders. This suggests that using slower speeds on conveying wheat gluten powder is inefficient as the total energy required for the wheat gluten powder to flow would be higher. This

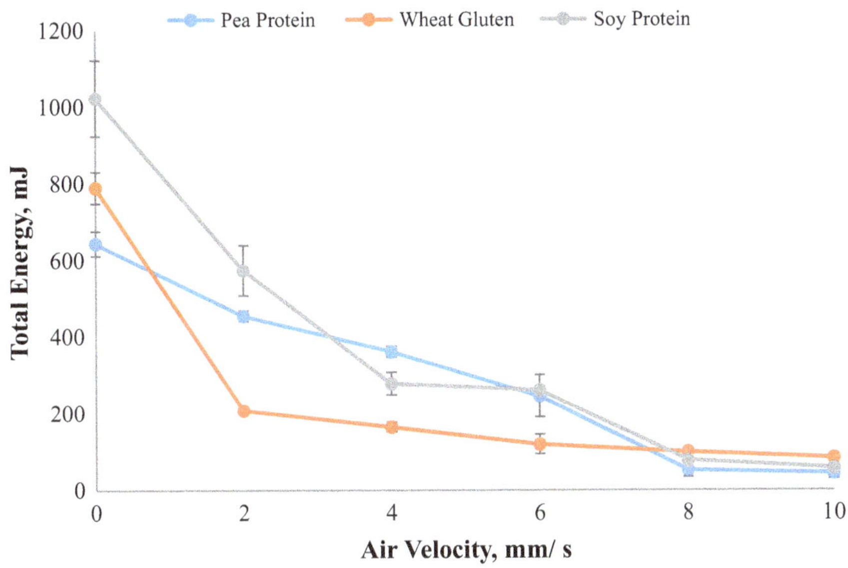

Fig. 9 Total energy of protein powders during aerated flow measured using the standard aeration test of the FT4™ Powder Rheometer. Blade tip speed during aeration tests was done at −60 mm/s for all air velocities

lower flowability demonstrates that particle interlocking could occur for wheat gluten powders if slower speeds are used to make it flow. Soy protein powders were also sensitive to slower flow rates to a lower extent compared to the wheat gluten powder. Pea protein is highly stable against changing flow rates as its FRI value is approximately 1. Lastly, consolidation (CE_{50}) energy values were measured for the protein powders after tapping for 50 times. Both wheat gluten and soy had higher CE_{50} values than the pea protein powder. This indicates that wheat gluten and soy would be less flowable than pea protein powder when consolidation occurs prior to flow.

Figure 9 shows the total energy required for the protein powders to flow under aerated conditions. This test works similarly to the stability and variable flow rate test, with the addition of aeration (0–10 mm/s) during testing to simulate fluidization of the powders. This test provides a method to determine the aerated flow properties of the powders (shown in Table 2). The basic flowability energy (BFE) is taken similarly as the stability and variable flow rate testing although the measurement is done at a blade tip speed of 70 mm/s. The aerated energy (AE_10) is the total energy required for the powder to flow at an aeration speed of 10 mm/s. Aeration ratio (AR) is the ratio of the BFE (at 0 mm/s) and the aerated flow energy (at 10 mm/s), which reflects the powders sensitivity to aeration. Normalized aeration sensitivity (NAS) is a measure of the powder sensitivity to slow aeration velocities. It is defined as the ratio of the highest difference in normalized test energy between consecutive tests and the air velocity [8].

Table 2
Aerated flow properties of protein powders from standard aeration FT4™ Powder Rheometer test

Protein powder	BFE, mJ	AE_10, mJ	AR	NAS, s/ mm
Pea	645.16 ± 32.28	44.61 ± 15.01	15.93 ± 6.59	0.17 ± 0.02
Wheat gluten	790.73 ± 40.79	85.24 ± 4.32	9.30 ± 0.87	0.37 ± 0.01
Soy	1024.46 ± 99.20	57.83 ± 19.41	18.58 ± 3.86	0.22 ± 0.02

BFE basic flowability energy, *AE* aeration energy at 10 mm/s air speed, *AR* aeration ratio, *NAS* normalized aeration sensitivity

The trends for the BFE values of the protein powders were similar to the stability and variable flow rate results. The soy protein powder was the least flowable among the protein powders as it had higher BFE and AE values. The AR values indicate that the soy protein powder was more sensitive to aeration followed by the pea and wheat gluten powders. In general, both AE and AR values are good indicators of powder fluidization [8]. However, the wheat gluten powder was more sensitive to aeration at slower air velocities given by its higher NAS values. This trend could be due to the higher sensitivity of the wheat gluten powder particles at slower flow rates, which was evidenced by the higher FRI values (1.85) observed in the stability and variable flow rate tests. Another objective of aeration testing is to find the minimum fluidization velocity for the powders. This is defined as the aeration velocity wherein the total energy recorded is close to 0 [8]. Based on Fig. 9, the protein powders were not fully fluidized at 10 mm/s as the AE_10 values were not close to 0 (44.61–85.24 mJ). Hence, it could be worthwhile to do aeration tests at faster velocities (>10 mm/s) to determine if better powder fluidization could be achieved.

Overall, evaluating the powder flow properties in a non-aerated state and different flow rates (stability and variable flow), confinement, and aerated states is crucial. The results of the dynamic flow tests indicate that powder flow changes under different process conditions. These changes are caused by different powder phenomena such as agglomeration, particle interlocking, characteristics, and conditions. These properties are especially relevant for processes that require powders to flow such as powder mixing, conveying, and transport.

3.5 Shear Flow Properties

The determination of the shear flow properties of powders is important in evaluating powder behavior in process steps involving consolidation and low flow rates. Results from shear flow testing could also be used in hopper designs to ensure mass flow during discharge. The FT4™ rheometer includes a shear cell accessory to allow the determination of the shear flow properties of the powder. The operating principle of shear flow testing involves the

Fig. 10 Summary of steps involved in shear cell flow testing using the FT4™ rheometer. (**a**) Assembled empty vessel with shear base. (**b**) Vessel with sample included. (**c**) Powder bed compression using vented piston. (**d**) Splitting of vessel and insertion of shear cell head. (**e**) Closed vessel after splitting. (**f**) Shear cell flow testing. The shear cell testing uses the vessel base with a grid (blue) that is included with the shear cell head accessory kit

Fig. 11 Summary of steps involved in wall friction testing against stainless-steel test material using FT4™ rheometer. (**a**) Assembled empty vessel. (**b**) Vessel with sample included. (**c**) Compression of powder bed using vented piston attachment. (**d**) Splitting of vessel and insertion of wall friction head attachment. (**e**) Closed vessel after attachment. (**f**) Wall friction testing. Testing was done against stainless steel (lower half of vessel)

application of shear (low speed) to the upper layer of the upper bed, while the lower layer is prevented from movement [9]. The applied shear force is gradually increased until it overcomes the powder shear strength as indicated by the movement of the upper layer against the lower layer of the powder bed. Figures 10 and 11 show the proper assembly and sequence of steps for the shear and wall friction flow properties testing, respectively. The following section discusses the steps needed to determine the shear and wall friction flow properties of protein powders. Please refer to **Notes 1–19** for tips in properly conducting the shear and wall friction tests.

3.5.1 Shear Cell

1. From the "Standard tests program" menu, select the "Shear Cell" test option, and follow the on-screen prompts for the testing procedure.
2. In the next screen, select the "50 mm" option as the vessel size.

3. Select consolidation pressure to be used for shear cell testing (3–15 kPa). Testing can be done at only one consolidation pressure at a time.
4. Screw the 48.00 mm blade into the rheometer hub (*see* **Note 6**).
5. Assemble the 50 × 85 mL vessel needed for testing, and screw the vessel into the rheometer platform. The vessel can be assembled with the white- or blue-gridded base included in the FT4™ accessory kit (*see* **Notes 4**, **7** and **8**).
6. In the screen controls, tick the check boxes shown on-screen for the 48.00 mm blade and 50 × 85 mL vessel requirements for testing.
7. Place the funnel at the top of the vessel, and tare the weighing scale from the on-screen controls. Fill the vessel with the powder sample until volume added is within ½ to ¾ of the vessel. Lower the safety cover of the rheometer (*see* **Notes 5** and **9**).
8. On-screen, select "Record sample mass" once sample weight has stabilized, and enter sample details. Press "Start Test."
9. The rheometer will proceed with the pre-conditioning step. Wait for this step to finish and follow the on-screen prompt.
10. Lift the safety cover, remove the 48 mm blade, and screw the vented piston attachment into the rheometer hub. On-screen, click "Okay." The test will proceed with compaction at the selected consolidation pressure.
11. After compaction, raise the safety cover, follow on-screen prompt to unscrew the vented piston attachment, and screw the shear cell head attachment.
12. Split the vessel, and catch excess powder using the scooper. Lower the safety cover. Click "OK" on the on-screen control panel.
13. The rheometer will then proceed with the shear cell testing. Once the test is done, save the results (refer to Subheading 3.2), and disassemble or clean the accessories attached to the rheometer to prepare for the next test. In case of failed tests, which is shown by an incomplete Mohr circle in the results section, (*see* **Notes 13–15**).

3.5.2 Wall Friction Testing

1. From the "Standard tests program" menu, select the "Wall Friction" test option, and follow the on-screen prompts for the testing procedure.
2. In the next screen, select the "50 mm" option as the vessel size.
3. Select consolidation pressure to be used for wall friction testing (3–15 kPa). Testing can be done at only one consolidation pressure at a time.

4. Screw the 48.00 mm blade into the rheometer hub (*see* **Note 6**).
5. Assemble the 50 mm × 85 mL vessel needed for testing. For this test, use the 50 × 85 mL stainless-steel cylinder for the lower half of the testing vessel and the white vessel base. Screw the vessel into the rheometer platform (*see* **Notes 4**, **7** and **8**).
6. In the on-screen controls, tick the check boxes for the 48.00 mm blade and 50 × 85 mL vessel requirements for testing.
7. Place the funnel at the top of the vessel, and tare the weighing scale from the on-screen controls. Fill the vessel with the powder sample until the volume added is within ½ to ¾ of the vessel. Lower the safety cover until it latches to the rheometer surface (*see* **Notes 5** and **9**).
8. On-screen, select "Record sample mass" once sample weight has stabilized. Enter sample details (name, run number, material, batch code, and reference no.) of test materials. Press "Start" once the details entered are correct.
9. Follow the on-screen prompts, the test will first proceed with pre-conditioning of the powder bed using the 48 mm blade.
10. After pre-conditioning, lift the safety cover, remove the 48 mm blade, and screw the vented piston attachment into the rheometer hub. On-screen, click "Okay." The test will proceed with compaction using the vented piston attachment at the selected consolidation pressure.
11. After compaction, raise the safety cover, and follow on-screen prompt to unscrew the vented piston attachment.
12. Remove the vented piston attachment head, and screw the wall friction attachment head into the spindle. Screw the spindle with the wall friction head into the rheometer hub.
13. Follow the on-screen prompts. Split the vessel and lower the safety cover. Click "OK" on the on-screen control panel.
14. The rheometer will then proceed with wall friction testing. Once the test is done, save the results (refer to Subheading 3.2), and disassemble or clean the accessories attached to the rheometer to prepare for the next test.

3.5.3 Sample Data Analysis for Shear Flow Property Testing

Evaluating the shear flow properties of powders is important to understand powder behavior under consolidated and low shear rate conditions. The Powder Rheometer uses a shear cell head that combines a straight line and rotational motion to identify the yield locus for the shear flow properties of powders. Using the combinations of normal and shear stress values, yield loci points are determined followed by Mohr circle analysis. Figure 12 shows the Mohr circle analysis of the shear flow properties of the protein

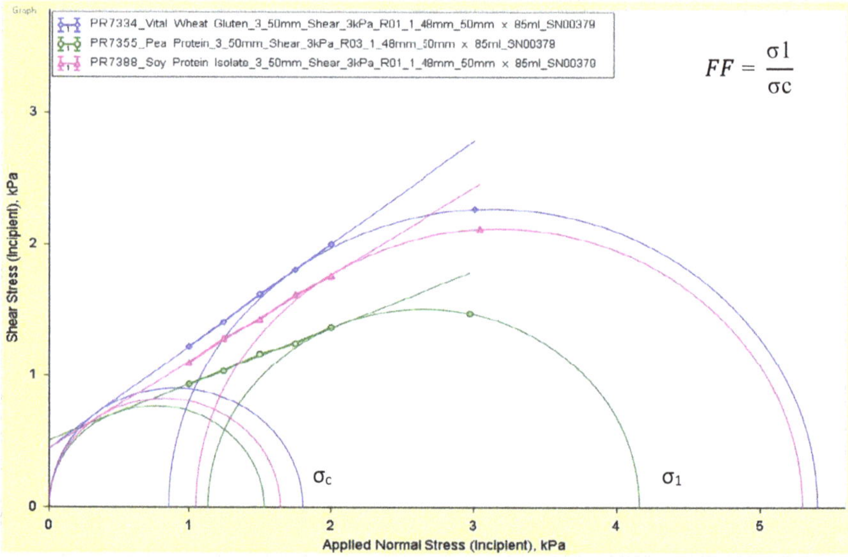

Fig. 12 Mohr circle analysis showing the shear flow properties of protein powders obtained from the output of the standard shear flow test (at 3 kPa) of the FT4™ Powder Rheometer

Table 3
Shear flow and wall friction properties of protein powders from standard shear cell and wall friction tests of the FT4™ Powder Rheometer

Protein powder	Cohesion	UYS, kPa	MPS, kPa	FF	AIF, °	WFA, °
Pea	0.51 ± 0.01	1.53 ± 0.01	4.15 ± 0.02	2.71 ± 0.01	22.75 ± 0.77	9.58 ± 0.58
Wheat gluten	0.35 ± 0.13	1.47 ± 0.48	5.31 ± 0.12	3.82 ± 1.17	39.00 ± 1.38	10.53 ± 0.53
Soy	0.44 ± 0.01	1.59 ± 0.08	5.18 ± 0.16	3.25 ± 0.05	32.09 ± 2.07	10.14 ± 0.13

UYS unconfined yield strength, *MPS* major principal stress, *FF* flow function (calculated as ratio of MPS and UYS), *AIF* angle of internal friction, *WFA* wall friction angle

powders. From this analysis, the shear flow properties (cohesion, UYS, MPS, FF, and AIF) values are determined (shown in Table 3). Cohesion values (intersection of yield locus) show that the pea protein powder is the most cohesive followed by soy and then wheat gluten protein powders. Cohesion is defined as the strength of the powder bed resulting from particle-particle interactions; higher values indicate greater internal resistance to powder flow. Hence, the pea protein powder develops greater internal resistance against shear flow compared to the other protein powders tested.

The unconfined yield strength (UYS; sigma c), major principal stress (MPS; sigma 1), and flow function (FF) values are other important shear flow property values for powders. The UYS value was highest for the soy protein powder followed by pea powder and wheat gluten. This value reflects the required stress applied to break a consolidated powder bed to initialize flow. The MPS value is defined as the maximum normal stress acting on the consolidated powder bed. This value was highest for wheat gluten followed by soy and pea protein. The UYS and MPS values are crucial as their ratio (MPS/UYS) gives the flow function value for the powder. Flow function (FF) is used as an index for comparative powder flow assessment [10]. This was first used by Jenike [11], wherein powders were classified according to their flow function. This includes FF values of <1, ≥ 1 to <2, ≥ 2 to <4, ≥ 4 to <10, and ≥ 10 corresponding to hardened, very cohesive, cohesive, easy-flowing, and free-flowing powder flowability characteristics, respectively. In this case, the FF values of all protein powders ranged from 2.71 to 3.82, which indicates that the protein powders tested all had cohesive powder flow characteristics. The angle of internal friction (AIF) can be regarded as a measure of the powder bed strength with increased normal stress applied. This powder property has implications in maintaining production efficiency as powders with high AIF values are less likely to flow readily, which can result in clogging and blockages when they are consolidated (e.g., in hoppers and silos). Table 3 shows that the wheat gluten powder had the highest AIF values followed by soy and pea protein. This indicates that with consolidation, the wheat gluten powder becomes more cohesive than the other powders; hence, it could be helpful to include measures in wheat gluten powder processing that will limit consolidation.

Wall friction is an indicator of the sliding resistance of the powder and the surface of the process equipment used. Hence, higher WFA values correspond to higher sliding resistance (measured as wall friction angle) of the powder against a particular surface. This is crucial in understanding the discharge behaviors of powders in hoppers and chutes during processing. In this book chapter, testing was done against a stainless-steel surface as this is a common material used in processing equipment. The wall friction angle (WFA) values for the protein powders are also shown in Table 3, while Fig. 13 shows the plot between applied normal stress and shear stress. The values ranged from 9.58° to 10.53°, which indicate that the sliding resistance of the protein powders is relatively similar against a stainless-steel surface. Evaluating the shear flow properties of powders is important as it provides an overview of the powder behavior during consolidation and flow rate conditions. These tests will be useful in evaluating protein powders during storage and on designing more efficient hopper/chute designs for the equipment used for these powders.

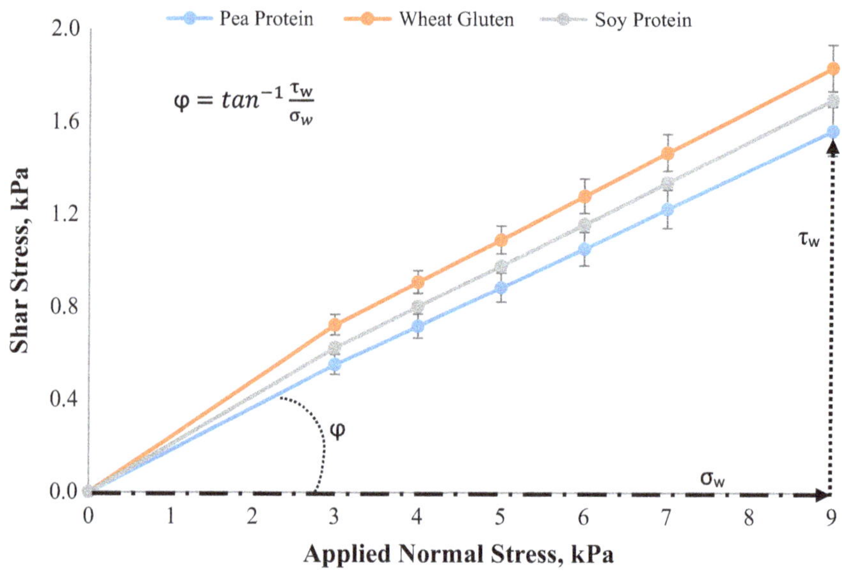

Fig. 13 Plot of applied normal stress vs. shear stress of the protein powders obtained from the standard wall friction test (at 9 kPa) of the FT4™ Powder Rheometer. The symbol "ϕ" corresponds to the wall friction angle. Test was done against a stainless-steel surface

4 Notes

1. Prepare the protein powders needed to be tested. A minimum of 500 g of protein powder sample is needed for conducting all the tests provided by the FT4™ Powder Rheometer.
2. A plastic or aluminum hand scooper is also recommended for dispensing the sample into the powder container for the powder tester.
3. Make sure that the appropriate softwares ("Powder Rheometer" and "Data Analysis v4") are installed in the computer connected to the FT4™ Powder Rheometer.
4. The parts of the vessels that need to be assembled are pre-labeled in the accessory kit. Refer to these labels when assembling the required vessel for each test.
5. When the FT4™ rheometer is operating, make sure that the safety cover is lowered as testing would not proceed otherwise.
6. When installing the attachments (blade, vented piston, shear cell head) to the rheometer, make sure that it is fully attached by checking that the notch in the rods/spindle is flushed against the rheometer hub. If not, rotate the spindle until it moves further inside the rheometer hub.
7. For the assembly of the vessel, please refer to the on-screen prompt to know proper assembly of the vessel.

8. During assembly of the vessel, make sure that the vessel can be split before being screwed into the rheometer platform by adjusting the position of the vessel splitter. This can be done by making sure that the attachments are level to the edge of the glass vessels to be used.
9. For improved consistency, putting powder samples at a consistent weight during testing is recommended.
10. When testing, an error message saying "problem moving into the vessel" could appear. This can be addressed by reducing the sample into the vessel by scooping out the top portion. Another method can be to remove the assembled vessel, re-position the vessel in the rheometer platform, and repeat testing.
11. Make sure to handle the accessories gently (48 mm blade, spindles, vented piston, wall friction head, and shear cell head) when using the FT4™ rheometer. Put accessories that are not in use inside their designated containers. Also, handle the accessories away from the blade/head ends when using.
12. If the same sample needs to be tested more than once, place test sample into a separate bag or container after each testing procedure. Do not use the same powder tested for a new test.
13. Failed test results can occur for shear cell testing. After each test, check the results of the Mohr circle analysis from the test. If the graph does not resemble the appearance shown in Fig. 12 (two Mohr circles, and a straight line is shown), please repeat the testing.
14. For shear flow testing, it is recommended to start testing with the lowest consolidation stress needed and increase it gradually after good analysis results are obtained. Good analysis results can be seen when the Mohr circle analysis is plotted correctly (refer to Fig. 12).
15. For both shear flow and wall friction testing, it is recommended to test at least two consolidation pressures if the results will be used for hopper design experiments.
16. To clean the rheometer, remove the spilled powders using a regular paint brush head or vacuum. In the case of sticky powders, surfaces and rheometer accessories can be wiped with a damp cloth or lint-free tissue.
17. Make sure that the vessels and attachments used prior to testing are cleaned properly before testing a new sample. Cleaning can be done by brushing the vessel. In the case of sticky powder samples, vessel parts can be disassembled, washed with running water, and dried before storing.

18. Place the rheometer attachments (blade, spindle, compressed piston, shear cell head, and wall friction head) in their proper locations inside the accessory kit box.
19. After testing is done, make sure to shut down rheometer when not in use, and lower the rheometer safety cover.

References

1. Intipunya P, Bhandari BR (2010) Chapter 22 – Chemical deterioration and physical instability of food powders. In: Skibsted LH, Risbo J, Andersen ML (eds) Chemical deterioration and physical instability of food and beverages. Woodhead Publishing, Cambridge
2. Grand View Research (2022) Plant based protein supplements market size, share, & trends analysis report by raw materials, by product (soy, spirulina, pumpkin seeds, wheat, hemp, rice, pea, and others), by distribution channel, by application, by region, and segment forecasts 2022–2030. Available from https://www.grandviewresearch.com/industry-analysis/plant-based-protein-supplements-market. Accessed 19 July 2023
3. Fu X, Huck D, Makein L, Armstrong B, Willen U, Freeman T (2012) Effect of particle shape and size on flow properties of lactose powders. Particuology 10(2):203–208. https://doi.org/10.1016/j.partic.2011.11.003
4. Babu KS, Siliveru K, Amamcharla JK, Vadlani PV, Ambrose RPK (2018) Influence of protein content and storage temperature on the particle morphology and flowability characteristics of milk protein concentrate powders. J Dairy Sci 101(8):7013–7026. https://doi.org/10.3168/jds.2018-14405
5. Divya S, Ganesh GNK (2019) Characterization of powder flowability using FT4-powder rheometer. J Pharm Sci Res 11(1):25–29
6. Freeman Technology (2021) Evaluating consolidation using the FT4 powder rheometer. Available from https://www.freemantech.co.uk/news/evaluating-consolidation-using-the-ft4-powder-rheometer. Accessed 19 July 2023
7. Bian Q, Sittipod S, Garg A, Ambrose RPK (2015) Bulk flow properties of hard and soft wheat flours. J Cereal Sci 63:88–94. https://doi.org/10.1016/j.jcs.2015.03.010
8. Irie KR, Petit J, Gnagne EH, Kouadio OK, Gaiani C, Scher J, Amani GN (2021) Effect of particle size on flow behaviour and physical properties of semi-ripe plantain (AAB *Musa* spp) powders. Int J Food Sci Technol 56(1):205–214. https://doi.org/10.1111/ijfs.14620
9. Freeman Technology (2023) Shear testing. Available from https://www.freemantech.co.uk/powder-testing/ft4-powder-rheometer-powder-flow-tester/shear-testing. Accessed 19 July 2023
10. Leung LY, Mao C, Srivastava I, Du P, Yang CY (2017) Flow function of pharmaceutical powders is predominantly governed by cohesion, not by friction coefficients. J Pharm Sci 106(7):1865–1873. https://doi.org/10.1016/j.xphs.2017.04.012
11. Jenike AW (1964) Storage and flow of solids, vol 53. University of Utah, Salt Lake City

Chapter 18

Molecular Size Distribution

Qin Zhang, Rongxue Zhou, Guo Huang, Yan Zhang, and Xiaonan Sui

Abstract

As one of the most basic structural parameters of proteins, molecular size distribution has an important effect on the properties of protein solutions, the processing properties, and the usability of protein. There are many methods for determining the molecular weight of proteins, such as size exclusion-high-performance liquid chromatography (SEC-HPLC), osmotic pressure, light scattering, ultracentrifugation, and polyacrylamide gel electrophoresis. Among these techniques, the SEC-HPLC technique has the advantages of faster, more accurate, better repeatability and automation. Therefore, this chapter provides a comprehensive guidance to the analysis of the molecular size distribution of proteins using SEC-HPLC.

Key words Molecular size distribution, Protein, SEC-HPLC

1 Introduction

The relative molecular weight of the protein is not the same because of the difference in the number of amino acid residues. Most of the proteins used in the food industry are mixtures with different relative molecular weights. The determination of protein molecular weight is helpful in ensuring the consistency, stability, and safety of protein products [1]. Additionally, the molecular weight information of proteins can help analyze their structure and function and provide a basis for the development and optimization of food [2].

Size exclusion chromatography (SEC) is a chromatographic technique that separates molecules based on their size. The principle is that the column is filled with gels with holes of a certain size (Fig. 1). Large molecular substances cannot enter the small hole so they are quickly eluded by the mobile phase, leading to short retention time and small elution volumes. Conversely, small molecules can penetrate into the gel holes, resulting in longer retention time and larger elution volume. Since 1964, size exclusion-high-

Yonghui Li (ed.), *Plant-Based Proteins: Production, Physicochemical, Functional, and Sensory Properties*, Methods and Protocols in Food Science, https://doi.org/10.1007/978-1-0716-4272-6_18,

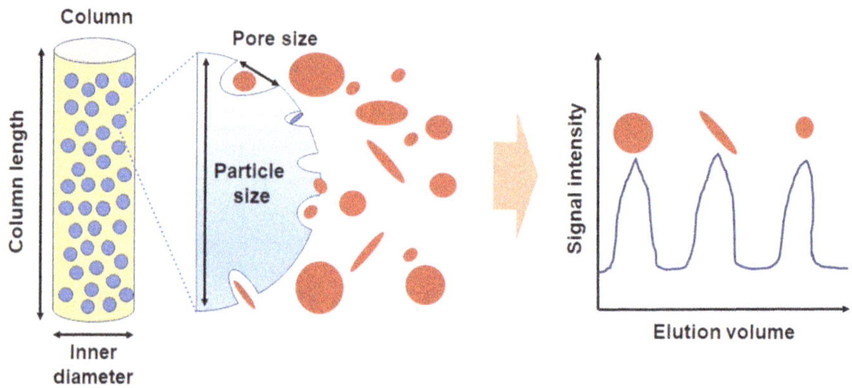

Fig. 1 Interpretation of an exclusion column: column with porous particles and relevant chromatogram of a size exclusion separation

performance liquid chromatography (SEC-HPLC) has become the most convenient, rapid, and widely used method for determining the molecular size distribution of polymers [3, 4]. Nowadays, SEC-HPLC has been instrumentalized and automated [5]. The gel chromatography can separate relative molecular weight ranging from millions to less than 100.

SEC-HPLC comprises liquid reservoir, pump, injector, column, detector, and recorder [6]. After the mobile phase was injected into the system with a high-pressure pump, the sample solution entered the chromatographic column through the sampler with the mobile phase. Due to the different distribution coefficients in the two phases, the components in the sample solution move relatively in the two phases and undergo repeated adsorption-desorption distribution processes. This leads to a large difference in the moving speed of each component in the sample, which is separated into a single component and flows out of the column in turn. When the sample is checked by the detector during the sample outflow process, the sample concentration will be converted into an electrical signal and transmitted to the recorder [7].

The relative molecular weight and its distribution of the polymer can be simultaneously determined by SEC-HPLC [8]. In addition, the elution time can be predicted, and the sample can be continuously loaded, resulting in a short time for gel chromatography to separate polymers [9]. According to the nature of the gel used, it can be categorized into gel filtration chromatography (GFC) using aqueous solution and gel permeation chromatography (GPC) using organic solvent [10]. GFC is suitable for the analysis of polypeptides, proteins, enzymes, and other biomolecules in aqueous solution [11]. GPC is mainly used for determination of molecular weight of organic polymer (such as polyethylene, polypropylene, polystyrene, polyvinyl chloride, polymethyl methacrylate) [12].

The purpose of this chapter is to provide guidance for the determination of the molecular size distribution of biomolecules, such as peptides, proteins, and biological enzymes in aqueous solutions. Therefore, this chapter will help standardize the methods used, so that it will facilitate the comparison of results.

2 Materials

The test will be performed on an HPLC system (Waters e2695, Milford, MA, USA) equipped with a diode array detector, an autosampler, and a vacuum degasser (Fig. 2). All solutions and samples are prepared with ultrapure water and chromatographic grade reagents:

1. An AdvanceBio SEC 300 × 4.6 mm column with a pore size of 2.7 μm (Agilent Technologies, Wilmington, DE, USA) (*see* **Note 1**).

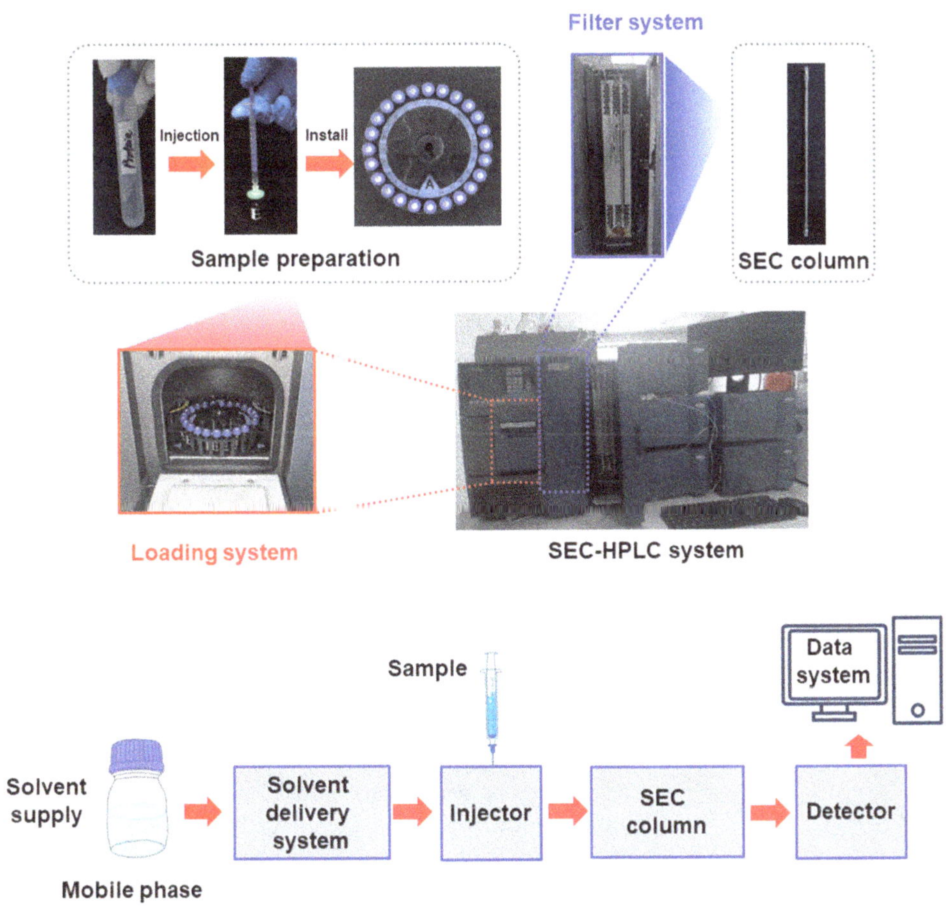

Fig. 2 Equipment diagram of protein molecular size distribution determination by SEC-HPLC

2. Injection needle, cellulose acetate membrane, and syringe (*see* **Note 2**).
3. 1 mg/mL protein solution. 20 mg protein was dissolved in 10 mL ultrapure water. The protein concentration in the sample is then measured by the Lowry protein detection kit, and the protein concentration is diluted to 1 mg/mL. Finally, the sample is filtered using a syringe with a 0.2 μL filter membrane.
4. The mobile phase. 100 mM sodium phosphate buffer (pH 7.0) and passed through a 0.22 μm organic membrane to obtain the mobile phase.
5. The protective solution of column (0.05 wt% NaN_3). 20 mL 1 wt% NaN_3 was mixed with 380 mL ultrapure water and stored in a protective solution bottle after passing through the membrane (*see* **Note 3**).
6. Select the appropriate standard molecular markers, and calculate the molecular size distribution range of the sample according to the standard calibration curve. Common standards are thyroglobulin (670 kDa), bovine γ-globulin (158 kDa), ovalbumin (43 kDa), cytochrome c (12.384 kDa), aprotinin (6.512 kDa), vitamin B12 (1.855 kDa), and glutathione (0.307 kDa).

3 Method

All procedures will be carried out at room temperature unless otherwise specified (*see* **Note 4**).

3.1 Sample Preparation for Analysis

1. The mobile phase was ultrasound for 15 min to remove the bubbles.
2. Shake the sample bottle gently to break the bubbles in the wall (*see* **Note 5**).

3.2 Determination of Protein Molecular Size Distribution

1. Firstly, turn on the computer and instruments, including the infusion pump, sampler, detector, and column temperature box. Then, open the OMNISEC 5.0 test software, and ensure that the software and hardware connection is normal (*see* **Note 6**). Regarding parameter settings, the flow rate of the mobile phase was set to 0.3 mL/min, the determination time was set to be 2–3 min greater than the peak time of the solvent, the column temperature was set to 30 °C, and the maximum column pressure was set to 8 MPa (*see* **Note 7**). The column eluent was monitored at a wavelength of 220 nm for 30 min.
2. Unscrew the exhaust valve counterclockwise, purge for 3 min, and close the valve to get rid of the bubbles. An alternative method is to use a syringe to remove air bubbles from the system (*see* **Note 8**).

3. Adjust the injection valve from "Load" to "Inject," stay for 1 min, and then clean the injection valve by injecting detergent into the injection port three times with the flat inlet injector.
4. After the end of **step 4**, run the instrument according to the method parameters until the baseline can be leveled to make gel filtration column balanced.
5. Next, balance the differential detector. Flush the reference pool, and eliminate bubbles in the pool. After the baseline is stable, check the balance value; if the value is more than 200, purge three times.
6. Set the sample name, sample ID, and data storage path in the software. Open the injection valve, inject 300 μL of sample into the injector using the injector, and close the injection valve 1 min later (*see* **Notes 9** and **10**).
7. Cease temperature control, and let the protection fluid into the column at a flow rate of 0.5 mL/min until the column temperature is stable. Remove the filling column from the instrument, plug both ends with plugs, and place it in the packaging box.
8. Shut down programs, computers, and devices in sequence.
9. We analyze the curve with "OMNISEC 5.0" software to obtain the molecular size distribution data using standard curve method (Fig. 3). Firstly, the method is established by selecting the light detection display, selecting the standard sample, and

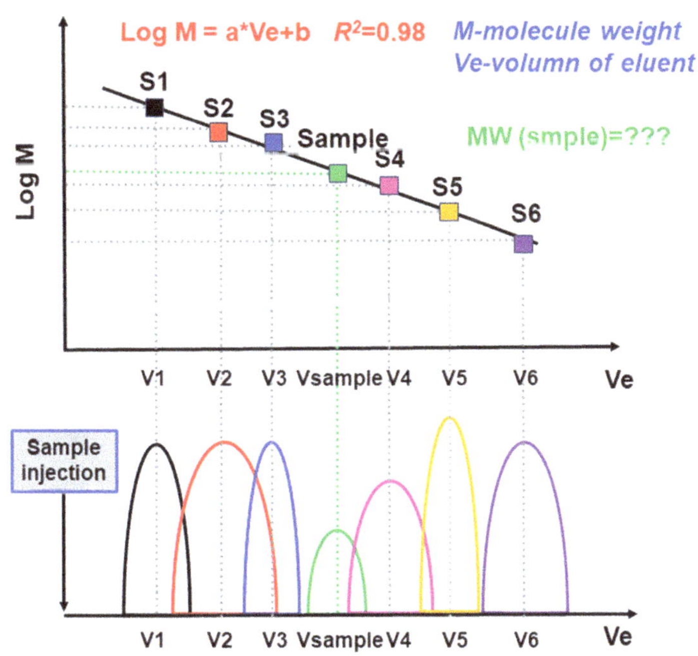

Fig. 3 The molecular size distribution of the sample to be measured was calculated by standard curve method

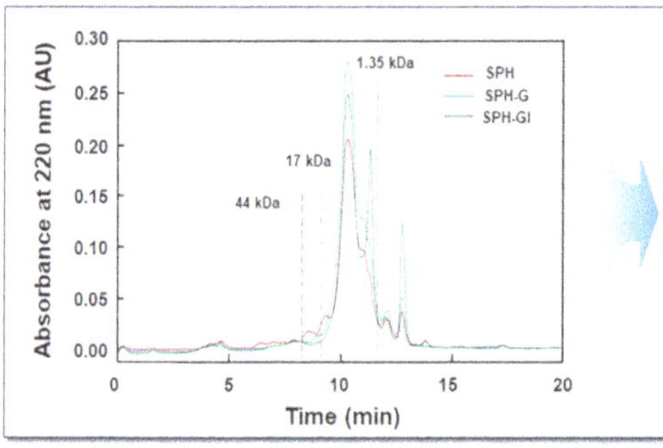

Sample	Molecular weight distribution (%)			
	>10 kDa	5-10 kDa	1-5 kDa	<1 kDa
SPH	27.83±0.09[a]	14.70±0.29[b]	43.81±1.03[b]	13.66±1.41[b]
SPH-G	11.00±0.72[b]	18.76±0.39[a]	53.17±0.04[a]	17.06±0.30[b]
SPH-GI	11.85±0.22[b]	15.32±0.19[b]	50.64±0.77[a]	22.19±1.17[a]

Fig. 4 The SEC-HPLC results. SPH represents soybean protein hydrolysate, SPH-G indicates gastric-digested SPH, and SPH-GI means gastrointestinal-digested SPH

entering the parameters of the mobile phase. Then, open the file containing the sample curve, adjust the peak shape after calibration baseline, and set the starting point and end point of the peak (*see* **Note 11**). Finally, the sample curve is calculated by using the established method, and the data report is printed (Fig. 4).

4 Notes

1. The gel column needs to be fully dissolved in the eluent for 1 day before using to ensure that the porous gel particles are full of water and evenly filled in the column tube. We should choose the appropriate eluent and flow rate to ensure the sample dissolution and flow and prevent sample adsorption or denaturation.
2. Clean all instrument used in the experiment with ultrapure water to avoid contamination.
3. NaN_3 is highly toxic, and all utensils and solutions that come into contact with NaN_3 must be dissolved with NaClO for proper disposal and safety.
4. Maintain the laboratory temperature within the range of 10–35 °C, and the humidity is less than 80%. The instrument should be placed in a place without corrosive gases, dust, and direct sunlight.
5. The sample should not be shaken violently to avoid breaking the polymer chain.
6. In order to prevent damage to the instrument, it's recommended to turn on all the hardware after opening the software.
7. The flow rate should be set in the range of 0.8–1.2 mL/min, and the maximum column pressure does not exceed 30 MPa.

8. Do not open the drain valve more than 180°; otherwise it is easy to introduce air bubbles into the infusion system during purge.
9. The volume of the sample ring is 100 μL, and the volume of the sample is preferably three times the volume of the obtained sample ring.
10. The sample needs to flow smoothly and evenly into the injector, and the valve needs to be opened and closed quickly.
11. If the peaks are not symmetrical, the columns need to be more uniform. Because if the column is too loose, it is easy to cause tailing, and if it is too tight, it will cause the front column to be dirty and regenerate the column.

References

1. Dornish M, Kaplan D, Skaugrud Ø (2001) Standards and guidelines for biopolymers in tissue-engineered medical products: ASTM alginate and chitosan standard guides. Ann N Y Acad Sci 944(1):388–397
2. de la Cruz L, Nguyen THD, Ozawa K, Shin J, Graham B, Huber T, Otting G (2011) Binding of low molecular weight inhibitors promotes large conformational changes in the dengue virus NS2B-NS3 protease: fold analysis by pseudocontact shifts. J Am Chem Soc 133(47):19205–19215
3. Kleist H (1964) Radiochromatographic studies on fulvic acids. Acta Biol Med Ger 13:949–951
4. Wang Z, Fingas M, Yang C, Christensen JH (1964) Crude oil and refined product fingerprinting: principles. In: Environmental forensics. Elsevier, pp 339–407
5. Ó'Fágáin C, Cummins PM, O'Connor BF (2017) Gel-filtration chromatography. In: Protein chromatography: methods and protocols. Humana Press, Springer, New York, pp 15–25
6. Kanu AB (2021) Recent developments in sample preparation techniques combined with high-performance liquid chromatography: a critical review. J Chromatogr A 1654:462444
7. Ó'Fágáin C, Cummins PM, O'Connor BF (2011) Gel-filtration chromatography. In: Protein chromatography: methods and protocols. Humana Press, New York, pp 25–33
8. Li M, Xing C, Liu L, Huang W, Meng Y (2022) Gel permeation chromatography-based method for assessing the properties of binders in reclaimed asphalt pavement mixtures. Constr Build Mater 316:126005
9. Gregges AR, Dowden BF, Barrall EM, Horikawa TT (1970) Modification of a gel permeation chromatograph for automatic sample injection and on-line computer data recording. Sep Sci 5(6):731–749
10. Viovy J, Lessec J (2005) Separation of macromolecules in gels: permeation chromatography and electrophoresis. Polym Anal Charact:1–41. https://doi.org/10.1007/BFb0008693
11. Stanton P (2004) Gel filtration chromatography. In: HPLC of peptides and proteins: methods and protocols. Humana Press, Totowa, pp 55–73
12. Neira-Velázquez MG, Rodríguez-Hernández MT, Hernández-Hernández E, Ruiz-Martínez AR (2013) Polymer molecular weight measurement. In: Handbook of polymer synthesis, characterization, and processing. Wiley, Chicester, pp 355–366

Chapter 19

Polyacrylamide Gel Electrophoresis

Changqi Liu, Jing Zhao, and Sahil Gupta

Abstract

Gel electrophoresis is a widely used analytical method for protein characterization. It also serves as a foundational step for various other analytical techniques such as Western blotting, N-terminal amino acid sequencing, and mass spectrometry. Within this chapter, we describe the protocols for vertical protein electrophoresis using nondenaturing, non-dissociating polyacrylamide gel electrophoresis (NDND-PAGE), and sodium dodecyl sulfate (SDS)-PAGE. Furthermore, procedures are outlined for sample preparation, gel preparation, and band visualization through Coomassie Brilliant Blue staining, silver staining, and glycoprotein staining.

Key words Electrophoresis, NDND-PAGE, SDS-PAGE, Protein separation

1 Introduction

Electrophoresis is a frequently employed method for separating and analyzing proteins, glycans, nucleic acids, and other charged particles or molecules. When subjected to an electric field, these charged molecules or particles move toward the electrode carrying an opposing charge. Due to their distinct charges and masses, different molecules and particles within a mixture migrate at varying speeds, resulting in their separation into different fractions. Electrophoresis can be carried out in free solutions (e.g., capillary electrophoresis) or with supporting media such as papers, films, and gels. Electrophoresis in a gel matrix stands as the most popular technique, employing a variety of gel materials such as starch gels, dextran gels, agarose gels, and polyacrylamide gels [1]. In this protocol, we will focus on protein electrophoresis in polyacrylamide gels.

Polyacrylamide gels are created through the chemical copolymerization of acrylamide monomers and a cross-linking reagent, *N*,*N'*-methylenebisacrylamide. The commencement of this reaction involves the utilization of ammonium persulfate as a catalyst, while *N*,*N*,*N'*,*N'*-tetramethylethylenediamine furnishes tertiary amino groups to initiate the release of radicals. The pore size of the gel

Yonghui Li (ed.), *Plant-Based Proteins: Production, Physicochemical, Functional, and Sensory Properties*, Methods and Protocols in Food Science, https://doi.org/10.1007/978-1-0716-4272-6_19,

can be precisely controlled by manipulating the total concentration of acrylamide (T) and the degree of cross-linking (C) [2], where T and C are defined as:

$$T = \frac{a + b}{V} \times 100\%$$

$$C = \frac{b}{a + b} \times 100\%$$

where a is the mass of acrylamide (g), b is the mass of N,N'-methylenebisacrylamide (g), and V is the volume (mL).

The polymerization process of these gels takes place within an inert setting to avoid any disruption caused by oxygen. This environment could involve a glass tube for cylindrical gels or the space between two glass plates for flat gels. Among these options, flat gels are more frequently employed and can be oriented either vertically or horizontally for running electrophoretic experiments. The instructions in this protocol are limited to vertical flat gels.

Proteins can be electrophoresed in their native form, as seen in techniques like isoelectric focusing (IEF) and nondenaturing, nondissociating polyacrylamide gel electrophoresis (NDND-PAGE). Alternatively, they can be transformed into sodium dodecyl sulfate (SDS)-protein micelles for uses in SDS-PAGE. In this protocol, we provide instructions for NDND-PAGE based on Andrew's method [3], as well as for SDS-PAGE following the Fling and Gregerson modification [4] of the Laemmli method [5].

2 Materials

2.1 Materials for Sample Preparation

2.1.1 Materials for Protein Solubilization

1. Equipment and supplies: analytical balance, centrifuge, pH meter, grinder, sieve, vortex mixer, magnetic stirrer (optional), graduated cylinders, volumetric flasks, beakers, magnetic bars, centrifuge tubes, microcentrifuge tubes.
2. Chemicals and reagents: boric acid, hydrochloric acid, sodium borate, sodium chloride, reagents for soluble protein assay (refer to Chapter 9).
3. Borate saline buffer: 0.1 M boric acid, 25 mM sodium borate, 75 mM NaCl, pH 8.45.

2.1.2 Materials for SDS-PAGE Sample Preparation

1. Equipment and supplies: analytical balance, pH meter, vortex mixer, magnetic stirrer, water bath or heat block, pipettes, graduated cylinders, volumetric flasks, beakers, magnetic bars, centrifuge tubes, microcentrifuge tubes.
2. Chemicals and reagents: bromophenol blue, glycerol, hydrochloric acid, β-mecaptoethanol, N-ethylmaleimide, sodium dodecyl sulfate (SDS), Tris base.

3. SDS-PAGE reducing sample buffer: 50 mM Tris–HCl, 1% w/v SDS, 0.01% w/v bromophenol blue, 30% w/v glycerol, 2% v/v β-mecaptoethanol (*see* **Note 1**), pH 6.8.
4. SDS-PAGE non-reducing sample buffer: 50 mM Tris–HCl, 1% w/v SDS, 0.01% w/v bromophenol blue, 30% w/v glycerol, 1 mM *N*-ethylmaleimide, pH 6.8.

2.1.3 Materials for NDND-PAGE Sample Preparation

1. Equipment and supplies: analytical balance, pH meter, vortex mixer, magnetic stirrer, pipettes, graduated cylinders, volumetric flasks, beakers, magnetic bars, centrifuge tubes, microcentrifuge tubes.
2. Chemicals and reagents: boric acid, bromophenol blue, glycerol, hydrochloric acid, sodium-ethylenediamine tetra acetic acid (Na-EDTA), Tris base.
3. NDND-PAGE sample buffer: 0.45 M Tris–HCl, 0.4 M boric acid, 12.5 mM Na-EDTA, 0.01% w/v bromophenol blue, 30% w/v glycerol, pH 8.5.

2.2 Materials for Gel Preparation

2.2.1 Materials for SDS-PAGE Gel Preparation

1. Equipment and supplies: analytical balance, pH meter, magnetic stirrer, glass plates, notched glass plates (or aluminum oxide ceramic plate), spacers, gel casting frames, gel casting stands with a rubber gasket bottom, combs, gradient maker (for gradient gel only), magnetic bars, peristaltic pump (optional, for gradient gel only), pipettes, graduated cylinders, volumetric flasks, beakers, 15 mL conical flasks, parafilm, gloves, Kimwipes.
2. Chemicals and reagents: acrylamide, ammonium persulfate (APS), hydrochloric acid, *N*,*N*′-methylenebisacrylamide (bisacrylamide), SDS, *N*,*N*,*N*′,*N*′-tetramethylethylenediamine (TEMED), Tris base, water-saturated *n*-butanol (*see* **Note 2**).
3. Acrylamide/bis stock (50%T/2.6%C): 50% w/v acrylamide, 1.35% w/v bis-acrylamide in distilled water.
4. Resolving buffer: 1.5 M Tris–HCl, 0.2% w/v SDS, pH 8.8.
5. Stacking buffer: 0.5 M Tris–HCl, 0.4% w/v SDS, pH 6.8.
6. 10% w/v APS solution.

2.2.2 Materials for NDND-PAGE Gel Preparation

1. Equipment and supplies: analytical balance, pH meter, magnetic stirrer, glass plates, notched glass plates (or aluminum oxide ceramic plate), spacers, gel casting frames, gel casting stands with a rubber gasket bottom, combs, gradient maker (for gradient gel only), magnetic bars, peristaltic pump (optional, for gradient gel only), pipettes, graduated cylinders, volumetric flasks, beakers, 15 mL conical flasks, parafilm, gloves, Kimwipes.

2. Chemicals and reagents: acrylamide, APS, bisacrylamide, boric acid, glycerol, hydrochloric acid, Na-EDTA, TEMED, Tris base, water-saturated *n*-butanol (*see* **Note 2**).
3. Acrylamide/bis stock (50%T/2.6%C): 50% w/v acrylamide, 1.35% w/v bis-acrylamide in distilled water.
4. Tris–borate buffer: 0.45 M Tris–HCl, 0.4 M boric acid, 12.5 mM Na-EDTA, pH 8.5.
5. 10% w/v APS solution.

2.3 Materials for Running the Gels

2.3.1 Materials for Running SDS-PAGE

1. Equipment and supplies: analytical balance, pH meter, gel electrophoresis system, power supply, vortex mixer, graduated cylinder, volumetric flask, syringe, or pipette.
2. Chemicals and reagents: glycine, SDS, Tris base.
3. SDS-PAGE running buffer stock: 0.5 M Tris–HCl, 1.9 M glycine, 1% w/v SDS, pH 8.7–8.8.

2.3.2 Chemicals and Reagents for Running NDND-PAGE

1. Equipment and supplies: analytical balance, gel electrophoresis system, power supply, vortex mixer, graduated cylinder, volumetric flask, syringe, or pipette.
2. Chemicals and reagents: boric acid, Na-EDTA, Tris base.
3. NDND-PAGE running buffer stock: 0.9 M Tris–HCl, 0.8 M boric acid, 25 mM Na-EDTA.

2.4 Materials for Band Visualization

2.4.1 Materials for Coomassie Brilliant Blue (CBB) Staining

1. Equipment and supplies: analytical balance, orbital shaker, graduated cylinder, volumetric flask, plastic trays with lids.
2. Chemicals and reagents: acetic acid, CBB R-250 dye, methanol.
3. CBB staining solution: 0.25% w/v CBB R-250, 50% v/v methanol, 10% v/v acetic acid in distilled water.
4. Destaining solution: 50% v/v methanol, 10% v/v acetic acid in distilled water.

2.4.2 Materials for Silver Staining

1. Equipment and supplies: analytical balance, orbital shaker, pipettes, graduated cylinder, volumetric flask, plastic trays with lids.
2. Chemicals and reagents: acetic acid, ethanol, formaldehyde (37% w/v), glutaraldehyde (25% w/v), glycine, silver nitrate, sodium acetate, sodium carbonate, sodium thiosulfate pentahydrate.
3. Fixing solution: 30% v/v ethanol, 10% v/v acetic acid in distilled water.
4. Sensitizer solution: dissolve 17 g of sodium acetate in distilled water, add 75 mL of ethanol, add 1.25 mL of glutaraldehyde (25% w/v), and 0.5 g of sodium thiosulfate pentahydrate before use and bring the volume to 250 mL.

5. Silvering solution: dissolve 0.5 g of silver nitrate in distilled water and bring the volume to 250 mL with distilled water, add 50 μL of formaldehyde (37% w/v) before use.
6. Developer solution: dissolve 7.5 g of sodium carbonate and 30 μL of formaldehyde (37% w/v) in distilled water and bring the volume to 300 mL with distilled water.
7. Stopping solution: 1% w/v glycine solution in distilled water.

2.4.3 Materials for Silver Staining

1. Equipment and supplies: analytical balance, orbital shaker, graduated cylinder, volumetric flask, plastic trays with lids.
2. Chemicals and reagents: acetic acid, methanol, periodic acid, Schiff's reagent.
3. Fixing solution: 50% v/v methanol, 10% v/v acetic acid in distilled water.
4. Washing solution: 3% v/v acetic acid.
5. Oxidizing solution: 1% w/v periodic acid in 3% v/v acetic acid.

3 Methods

3.1 Sample Preparation

3.1.1 Protein Solubilization

Isolated plant proteins (e.g., soy protein isolate) can be suspended in distilled water or appropriate buffers [6]. Soluble proteins can also be extracted from plant tissues. The extraction conditions (e.g., solvent, pH, ionic strength, temperature, time) need to be optimized for protein extraction efficiency [7]. We will use protein extraction from plant seeds as an example [8].

1. Grind plant seeds in a grinder to pass a twenty-mesh sieve (*see* **Note 3**). Defat the seed flour as needed (*see* **Note 4**).
2. Disperse the seed flour in 10 volumes of borate saline buffer with continuous vortex mixing or magnetic stirring for 1 h at room temperature.
3. Centrifuge the suspension at 16,000 *g* for 15 min at room temperature.
4. Collect the supernatant and determine the soluble protein concentration using methods as described in Chapter 9.

3.1.2 Electrophoresis Sample Preparation

Different sample preparation protocols are employed for SDS-PAGE and NDND-PAGE. Refer to subheadings "SDS-PAGE Sample Preparation" and "NDND-PAGE Sample Preparation". Within SDS-PAGE, proteins can be electrophoresed under either reducing or non-reducing conditions. In the reducing condition, a reducing agent (e.g., β-mercaptoethanol, dithiothreitol) is introduced to break disulfide bonds. In the non-reducing condition, *N*-ethylmaleimide can be added to prevent formation of disulfide artifacts during sample preparation [9].

SDS-PAGE Sample Preparation

1. Dilute the soluble protein suspension to an appropriate concentration (1–3 mg/mL) in a microcentrifuge tube (*see* **Note 5**).
2. Mix the diluted protein suspension with an equal volume of reducing or non-reducing SDS-PAGE sample buffer using a vortex mixer.
3. Heat the samples in a bath or heat block between 60–85 °C for 5 min (*see* **Note 6**).
4. For reduced samples, add 1 μL β-mercaptoethanol per 100 μL sample solution after cooling (*see* **Note 7**).

NDND-PAGE Sample Preparation

1. Dilute the soluble protein suspension to an appropriate concentration (e.g., 2 mg/mL) in a microcentrifuge tube.
2. Mix the diluted protein suspension with an equal volume of NDND-PAGE sample buffer using a vortex mixer.

3.2 Casting the Gels

3.2.1 Assembling Gel Cassettes

1. Assemble gel cassettes including a glass plate, a notched glass plate (or aluminum oxide ceramic plate; *see* **Note 8**), and two spacers as illustrated in Fig. 1.
2. Fasten the gel cassettes in gel casting frames (*see* **Note 9**).
3. Place the assembled gel cassettes on gel casting stands (*see* **Note 10**).

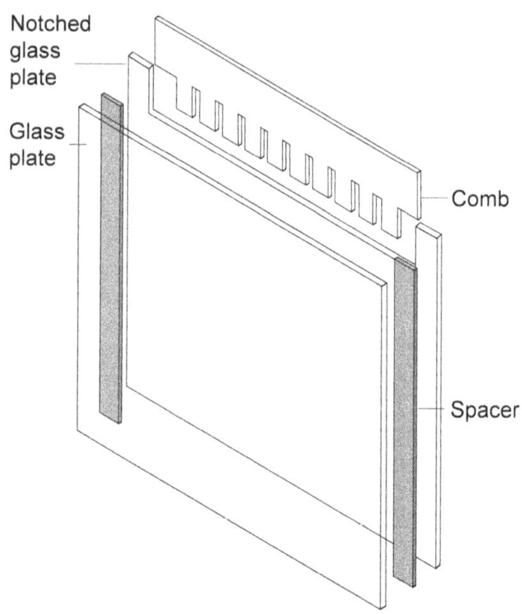

Fig. 1 Assembly of a gel cassette

3.2.2 Preparation of Resolving Gels

The polyacrylamide concentration can be manipulated to regulate the pore size of the gels. Typically, low-concentration gels are used to separate large proteins, while high-concentration gels are used to separate small proteins. By continuously altering the acrylamide concentration in the polymerization solution, a gradient gel can be created. These gradient gels allow for the separation of proteins spanning a broader range of sizes [1]. For details on SDS-PAGE monomer gel preparation, refer to subheading "Preparation of SDS-PAGE Monomer Resolving Gels", for SDS-PAGE gradient gel preparation, proceed to subheading "Preparation of SDS-PAGE Gradient Resolving Gels", and for NDND-PAGE gradient gel preparation, proceed to subheading "Preparation of NDND-PAGE Gradient Resolving Gels".

Preparation of SDS-PAGE Monomer Resolving Gels

1. To prepare SDS-PAGE monomer gels, mix the resolving buffer, acrylamide stock, distilled water, 10% (w/v) APS solution, and TEMED in conical flasks in accordance with the specified volumes indicated in Table 1 (*see* **Note 11**).
2. Seal the conical flasks with parafilm and mix well. Make sure that no solution is spilled.
3. Pipette the mixed resolving gel solution into the gel cassettes (*see* **Note 12**).
4. Overlay the resolving gel solution with a thin layer of water saturated butanol to prevent oxygen from diffusing into the solution (*see* **Note 13**).
5. Let the resolving gels polymerize for at least an hour before proceeding to Subheading 3.2.3.

Preparation of SDS-PAGE Gradient Resolving Gels

1. To prepare SDS-PAGE gradient gels, mix the resolving buffer, acrylamide stock, distilled water, 10% (w/v) APS solution, and TEMED in conical flasks in accordance with the specified volumes indicated in Table 2 (*see* **Note 11**).

Table 1
Recipes for preparing SDS-PAGE monomer gels (sufficient for making two 8.6 × 6.7 × 0.075 cm gels)

	Resolving gel (12%T/2.6%C)	Stacking gel (4%T/2.6%C)
Resolving buffer	5 mL	–
Stacking buffer	–	2.5 mL
Acrylamide/bis stock (50%T/2.6%C)	2.4 mL	0.8 mL
Distilled water	2.6 mL	6.7 mL
APS solution (10%)	50 μL	50 μL
TEMED	10 μL	10 μL

Table 2
Recipes for preparing SDS-PAGE gradient gels (sufficient for making an 18 × 16 × 0.15 cm gel)

	Light solution (8%T/2.6%C)	Dense solution (25%T/2.6%C)	Stacking gel (4%T/2.6%C)
Resolving buffer	7.5 mL	7.5 mL	–
Stacking buffer	–	–	2.5 mL
Acrylamide/bis stock (50%T/2.6%C)	2.4 mL	7.5 mL	0.8 mL
Distilled water	5.1 mL	–	6.7 mL
APS solution (10%)	50 μL	50 μL	50 μL
TEMED	10 μL	10 μL	10 μL

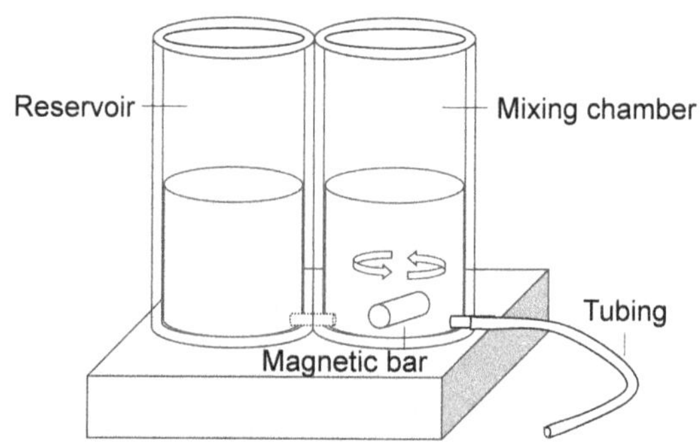

Fig. 2 Gradient maker

2. Seal the conical flasks with parafilm and mix well. Make sure that no solution is spilled.
3. Simultaneously, pour the light solution into the reservoir, and the dense solution into the mixing chamber of the gradient maker (Fig. 2).
4. Stir the solution in the mixing chamber using a magnetic stirrer (*see* **Note 14**).
5. Place the tip of the tubing into the cassette.
6. Use gravity or a peristaltic pump to deliver the solution into the cassette (*see* **Note 15**).
7. Rinse the gradient maker with distilled water immediately to avoid clogging of the tubes and connections.
8. Overlay the resolving gel solution with a thin layer of water saturated butanol (*see* **Note 13**).
9. Let the resolving gels polymerize for at least 1 h before proceeding to Subheading 3.2.3.

Table 3
Recipes for preparing NDND-PAGE gradient gels (sufficient for making an 18 × 16 × 0.15 cm gel)

	Light solution (3%T/2.6%C)	Dense solution (30%T/2.6%C)	Stacking gel (3%T/2.6%C)
Tris–borate buffer	3.0 mL	3.0 mL	2.0 mL
Acrylamide/bis stock (50%T/2.6%C)	0.9 mL	9.0 mL	0.6 mL
Distilled water	10.7 mL	2.6 mL	7.4 mL
Glycerol	0.4 mL	0.4 mL	–
APS solution (10%)	50 μL	50 μL	50 μL
TEMED	10 μL	10 μL	10 μL

Preparation of NDND-PAGE Gradient Resolving Gels

1. To prepare NDND-PAGE gradient gels, mix the Tris–borate buffer, acrylamide stock, distilled water, glycerol, 10% (w/v) APS solution, and TEMED in conical flasks in accordance with the specified volumes indicated in Table 3 (*see* **Note 11**).
2. Seal the conical flasks with parafilm and mix well. Make sure that no solution is spilled.
3. Simultaneously, pour the light solution into the reservoir, and the dense solution into the mixing chamber of the gradient maker (Fig. 2).
4. Stir the solution in the mixing chamber using a magnetic stirrer (*see* **Note 14**).
5. Place the tip of the tubing into the cassette.
6. Use gravity or a peristaltic pump to deliver the solution into the cassette (*see* **Note 15**).
7. Rinse the gradient maker with distilled water immediately to avoid clogging of the tubes and connections.
8. Overlay the resolving gel solution with a thin layer of water saturated butanol (*see* **Note 13**).
9. Let the resolving gels polymerize for at least 1 h before proceeding to Subheading 3.2.3.

3.2.3 Preparation of Stacking Gels

1. Rinse the polymerized resolving gels several times with distilled water and dry the inside of the glass plates with Kimwipes.
2. Mix the stacking buffer (SDS-PAGE) or Tris–borate buffer (NDND-PAGE), acrylamide stock, distilled water, 10% (w/v) APS solution, and TEMED in accordance with the specified volumes indicated in Tables 1, 2, and 3 (*see* **Note 11**).
3. Seal the conical flasks with parafilm and mix well. Make sure that no solution is spilled.

4. Pipette the mixed stacking gel solution into the gel cassettes (*see* **Note 12**).
5. Insert the combs into the gel cassettes (Fig. 1, *see* **Note 16**).
6. Let the gels to polymerize overnight (*see* **Note 17**).

3.3 Running the Gels

1. Assemble the electrophoresis system.
2. Remove the combs from the gel cassettes.
3. Dilute the running buffer stock ten-fold with distilled water, mix, and pour into the chamber (*see* **Note 18**).
4. Vortex mix the protein samples in microcentrifuge tubes.
5. Load the samples and molecular weight standards using a syringe or pipette (*see* **Note 19**). Push the samples slowly into the wells (*see* **Note 20**).
6. Connect the system to the power supply.
7. Run the gels at a constant current (10 mA/gel for stacking gel and 20 mA/gel for resolving gel) until the tracking dye reached the gel edge (*see* **Note 21**).
8. Turn off the power supply.
9. Pour the running buffer and remove the gel cassettes.

3.4 Band Visualization

1. Open the gel cassettes with a plastic wedge (*see* **Note 22**).
2. Carefully remove the gels and subject them to staining.
3. Clean and dry the glass and ceramic plates.

Various staining methods allow for the visualization of proteins and polypeptides on gels [1]. Here, three examples are provided: CBB staining [10], silver staining [11], and glycoprotein staining [12]. CBB staining stands out as the most popular approach. Silver staining delivers high sensitivity (100 pg/band). Glycoprotein staining allows for identification of glycoproteins. Proceed to Subheading 3.4.1 for CBBR staining, Subheading 3.4.2 for silver staining, and Subheading 3.4.3 for glycoprotein staining. Examples of gels stained by different methods are illustrated in Fig. 3.

3.4.1 CBB Staining (See Note 23)

1. Fill the plastic tray with CBB staining solution (*see* **Note 24**). Carefully place the gel in the tray. Make sure there is enough staining solution to cover the gel.
2. Gently shake the tray on an orbital shaker overnight.
3. Remove the staining solution and rinse the gel with distilled water.
4. Add enough destaining solution to cover the gel and shake for 2–4 h.

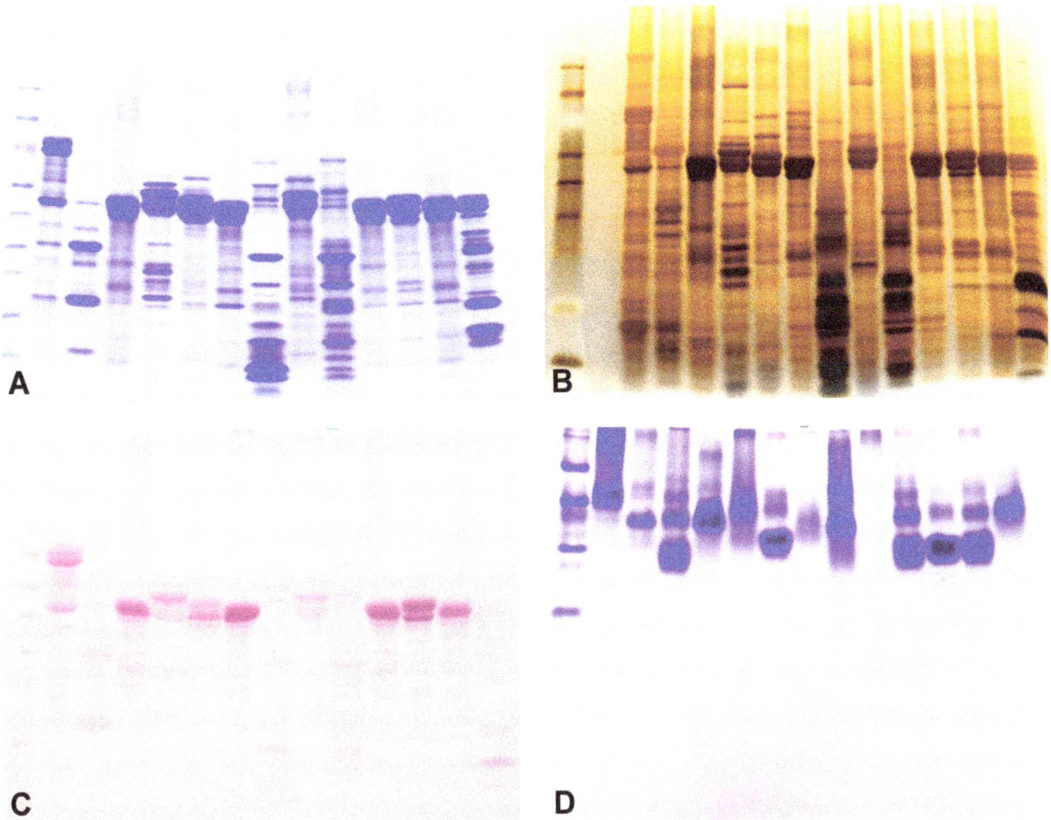

Fig. 3 (**a**) CBB staining of SDS-PAGE reducing gel. (**b**) Silver staining of SDS-PAGE non-reducing gel. (**c**) Glycoprotein staining of SDS-PAGE reducing gel. (**d**) Glycoprotein staining followed by CBB staining of NDND-PAGE gel. Protein load: 80 μg/lane for **a**, **c**, and **d**, 20 μg/lane for **b**

5. Remove the destaining solution and add 10× diluted destaining solution to cover the gel.
6. Gently shake the tray on an orbital shaker until the blue background is removed. Replace the 10× diluted destaining solution frequently.

3.4.2 Silver Staining

1. Add 250 mL of fixing solution into a plastic tray (*see* **Note 24**).
2. Carefully place the gel in the tray.
3. Gently shake the tray on an orbital shaker for 15 min.
4. Remove the fixing solution and repeat **step 2**.
5. Remove the fixing solution and add 250 mL of sensitizer solution.
6. Gently shake the tray on an orbital shaker for 30 min or overnight.
7. Remove the sensitizer solution and wash the gel with 250 mL of water for 5 min.

8. Repeat **step 6** for 2 more times.
9. Remove the water and add 250 mL of silvering solution.
10. Gently shake the tray on an orbital shaker for 20 min.
11. Remove the silvering solution and rinse with distilled water.
12. Add 100 mL of developer solution and shake for 10 s.
13. Remove the developer solution and rinse with water.
14. Add 200 mL of developer solution and shake for 1 min.
15. Remove the developer solution and allow the color to develop for 5–10 min.
16. Add the 250 mL of stopping solution and shake for 10 min.
17. Remove the stopping solution and wash the gel with 250 mL of water for 5 min.
18. Repeat **step 16** for 2 more times.

3.4.3 Glycoprotein Staining

1. Add 250 mL of fixing solution into a plastic tray (*see* **Note 24**).
2. Carefully place the gel in the tray.
3. Gently shake the tray on an orbital shaker for 60 min.
4. Remove the fixing solution and wash with 250 mL of washing solution for 15 min.
5. Repeat **step 3** one more time.
6. Remove the washing solution and add 250 mL of oxidizing solution.
7. Gently shake the tray on an orbital shaker for 60 min.
8. Remove the oxidizing solution and wash with 250 mL of washing solution for 15 min.
9. Repeat **step 7** for 3 more times.
10. Remove the washing solution and add 15 mL of Schiff's reagent.
11. Gently shake the tray on an orbital shaker for 15–30 min.
12. Remove the Schiff's reagent and wash with 250 mL of washing solution for 10 min.
13. Repeat **step 11** for 2 more times (*see* **Note 25**).

3.5 Result Documentation and Analysis

Following band visualization, the gel image can be documented by a camera, scanner, or gel imaging system. Programs are available for analysis of relative migration distance, molecular weight (if run with molecular weight standards), and band intensity of visualized polypeptide bands. Gels can be preserved by a gel dryer or subjected to further analysis using techniques such as Western blotting [13], 2D electrophoresis [14], N-terminal amino acid sequencing [15], and mass spectrometry [15].

4 Notes

1. The 2% v/v β-mercaptoethanol can be substituted by 5% v/v dithiothreitol. Dithiothreitol offers the advantage of reduced volatility and eliminates the risk of diffusing into lanes containing non-reduced proteins. Nonetheless, β-mercaptoethanol is needed for proteins with densely packed tertiary structures [1].
2. Monomeric acrylamide and bis are toxic. Avoid skin contact when handling the chemicals and reagents. Polymerize the remains with an excess of APS and follow laboratory safety guidelines for disposal.
3. A fine powder helps improve the protein extractability. We have found 20 mesh to be a desirable particle size. However, the seed flour can be ground to a smaller particle size (e.g., 40 mesh) to further improve protein extraction efficiency.
4. For samples with a high fat content, the defatting process is recommended as fat interferes with protein extraction. Defatting also helps reduce the particle size, which improves the wetting of the powder and extraction of the protein.
5. Protein concentration is dependent on the size of the well and desired protein load in each well. Make a more concentrated sample if the well cannot accommodate a large volume.
6. To prevent protein fragmentation, it is recommended to refrain from boiling, which could also help prevent the tubes from popping and causing sample spillage. To prevent evaporation, ensure the lids of the microcentrifuge tubes are kept closed.
7. The reducing agent might have been oxidized upon heating. Reintroducing the reducing agent can prevent the formation of disulfide bonds.
8. For certain set-ups, two glass plates with even upper edges are used.
9. Make sure the glasses are clean and free of chips and scratches.
10. It is recommended to check for leaks of the assembled gel cassettes by adding water into the cassettes followed by drying. Make sure it is well dried as any residual water will interfere with gel formation.
11. Polymerization of acrylamide commences upon the addition of APS and TEMED. Introduce these two reagents last and act fast once they are added.
12. Avoid the bubbles on top of the solutions.

13. Make sure to use water-saturated butanol to avoid absorbing water from the solution.
14. Use a moderate speed to avoid generating air bubbles.
15. For reproducible results, use the same pump speed or place the gradient maker at the same level above the gel cassette.
16. Be careful not to trap air bubbles.
17. To prevent drying of the gels, it is advisable to remove the gel cassettes from the casting stand after 1 h of polymerization. Place them within a plastic bag containing a few milliliters of gel buffer diluted four times with water. Seal the plastic bags and leave the gels at room temperature if they are to be used the following day. For extended storage, keep the bags in a refrigerator for a couple of weeks [1].
18. Make sure that the chamber is not leaking. The leak test can be performed with distilled water.
19. To prevent the diffusion of the reducing agent from reduced samples to non-reduced samples, it is advised to avoid running both the reduced and non-reduced samples within the same gel. If it is necessary to run them together, it is recommended to maintain an empty lane between the reduced and non-reduced samples. Be aware that the molecular weight standard may also contain reducing agent.
20. Do not overfill the wells.
21. It is recommended to cool the gels with a thermostatic circulator during electrophoresis.
22. Opening the cassettes with a spatula or knife may damage the glass and ceramic plates.
23. Commercial protein staining system is available to allow faster and automated gel staining.
24. Keep the tray filled with solution before adding the gel to prevent gel damage.
25. The gels can be subjected to CBB staining following the glycoprotein staining.

References

1. Westermeier R (2016) Electrophoresis in practice: a guide to methods and applications of DNA and protein separations. John Wiley & Sons, Weinheim, Germany
2. Hjerten S, Mosbach R (1962) "Molecular-sieve" chromatography of proteins on columns of cross-linked polyacrylamide. Anal Biochem 3:109–118
3. Andrews AT (1986) Electrophoresis: theory, techniques, and biochemical and clinical applications. Clarendon Press, Oxford, England
4. Fling SP, Gregerson DS (1986) Peptide and protein molecular weight determination by electrophoresis using a high-molarity tris buffer system without urea. Anal Biochem 155:83–88

5. Laemmli UK (1970) Cleavage of structural proteins during the assembly of the head of bacteriophage T4. Nature 227:680–685
6. Wang Y, Ma T, Liu C et al (2021) L-Histidine improves solubility and emulsifying properties of soy proteins under various ionic strengths. LWT Food Sci Technol 152:112382
7. Liu C, Zaffran VD, Gupta S et al (2019) Pecan (*Carya illinoinensis*) detection using a monoclonal antibody-based direct sandwich enzyme-linked immunosorbent assay. LWT Food Sci Technol 116:108516
8. Su M, Venkatachalam M, Gradziel TM et al (2015) Application of mouse monoclonal antibody (mAb) 4C10-based enzyme-linked immunosorbent assay (ELISA) for amandin detection in almond (*Prunus dulcis* L.) genotypes and hybrids. LWT Food Sci Technol 60: 535–543
9. Liu C, Xiong YL (2015) Oxidation-initiated myosin subfragment cross-linking and structural instability differences between white and red muscle fiber types. J Food Sci 80:C288–C297
10. Groth SFDS, Webster R, Datyner A (1963) Two new staining procedures for quantitative estimation of proteins on electrophoretic strips. Biochim Biophys Acta 71:377–391
11. Heukeshoven J, Dernick R (1985) Simplified method for silver staining of proteins in polyacrylamide gels and the mechanism of silver staining. Electrophoresis 6:103–112
12. Dubray G, Bezard G (1982) A highly sensitive periodic acid-silver stain for 1,2-diol groups of glycoproteins and polysaccharides in polyacrylamide gels. Anal Biochem 119:325–329
13. Gupta S, Sathe SK, Su M et al (2021) Germination reduces black gram (*Vigna mungo*) and mung bean (*Vigna radiata*) vicilin immunoreactivity. LWT Food Sci Technol 135:110217
14. Sathe SK, Kshirsagar HH, Sharma GM (2012) Solubilization, fractionation, and electrophoretic characterization of Inca peanut (*Plukenetia volubilis* L.) proteins. Plant Foods Hum Nutr 67:247–255
15. Sharma GM, Irsigler A, Dhanarajan P et al (2011) Cloning and characterization of an 11S legumin, Car i 4, a major allergen in pecan. J Agric Food Chem 59:9542–9552

Chapter 20

Secondary Structural Analysis Using Fourier Transform Infrared and Circular Dichroism Spectroscopy

Guo Huang, Qin Zhang, Xiaoshuai Wang, Yan Zhang, and Xiaonan Sui

Abstract

Secondary structures, such as α-helices or β-sheets, play a crucial role in maintaining protein stability and enhancing functional properties, resulting in the resource of particular interest to structure—function relationships analysis. Fourier transform infrared (FT-IR) spectroscopy is a nondestructive technique for structural characterization of proteins and polypeptides, which is a widely utilized tool for determining secondary structures by analyzing the amide I band. FT-IR spectroscopy exhibits exceptional versatility and adaptability when applied to powdered samples with high absorption, as it remains unaffected by artifacts resulting from the scattering of radiation by sample particles. In contrast, circular dichroism (CD) spectroscopy, a traditional technique, is employed for the characterization of protein secondary structures in solution. This chapter provides an overview of the fundamental procedures involved in obtaining and interpreting FT-IR and CD data, along with techniques for analyzing spectra to estimate the composition of protein secondary structures.

Key words Secondary structure, Fourier transform infrared, Circular dichroism

1 Introduction

Primary structure is defined by the amino acid sequence located in the protein structure, whereas the more complex structures are dictated by the folding and intra-molecular interactions between the constituent amino acid residues. Proteins adopt diverse and intricate structure, which confer on them a wide variety of molecular functions. Secondary structure in proteins occupies a pivotal position in the hierarchical classification of structure, the research of which exhibited an explosive growth in food and biomedicine field with the number of publications about secondary structures of proteins added to the Web of Science currently growing at an ever-increasing rate. The peptide backbone is created by the patterns of hydrogen bonds between amine hydrogen and carbonyl oxygen atoms, which are crucial for the formation of the α-helix and β-sheet, the two primary types of secondary structure

Yonghui Li (ed.), *Plant-Based Proteins: Production, Physicochemical, Functional, and Sensory Properties*,
Methods and Protocols in Food Science, https://doi.org/10.1007/978-1-0716-4272-6_20,

[1]. Knowledge of secondary structures of proteins is essential for getting an insight into protein conformation and stability. For example, temperature dependent analysis of the secondary structure is critical in enhancing digested conditions for polyphenols-rich soy protein foods [2]. Apart from this, this is also currently being done for a better understanding of the structure-activity relationships [3–5]. Protein conformation changes and protein secondary structures have been detected using a range of biophysical methods including X-ray crystallography and nuclear magnetic resonance (NMR) in the past few years. Nevertheless, these methods necessitate a significant amount of time, demand sample preparation as well as specific reagents and equipment, consequently leading to high expenses [6].

The Fourier transform infrared (FT-IR) spectroscopy is extensively employed to examine the secondary structure of proteins and observe alterations in local conformation by absorbing infrared light and stimulating vibrational excitations in molecules. The amide bands are nine characteristic group frequencies in protein that arise from the vibrations of the polypeptide repeat units. The amide I (1700–1600 cm^{-1}) and amide II (1600–1500 cm^{-1}) bands are the two most significant bands related to protein secondary structure among them. Amide I refers to the absorptions related to C=O stretching, whereas amide II corresponds to the absorptions associated with N-H bending. Reports indicate that the C=O and N-H bonds play a role in maintaining the various components of secondary structure through hydrogen bonding. This sensitivity to the protein's secondary structure composition is observed in both the amide I and amide II bands. In contrast, the amide II band exhibits lower sensitivity to structure and is less useful in predicting the secondary structure of proteins, leading to greater emphasis on the amide I band [7]. The amide I contains overlapping spectral bands that are unique to each type of secondary structure. In this particular framework, mathematical methods like band curve-fitting and second represents as effective approaches to overcome these constraints and accurately measure the secondary configuration of proteins with exceptional spectral precision (Fig. 1) [8]. On the other hand, FT-IR spectroscopic tests can be utilized on a diverse range of samples regardless of their physical state.

Outside of the FT-IR spectroscopy, circular dichroism (CD) spectroscopy is another potential technique for detection of protein conformation and estimation of secondary structure in solution, which measures the differential absorption between left and right circularly polarized light in the far ultraviolet region, mainly from 250 to 180 nm [10]. When the peptide bonds (symmetric chromophores) and amino acid sidechains of proteins are aligned in this region, exciton splitting and optical transitions were occurred, resulting in the characteristic CD spectra of different structural

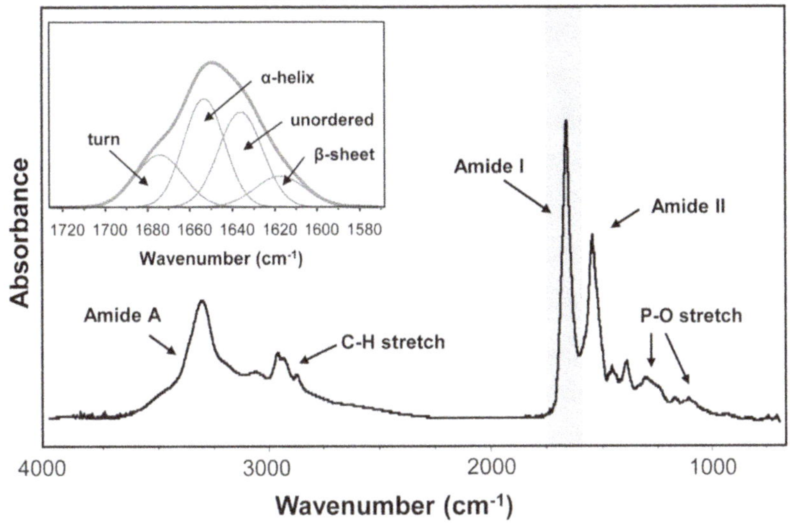

Fig. 1 In the typical FT-IR spectrum of protein, the bands recognized and allocated include the Amide I and Amide II bands, which are situated at approximately 1650 cm^{-1} and 1540 cm^{-1}, correspondingly. The inset displays an enlarged perspective of the Amide I peak, which can be separated into its individual components and fitted to a curve for studying the composition of the secondary structure [9]

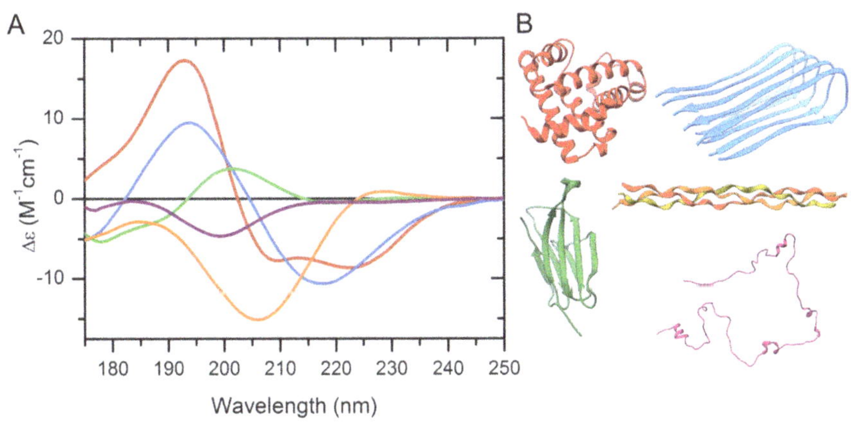

Fig. 2 Circular dichroism spectra of various protein structures in the far ultraviolet range. Red, blue, green, orange, and purple solid lines represent α-helix, parallel β-sheet, antiparallel β-sheet, polyproline-helix, and disordered chain, respectively [12]

elements (α-helical, β-sheet, turn, and random coil) formed [11]. Deconvolution is a more intricate procedure in contrast to assigning a band to each secondary structure, even though the various forms of regular secondary structures exhibit the distinctive CD spectra (Fig. 2). Conducting secondary structure analyses requires curve fitting methods that rely on a collection of reference spectra comprising recognized secondary structure elements. Regression analysis allows for the estimation of the constituents in a spectrum of an unidentified or sample nature [10]. The method is

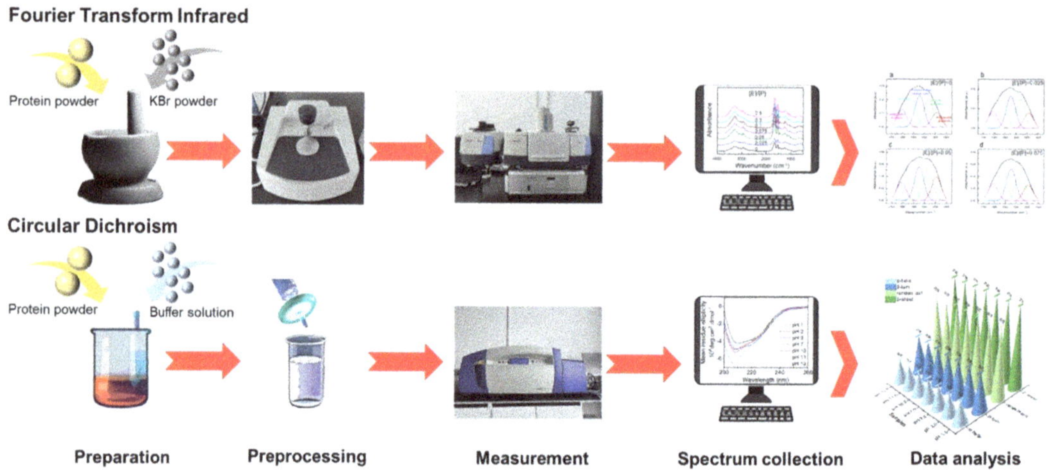

Fig. 3 Flow diagram of the measurement process of protein secondary structure with Fourier transform infrared and circular dichroism methods

complicated by the necessity of optically clean solutions, as any scattering components can impact the results, and the requirement for highly accurate determinations of protein concentration. Nevertheless, the advantage of only requiring small volumes of dilute, aqueous protein solutions is noteworthy.

Therefore, in this study, we describe the procedure and analytical methods of circular dichroism and infrared spectroscopy for the dry and wet states of the samples, with a view to providing insights into the analysis of the secondary structure of proteins. A flow diagram (Fig. 3) shows the flow chart of the measurement of secondary structure of protein.

2 Materials

2.1 Sample Preparation for Analysis

1. The protein powder can be dialyzed before use to remove salt ions that may affect the accuracy of the measurement results. After dialysis, freeze-dry storage.
2. Protein powder should be stored in a desiccator to minimize the water content to reduce the interference of water vapor on the measured spectral curve results. In addition, ensure that the protein purity is guaranteed to be above 95% [13].
3. Meanwhile, KBr pellets should be prepared and stored dry.
4. Use CaF_2 to remove water to minimize the interference of water vapor on protein spectra from the sample detection chamber of the instrument.
5. 0.01 M, pH 7.4 phosphate buffer solution (PBS): Prepare 800 mL of distilled water in a suitable container. Next, add 0.01 M phosphate buffered saline powder (Boster Biological

Technology, Pleasanton CA, USA, Catalog # AR0030-E) to the solution. Finally, add distilled water until the volume is 1 L.

6. 1 mg/mL protein solution: Dissolve 15 mg of protein powder in 10 mL of 0.01 M, pH 7.4 PBS and then filter using a 0.22 μm syringe filter (Millipore). The protein content of the solution was determined by means of the micro-*Kjeldahl* Nitrogen method. Next, dilute to a final concentration of 1 mg/mL.

3 Methods

3.1 FT-IR Spectroscopy

3.1.1 Sample Determination and Spectrum Acquisition

The FT-IR spectra was recorded by a Nicolet 6700 FT-IR spectrometer (Thermo Scientific, WA, USA), for which the protein powder was dried, ground in an agate mortar, thoroughly mixed with KBr at a ratio of 1:100 in weight pressed into small discs at room temperature (*see* **Note 1**). Measurements were performed in the spectral range from 4000 to 500 cm^{-1} with a spectrometer resolution of 4 cm^{-1} using an accumulation of 32 scans [14–18]. The spectra of the KBr pellets blank and the machine background were also measured. The FT-IR spectral analysis was conducted using the PeakFit Version 4.12 software (SPSS Inc., Chicago, IL, USA).

3.1.2 Spectral Peak Analysis

1. Subtract reference spectra: The transmittance of its spectral data is converted into absorbance for further analysis. The background spectra corresponding to the KBr pellets and machine were subtracted from all the FT-IR spectra of protein using the infrared spectroscopy analysis software package OMNIC version 8.2 (Thermo Scientific, WA, USA).
2. Spectral smoothing: The resulting protein difference spectra were smoothened using a seven-point Savizky-Golay function to remove any possible white noise (*see* **Note 2**). After completion, save its data in CVS format.
3. Perform baseline correction: Import its data into PeakFit Version 4.12 software, select the amide band I (1700–1600 cm^{-1}) for baseline correction by the second derivative function.
4. Normalized: The area of the amide band I was normalized to remove concentration differences.
5. Self-deconvolution: To get information about protein secondary structure, the spectra in Amide I region were further analyzed using the pre-installed AutoFit function by Gaussian deconvolution fitting (bandwidth 50 cm^{-1}, enhancement 2) [19].
6. Second-derivative analysis: To enhance the spectral resolution in the range of 1700–1600 cm^{-1}, the application of second-derivative analysis techniques was implemented. A point

convolution of either 11 or 13 was utilized to generate the second derivatives. The increase in resolution caused by the second derivative determined both the quantity and location of the bands to be fitted.

7. Peak fitting: To measure the size of the various parts of the amide I contour, we utilized least-squares iterative curve fitting to match the Lorentzian line shapes to the spectra within the range of 1700–1600 cm^{-1}. This process was repeated multiple times until the value of R^2 was equal to or greater than 0.99.

3.1.3 Secondary Structure Content Characterization

The identification and calculation of the secondary structure content could be accomplished by analyzing the frequency of the Lorentzian band maximum associated with the separated independent sub-peaks, in which its frequency was assigned as: α-helix (1646–1664 cm^{-1}), β-sheet (1615–1637 and 1682–1700 cm^{-1}), β-turns (1664–1681 cm^{-1}), and random coil (1637–1645 cm^{-1}) [20]. The summation of the areas of all component bands allocated to a specific conformation is subsequently divided by the total area. The resulting value is regarded as the proportion of the polypeptide chain adopting that conformation [21]. Figure 4 showed an example of FT-IR spectra of soy protein isolate (SPI) and SPI-anthocyanins conjugates. Subsequently, the amide I band was used for secondary structures analysis and collected in Table 1.

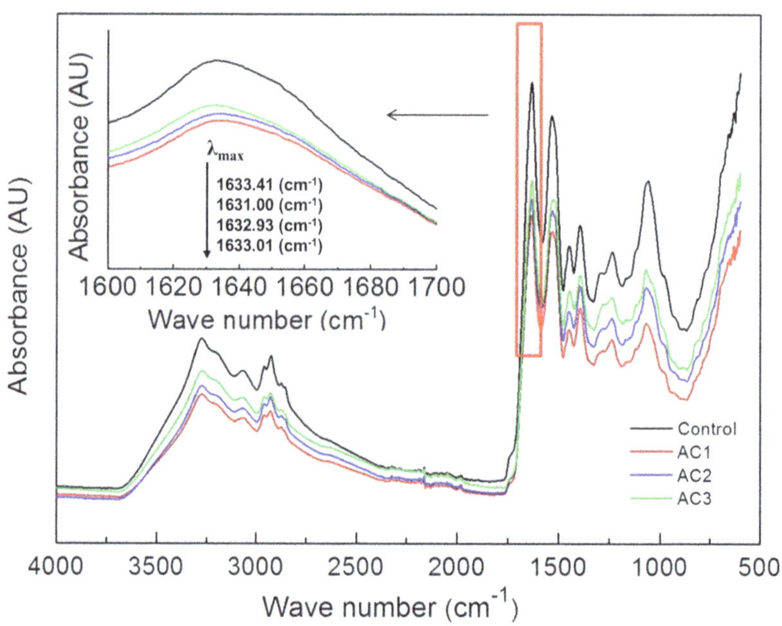

Fig. 4 FT-IR spectra of SPI (Control) and AC1 (SPI: anthocyanins, w/w, 60:1), AC2 (SPI: anthocyanins, w/w, 40:1) and AC3 (SPI: anthocyanins, w/w, 20:1). The inserted plot represents FT-IR spectra at amide I band (1600–1700 cm^{-1}) [15]

Table 1
Secondary structure contents (%) of SPI and SPI-anthocyanins conjugates [15]

Samples	α-helix	β-sheet	β-turn	Random coil
Control	17.09 ± 0.13^{a}	44.19 ± 1.96^{a}	21.47 ± 1.99^{b}	17.24 ± 0.11^{c}
AC1	12.73 ± 0.24^{c}	38.40 ± 0.68^{b}	26.50 ± 1.37^{a}	22.37 ± 1.27^{b}
AC2	13.00 ± 0.16^{bc}	35.85 ± 0.32^{bc}	28.26 ± 0.86^{a}	22.89 ± 0.61^{b}
AC3	13.89 ± 0.50^{b}	30.98 ± 2.40^{c}	26.47 ± 0.69^{a}	28.66 ± 2.30^{a}

Mean values in the same column followed by the difference superscript letters are significantly different ($p < 0.05$)

3.2 Circular Dichroism (CD) Spectroscopy

3.2.1 Sample Determination and Spectrum Acquisition

1. Instrument startup: During instrument start-up and operation, the CD machines must be purged with nitrogen to remove oxygen. Before starting the machine, turn on the nitrogen and flush the optical chamber at about 10–15 L/min for 15–20 min (see manufacturer's recommended times). Next, a water-circulating heating pad was used to maintain test temperature. After that, the light source was turned on for at least 30 min to warm up before the measurements (see manufacturer's recommended times). Afterward, other programs on the CD machines are opened one after another and connected to the computer waiting for testing. If you are new to CD and there is no one to teach you the operation of your specific machine, call the manufacturer and ask for their start up protocol (*see* **Note 3**).
2. Parameters settings: Start the CD collection program and then set up the monitoring database to store the data and document changes. Thereafter, control of temperature was set to test temperature. Set the desired equilibration time (*see* **Note 4**), the half-band width (*see* **Note 5**) and the wavelength range (*see* **Note 6**), the wavelength interval (*see* **Note 7**), the time for data collection at each point (*see* **Note 8**), and the instrument time constant (*see* **Note 9**). At the same time, the absorbance and ellipticity were electronically logged, as were the photomultiplier tube (PMT) voltage (*see* **Note 10**), if available. In this way, detailed information on procedures with references is provided in the relevant manufacturers web sites (https://jascoinc.com/applications/cd-measurement-of-protein-water-solution-in-vacuum-uv-region/?print=pdf or https://jascoinc.com/training-video/spectra-measurement-parameters/) and publications [22].
3. Instrument calibration: To assess the success of the calibration adjustment procedure, calibration was compared to that of actual determined values using external standards ((1S)-(+)-

Camphor-10-sulfonic acid (CSA) (Aldrich, C2107)). Please refer to the manufacturer's instructions for your particular instrument.

4. The water background determination: Spectra was acquired with water that filled the cell. The spectrum of the water should be relatively flat and have no ellipticity to ensure the cell was clean.
5. The buffer solution background determination: The cell contained the buffer solution instead of water and then the buffer solution spectrum collected was used as a background for the following protein spectra. The collected spectrum of the buffer solution was then utilized as a background for the subsequent protein spectra. Ensure that the buffer spectrum does not exhibit any ellipticity caused by dichroic components or high absorbance, as this may result in a weak signal-to-noise ratio and potentially misleading peaks. The buffer and water spectra should overlap, within the margin of error, but the buffer spectrum typically exhibits a lower signal to noise ratio than the water spectrum at shorter wavelengths. For details of the buffer solution, please see the relevant sections of the publishes [10, 23].
6. The protein solution determination: For typical measurements in a 0.1 cm cell (260–185 nm), depending on the buffer, making solutions of 0.05–0.2 mg/mL protein. For measurements in 0.01–0.02 cm cell (260–178 nm) use 0.2–1 mg/mL protein, and for 1 cm cell (260–200 nm) use 0.005–0.01 mg/mL protein. The protein solution was sealed in a cell with specific measuring range, and then the total absorbance of the protein solution was kept at or below 1.2 absorbance units to achieve great measurements [23]. In order to meet the absorbance criterion, the vast majority of these adaptations involve adjusting either the concentration of protein or the cell path-length or both. Next, three scans total were collected and averaged for each final CD spectrum, and spectra were collected on independent duplicate samples to ensure data reproducibility [2]. Save the raw data on the same folder as your working directory. After these measure sessions, the cells were rinsed and dried between runs. For a detailed description, please refer to the manufacturer's instructions for your particular instrument. Whereas, if the outcome is bad, you can change these parameters and troubleshoot technical issues as reported in the literature [23, 24].

3.2.2 Spectral Data Preprocessing

1. Smooth the spectra of the sample: After data acquisition, spectroscopic data pre-processing was performed. First, the baseline was subtracted from the spectra of the sample, followed by subtracting the spectrum of the blank. The sample's spectra

were smoothed using the CD software's built-in procedures, some of which will automatically select the optimal smoothing parameters as described in the manual. For further details, we refer the reader to manufacturer's manual.

2. Data transformation: Utilize the integrated software programs along with the relevant protein details to convert the data into mean residue ellipticity or $\Delta\varepsilon$. All data were saved for subsequent data analysis at the lead researchers.

3.2.3 Computational Study of Protein Secondary Structure

1. Calculated data: CD instruments are supplied with their own data processing software packages, which enable basic processing procedures. The secondary structures of protein were conducted on built-in analyses program for convenience. In practice of course, some software used for calculating protein secondary structure content and other computational analyses are available at request. Detailed information about software and methods used are available online in the Chemical Society Reviews [23]. Examine the data utilizing suitable techniques. For the most precise outcomes, it is advisable to utilize a wide range of techniques, such as CONTIN, CONTINLL, SELCON3, and CDSSTR, which are available in the online version of the paper [10, 25]. Some of the analysis programs output the fitted curve to the raw data. Recognizing that the programs producing the most precise matches to the data might not necessarily offer the most dependable predictions of protein structure is crucial. Incorporating a larger number of variables can lead to better fits. As a result, the use of CONTIN, a software that utilizes multiple reference spectra to match protein CD data, consistently produces an excellent match to the original data when compared to approaches that use a smaller set of reference sets. In Fig. 5, we provide an illustrative case for

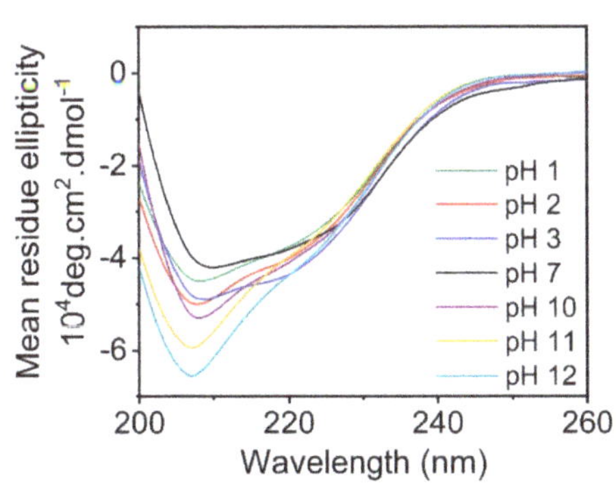

Fig. 5 Far-UV CD spectra of the natural state of SPI (pH 7.0) and the molten globule state generated by SPI (SMG) in extreme pH solutions after different pH induction [26]

Table 2
Secondary structure content of SMG after different pH induction [26]

Samples	Secondary structure content (%)			
	α-helix	β-sheet	β-turn	Random coil
pH 1.0	11.29 ± 0.11^{a}	37.80 ± 0.20^{abc}	17.76 ± 0.12^{bc}	33.15 ± 0.05^{a}
pH 2.0	11.21 ± 0.07^{a}	37.74 ± 0.14^{abc}	17.87 ± 0.06^{ab}	33.18 ± 0.08^{a}
pH 3.0	11.32 ± 0.08^{a}	37.68 ± 0.19^{bc}	17.79 ± 0.09^{bc}	33.21 ± 0.56^{a}
pH 7.0	11.31 ± 0.17^{a}	37.92 ± 0.10^{ab}	17.68 ± 0.08^{c}	33.09 ± 0.10^{a}
pH 10.0	11.31 ± 0.11^{a}	37.71 ± 0.09^{abc}	$17.81 \pm 0.17b^{c}$	33.17 ± 0.06^{a}
pH 11.0	11.30 ± 0.87^{a}	37.56 ± 0.12^{c}	17.96 ± 0.05^{a}	33.18 ± 0.13^{a}
pH 12.0	10.97 ± 0.11^{b}	37.99 ± 0.16^{a}	17.95 ± 0.07^{a}	33.09 ± 0.01^{a}

Values are means ± SD. Values in the same column with different superscript letters (a–g) are significantly different, $n = 3$ ($p < 0.05$)

CD spectra of SPI by CONTIN and then the secondary structures of SPI presented in Table 2.

2. When evaluating predictions of the secondary structure for a particular protein, it is vital to recognize the presence of numerous algorithms that employ different computational methods and utilize references acquired from various spectral ranges and protein datasets. Consequently, establishing universally applicable guidelines becomes challenging. However, the analysis's dependability can be assessed by taking into account these factors. (1) The NRMSD parameter, also referred to as the 'goodness of fit,' is a useful tool for evaluating the level of agreement between the experimental data and the theoretically calculated CD spectrum. It is based on the derived secondary structure composition and covers the entire range of wavelengths of interest. Typically, an NRMSD value exceeding 0.25 is considered indicative of an error in the analysis procedure. In practical terms, striving for NRMSD values below 0.1, and ideally 0.05 or lower, is recommended. (2) The availability of the R value indicates the suitability of the secondary structure composition obtained from CD. It aids in determining the most suitable algorithm and reference dataset for a specific case. (3) It is essential to compare the outcomes obtained using various algorithms. The reliability of an analysis can be assessed when different algorithms yield highly similar estimates for the secondary structural characteristics.

4 Notes

1. The operation conditions used to operate the FT-IR instrument were as recommended by the manufacturer (https://www.thermofisher.cn/cn/en/home/industrial/spectroscopy-elemental-isotope-analysis/molecular-spectroscopy/fourier-transform-infrared-spectroscopy.html).
2. The noise level is defined as the difference between the smoothed and the raw spectrum. An appropriate smoothing parameter is applied to avoid over-smoothing and decrease possible noise disturbance.
3. CD instrument specific parameters and further method details can be found in the official JASCO website (https://jascoinc.com/learning-center/theory/spectroscopy/circular-dichroism-spectroscopy/) or Applied Photophysics Ltd. website (https://photophysics.com/products/circular-dichroism/).
4. Usually globular proteins reach equilibrium within 2 min.
5. Adjust the bandwidth to be within the range of 1–1.5 nm. The spectra obtained from these values have satisfactory signal to noise ratios and sufficient spectral resolution due to the broad UV bands.
6. The wavelength range for protein samples in transparent buffers is 260 to 185 nm for concentrations of 0.1–0.2 mg/mL in cells with a thickness of 0.1 cm. For concentrations of 0.2–0.8 mg/mL in cells with a thickness of 0.01 cm, the range is from 260 to 178 nm. Lastly, for concentrations of 0.01–0.02 mg/mL in cells with a thickness of 1 cm, the range is from 260 to 200 nm.
7. For samples with signal to noise >20, set the wavelength interval to 0.5 nm, while for samples with low ellipticity, set it to either 0.1, 0.2, or 0.25 nm. Accurate spectra can be acquired by gathering information at intervals of 0.10–0.5 nm, using half-band widths that vary from 1 to 1.5 nm.
8. For samples with a concentration of around 0.1 mg/mL in a clear buffer, a collection time of 1 second per data point should be enough. Nevertheless, when acquiring duplicate spectra with intervals varying from 0.1 to 0.5 nm, it is advisable to consider a longer collection time of 0.5 s per data point as satisfactory. In instances where the protein concentration is low or the buffer exhibits high absorbance, it is advisable to augment the averaging time. The relative signal to noise ratio will increase proportionally to the square root of the signal averaging time.

9. For routine collection of CD spectra, it is recommended to use a time constant of 100 ms. Nevertheless, when there is a need for swift data gathering, like in the assessment of halted stream CD, it is crucial to decrease the time constant to 100 μs or below, guaranteeing that it does not surpass 1/10 of the data averaging time at every point.
10. In the majority of CD machines, the preservation of a consistent current is attained by elevating the voltage in correspondence with the reduction in light intensity. When the machine scans towards shorter wavelengths, the absorbance levels increase, leading to a rise in the PMT voltage. Nevertheless, it is crucial to acknowledge that once the dynode voltage exceeds 500 volts, the ratio of signal to noise decreases considerably, frequently resulting in data that is both noisy and unreliable.

References

1. Weidner T, Apte JS, Gamble LJ, Castner DG (2010) Probing the orientation and conformation of α-helix and β-strand model peptides on self-assembled monolayers using sum frequency generation and NEXAFS spectroscopy. Langmuir 26(5):3433–3440
2. Zhang Y, Chen S, Qi BK, Sui XN, Jiang LZ (2018) Complexation of thermally-denatured soybean protein isolate with anthocyanins and its effect on the protein structure and in vitro digestibility. Food Res Int 106:619–625
3. Qi BK, Ding J, Wang ZJ, Li Y, Ma CG, Chen FS, Sui XN, Jiang LZ (2017) Deciphering the characteristics of soybean oleosome-associated protein in maintaining the stability of oleosomes as affected by pH. Food Res Int 100: 551–557
4. Tian R, Feng JR, Huang G, Tian B, Zhang Y, Jiang LJ, Sui XN (2020) Ultrasound driven conformational and physicochemical changes of soy protein hydrolysates. Ultrason Sonochem 68:105202
5. Dong YB, Lan T, Wang X, Zhang Y, Jiang LJ, Sui XN (2020) Preparation and characterization of soy protein microspheres using amorphous calcium carbonate cores. Food Hydrocoll 107:105953
6. Fevzioglu M, Ozturk OK, Hamaker BR, Campanella OH (2020) Quantitative approach to study secondary structure of proteins by FT-IR spectroscopy, using a model wheat gluten system. Int J Biol Macromol 164:2753–2760
7. DeFlores LP, Ganim Z, Nicodemus RA, Tokmakoff A (2009) Amide I′-II′ 2D IR spectroscopy provides enhanced protein secondary structural sensitivity. J Am Chem Soc 131(9):3385–3391
8. Kong JL, Yu SN (2007) Fourier transform infrared spectroscopic analysis of protein secondary structures. Acta Bioch Bioph Sin 39(8): 549–559
9. Miller LM, Bourassa MW, Smith RJ (2013) FTIR spectroscopic imaging of protein aggregation in living cells. BBA-Biomembranes 1828(10):2339–2346
10. Greenfield NJ (2006) Using circular dichroism spectra to estimate protein secondary structure. Nat Protoc 1(6):2876–2890
11. Miles AJ, Ramalli SG, Wallace BA (2022) DichroWeb, a website for calculating protein secondary structure from circular dichroism spectroscopic data. Protein Sci 31(1):37–46
12. Micsonai A, Bulyáki É, Kardos J (2021) BeStSel: from secondary structure analysis to protein fold prediction by circular dichroism spectroscopy. In: Chen YW, Yiu C-PB (eds) Structural genomics: general applications. Springer US, New York, pp 175–189
13. Yang HY, Yang SN, Kong JL, Dong AC, Yu SN (2015) Obtaining information about protein secondary structures in aqueous solution using Fourier transform IR spectroscopy. Nat Protoc 10(3):382–396
14. Lan T, Dong YB, Zheng M, Jiang LJ, Zhang Y, Sui XN (2020) Complexation between soy peptides and epigallocatechin-3-gallate (EGCG): formation mechanism and morphological characterization. LWT 134:109990

15. Jiang LJ, Liu YJ, Li L, Qi BK, Ju MN, Xu Y, Zhang Y, Sui XN (2019) Covalent conjugates of anthocyanins to soy protein: unravelling their structure features and in vitro gastrointestinal digestion fate. Food Res Int 120:603–609
16. Xu ZJ, Hao NR, Li LW, Zhang Y, Yu LL, Jiang LZ, Sui XN (2019) Valorization of soy whey wastewater: how epigallocatechin-3-gallate regulates protein precipitation. ACS Sustain Chem Eng 7(18):15504–15513
17. Xu ZJ, Shan GC, Hao NR, Li LW, Lan T, Dong YB, Wen JY, Tian R, Zhang Y, Jiang LJ, Sui XN (2022) Structure remodeling of soy protein-derived amyloid fibrils mediated by epigallocatechin-3-gallate. Biomaterials 283:121455
18. Lan T, Dong YB, Xu ZJ, Zhang Y, Jiang LJ, Zhou WB, Sui XN (2023) Quercetin directed transformation of calcium carbonate into porous calcite and their application as delivery system for future foods. Biomaterials 301: 122216
19. Briggs J, Woys AM (2016) Simple method for monitoring protein secondary structure during thermal unfolding and aggregation. Spectroscopy 31(4):47–47
20. Dou W, Zhang X, Zhao Y, Zhang Y, Jiang LZ, Sui XA (2022) High moisture extrusion cooking on soy proteins: importance influence of gums on promoting the fiber formation. Food Res Int 156:111189
21. Ouameur AA, Mangier E, Diamantoglou S, Rouillon R, Carpentier R, Tajmir-Riahi H (2004) Effects of organic and inorganic polyamine cations on the structure of human serum albumin. Biopolymers 73(4):503–509
22. Banerjee R, Sheet T (2017) Ratio of ellipticities between 192 and 208 nm (R_1): an effective electronic circular dichroism parameter for characterization of the helical components of proteins and peptides. Proteins 85(11): 1975–1982
23. Miles AJ, Janes RW, Wallace BA (2021) Tools and methods for circular dichroism spectroscopy of proteins: a tutorial review. Chem Soc Rev 50(15):8400–8413
24. Kelly SM, Jess TJ, Price NC (2005) How to study proteins by circular dichroism. Biochim Biophys Acta 1751(2):119–139
25. Greenfield NJ (2006) Using circular dichroism collected as a function of temperature to determine the thermodynamics of protein unfolding and binding interactions. Nat Protoc 1(6): 2527–2535
26. Wang L, Wen JY, Wang LY, Jiang LZ, Zhang Y, Sui XN (2023) Characterization of the extreme pH-induced molten globule state of soy protein isolate and its influence on functional properties. Food Hydrocoll 144:109040

Chapter 21

Surface Hydrophobicity

Yue Zhang and Zile Xiao

Abstract

Surface hydrophobicity is a significant indicator for evaluating the surface properties of plant proteins. It plays a crucial role in determining the functionality of proteins, such as emulsification and foam stabilization abilities. This chapter provides a comprehensive analysis guideline for measuring surface hydrophobicity using two different methods: spectrofluorimetric assessment with the fluorescence probe 1-anilinonaphthalene-8-sulfonic acid (ANS), and hydrophobic-interaction capacity measurement with bromophenol blue (BPB). Therefore, the surface hydrophobicity of plant proteins can be expressed by a relative index value (S_0) and the bound BMP value, respectively.

Key words Surface hydrophobicity, Fluorescence probe, ANS, Bromophenol blue, Hydrophobic interaction

1 Introduction

Hydrophobicity refers to the tendency of non-polar solutes to stick together in water-based environments [1]. It is the main driving force in the folding of proteins, influencing their structure and functions. Surface hydrophobicity reflects the distribution of hydrophobic residues on the surface of protein, which provides insight into the changes occurring on the protein surface, and therefore, is crucial in determining the functionality of food proteins including emulsification and foam stabilization [2]. Many plant proteins, such as zein and legume proteins, exhibit higher surface hydrophobicity compared to commercial animal-sourced protein ingredients like whey protein and caseinate. However, they still hold promise as emulsifiers for stabilizing interfaces via the Pickering particle- stabilized mechanism [3]. In this regard, surface hydrophobicity is always measured as a key indicator for predicating and evaluating the surface properties and functionality of plant proteins.

Yonghui Li (ed.), *Plant-Based Proteins: Production, Physicochemical, Functional, and Sensory Properties*, Methods and Protocols in Food Science, https://doi.org/10.1007/978-1-0716-4272-6_21,

The most popular method for measuring the surface hydrophobicity is spectrofluorimetric assessment using fluorescence probes. This method is preferred due to its relatively rapid, non-destructive, and simple nature [4]. Several fluorescence probes including uncharged probes like 6-propionyl-2-(N,N-dimethylamino)-naphthalene (prodan), anionic aliphatic probes such as cis-parinaric acid (CPA), and aromatic probes like 1-anilinonaphthalene-8-sulfonic acid (ANS) have been applied for the assessment of surface hydrophobicity in proteins [5]. The mechanism behind the use of fluorescence probes is that these probes have low quantum yields in aqueous solutions. However, when bound to accessible hydrophobic regions of proteins, the fluorescence increases, which is dependent on the hydrophobic/hydrophilic circumstances and can therefore be used as an indicator of protein surface hydrophobicity [6].

The drawback for fluorescence probe assessment is the need of spectrofluorometer, which may not be existed in many laboratories. There are also other alternative simple methods such as hydrophobic-interaction capacity measurement by using bromophenol blue (BPB) [7], Coomassie brilliant blue G-250 (CBBG) [8] or detergent molecules [4]. The mechanism involves the non-covalent binding (mainly hydrophobic interaction) of BPB and CBBG to aromatic and basic amino acid residues on the surface of the detected proteins. This binding results in the formation of insoluble complexes between the protein and BPB/CBBG, which can be precipitated by centrifugation. As a consequence, the absorbance attributed to free BPB/CBBG molecules in the supernatant decreases. Therefore, the amount of protein-bound BPB/CBBG can serve as an indicator of protein surface hydrophobicity. This assessment can be performed using a simple absorbance measurement, which shows promise as an alternative method to complement fluorescence probe assessment.

The aim of this chapter is to thoroughly describe the analytical procedures for determining surface hydrophobicity of plant proteins. Two surface hydrophobicity measurements including spectrofluorimetric assessment and hydrophobic-interaction capacity measurement are described here, by using fluorescence probe ANS and bromophenol blue (BPB), respectively. In this particular study, soy protein isolate (SPI) was used as the protein sample for performing both the spectrofluorimetric assessment and hydrophobic-interaction capacity measurement. The purpose of this chapter is to help standardize the methodologies employed in these tests, and it will be easier for readers to compare and interpret the results obtained using different methods.

2 Materials

Prepare all solutions using distilled water and/or analytical grade reagents. Prepare and store all reagents at room temperature (unless indicated otherwise).

Protein solution to be tested. In this case, 0.1% w/v SPI and bamboo shoot protein (BSP) dissolved in 0.01 M phosphate buffer (pH 7.0) distilled water were used as examples.

To further illustrate the change in surface hydrophobicity, another set of experiments was conducted. In this set, the SPI solution was heated in a water bath at 95 °C for 10 min and then cooled to room temperature, resulting in a heated SPI sample.

2.1 Additional Materials for Spectrofluorimetric Assessment with ANS Probe

8.0 mM 1-anilinonaphthalene-8-sulfonic acid (ANS) solution: Dissolve 0.0128 g of ANS in 5 mL of 0.01 M phosphate buffer (pH 7.0).

The test will be performed using fluorescence spectrophotometer (F-4700, Hitachi, Japan) with quartz cuvette (4 clear sides).

2.2 Additional Materials for BPB Binding Measurement

1 mg/mL Bromophenol blue (BPB): Weigh 0.01 g BPB and dissolve in 10 mL distilled water.

Centrifugal Machine.

The absorbance will be measured using Uv-vis Spectrophotometer (5100, Metash

Instruments, Shanghai, China) with two clear sides cuvette (glass/quartz).

3 Methods

Carry out all procedures at room temperature unless otherwise specified.

3.1 Spectrofluorimetric Assessment with Fluorescent Probe ANS

The ANS probe is used as an example for spectrofluorimetric assessment with fluorescent probe. Figure 1 shows the steps of this fluorescent probe method in a simplified and schematic way.

1. The 0.1%w/v SPI and BSP protein solutions are prepared in 0.01 M phosphate buffer (pH 7.0) (*see* **Note 1**).
2. Prepare serial dilutions of the protein solution with concentrations of 0.02%, 0.04%, 0.06%, 0.08%, and 0.1% (w/v) in 0.01 M phosphate buffer, pH 7.0 (*see* **Note 2**).
3. Mix an aliquot of 2 mL of each dilution solution with 10 μL of ANS (8.0 mM), vortex and keep the mixtures in the dark for at least 15 min. Add 10 μL of the solvent used to prepare the ANS solution to the other tube of each dilution solution as a control (blank).

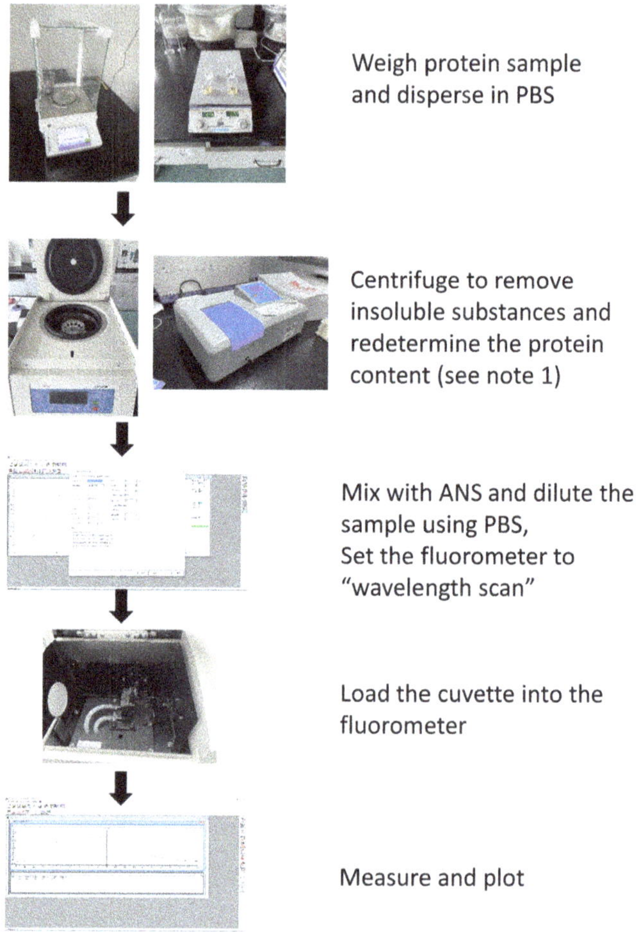

Fig. 1 Schematic and simplified representation of the steps of the spectrofluorimetric assessment

4. To standardize the spectrofluorometer and ensure that the fluorescence measurements are readable in the following test, adjusting the reading for equivalent concentrations of ANS in methanol (e.g., 8 mM ANS in methanol) to an arbitrary value is recommended.
5. Set the excitation wavelength (λ_{ex}) to 390 nm, set the emission wavelength (λ_{em}) to 400–600 nm or 470 nm (as the maximum emission peak for ANS) directly.
6. Add 2 mL of sample or blank into the cuvette and load it into the instrument, measure the relatively fluorescence intensity and record the maximum value of emission or the value at $\lambda_{em} = 470$ nm directly (*see* **Note 3**).
7. Subtract the intensity value of the corresponding blank from each sample intensity value. The resulting net intensity value can be used for further analysis.

8. Plot net intensity value versus protein concentration (%), and find the slope of linear regression (as shown in Fig. 2). The slope is expressed as the index of surface hydrophobicity (S_0). In this case, the values of S_0 for SPI, BSP, and heated SPI at pH 7.0 are determined as 779.91, 492.64, and 1708.62, respectively (*see* **Note 4**).

3.2 Bromophenol Blue (BPB) Binding Measurement

BPB is used as an example for hydrophobic-interaction capacity measurement. In this case, the absorbance at 595 nm will be measured while that will be set to 585 nm for CBBG [8]. Figure 3 shows the steps of this BPB binding method in a simplified and schematic way.

1. Mix 200 μL of 1 mg/mL BPB with 1 mL of the prepared protein sample and stir for 10 min. A control is made by mixing 200 μL of 1 mg/mL BPB with 1 mL PBS (pH 7.0).
2. Centrifuge at 2000× *g* for 15 min to get the supernatant and mix the supernatant with PBS in a ratio of 1 part supernatant to 9 parts PBS.
3. Measure the absorbance (Abs) value of both sample and control at 595 nm.
4. The amount of bound BPB can be obtained by the equation below:

$$\text{Bound BPB } (\mu\text{g}) = \frac{\text{Abs of control} - \text{Abs of sample}}{\text{Abs of control}} \times 200\mu\text{g}$$

The results are expressed as the weight value of bound BPB with a unit of μg. In this case, the bound BPB values of SPI, BMP, and heated SPI at pH 7.0 are determined to be 23.76, 6.59, and 56.60 μg, respectively. This value is used as an indicator of the surface hydrophobicity of the protein, with a higher value suggesting a greater surface hydrophobicity. Despite the different units used in the two methods, the comparison in this case reveals the same trend, indicating that both methods are reliable for determining surface hydrophobicity.

4 Notes

1. To ensure accurate fluorescence measurements, it is important to ensure that the protein solutes are completely dispersed. In the case of using stock SPI dispersion, any insoluble aggregates were removed by centrifugation at 10,000 g for 15 min. This step was taken to prevent any disruption or interference with the fluorescence measurement. It is advisable to redetermine the protein content using a reliable method such as the Lowry Assay to ensure the accuracy of the protein content in the following tests.

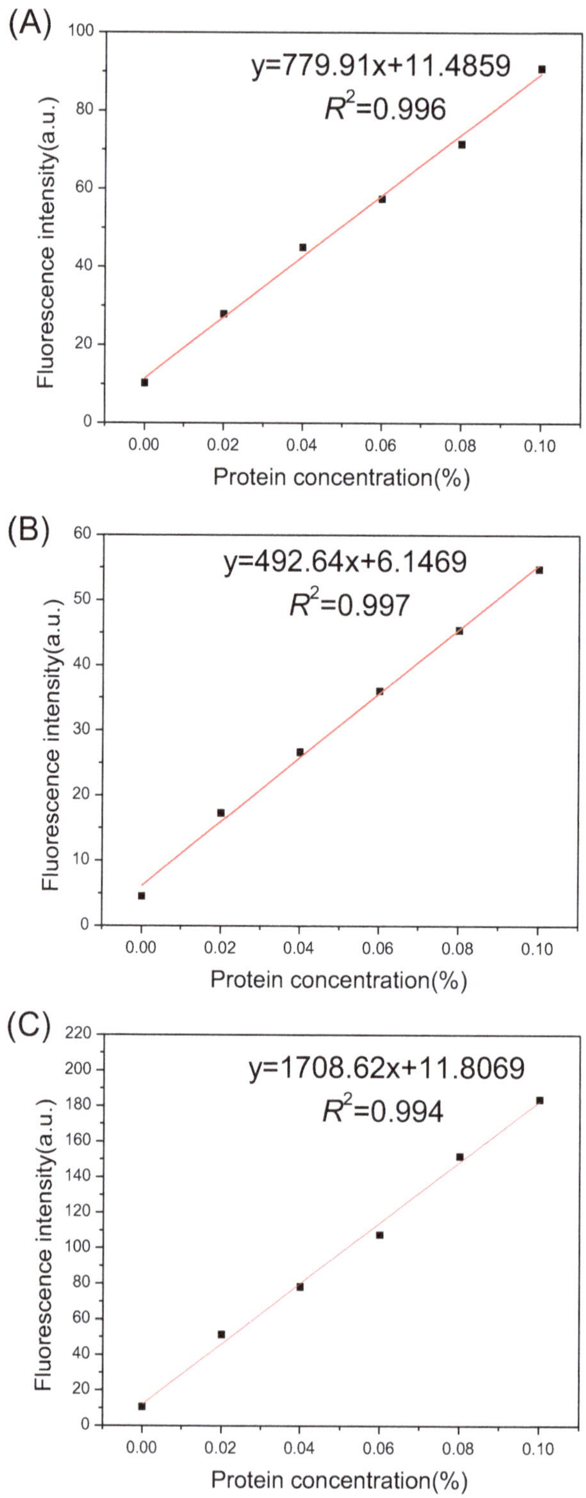

Fig. 2 Determine the index of surface hydrophobicity (S_0) by plotting the net fluorescence intensity versus protein concentration. (**a**), (**b**), and (**c**) represent SPI, BSP, and heated SPI, respectively

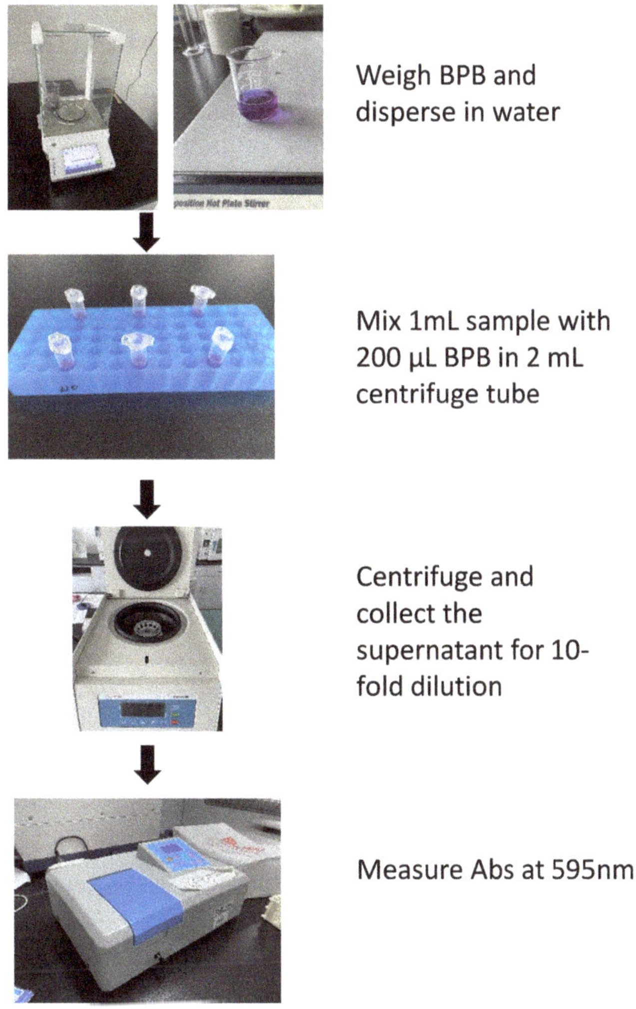

Fig. 3 Schematic and simplified representation of the steps of BPB binding method

2. In order to ensure accurate fluorescence observation, it may be necessary to adjust the protein concentration used in this measurement based on factors such as protein solubility and the amount of fluorophores (e.g., Tryptophan) in the protein samples.
3. The maximum emission peak of a fluorophore like ANS can be influenced by various factors, including the solution environment, pH, temperature, and molecular interactions. (e.g., in another case, the maximum emission of ANS was 485 nm [9]). To accurately determine the maximum emission wavelength for plotting, it is recommended to read the entire emission spectra and identify the wavelength at which the fluorescence intensity is highest.

4. The value of S_0 is a relative value, so it is better to compare the S_0 of the sample with other standard proteins in practice. In this case, SPI showed a higher S_0 value compared to BSP, indicating relatively higher surface hydrophobicity under the same measuring conditions. After heating, SPI demonstrated a significant increase in S_0, suggesting an enhanced surface hydrophobicity. This can be attributed to the denaturation induced by the heat, which exposes more hydrophobic residues on the molecular surface [10].

References

1. Murphy KP, Privalov PL, Gill SJ (1990) Common features of protein unfolding and dissolution of hydrophobic compounds. Science 247(4942):559–561
2. Burger TG et al (2022) The impact of spray drying conditions on the physicochemical and emulsification properties of pea protein isolate. LWT 153:112495
3. Kim W, Wang Y, Selomulya C (2020) Dairy and plant proteins as natural food emulsifiers. Trends Food Sci Technol 105:261–272
4. Wrolstad RE et al (2005) Handbook of food analytical chemistry, volume 1: water, proteins, enzymes, lipids, and carbohydrates. Wiley
5. Alizadeh-Pasdar N, Li-Chan EC (2000) Comparison of protein surface hydrophobicity measured at various pH values using three different fluorescent probes. J Agric Food Chem 48(2):328–334
6. Cardamone M, Puri N (1992) Spectrofluorimetric assessment of the surface hydrophobicity of proteins. Biochem J 282(2):589–593
7. Chelh I, Gatellier P, Santé-Lhoutellier V (2006) A simplified procedure for myofibril hydrophobicity determination. Meat Sci 74(4):681–683
8. Cao Y, Zhao J, Xiong YL (2016) Coomassie Brilliant Blue-binding: a simple and effective method for the determination of water-insoluble protein surface hydrophobicity. Anal Methods 8(4):790–795
9. Wang L, Zhang Y (2017) Eugenol nanoemulsion stabilized with zein and sodium caseinate by self-assembly. J Agric Food Chem 65(14):2990–2998
10. Wang Z et al (2014) Relationship between secondary structure and surface hydrophobicity of soybean protein isolate subjected to heat treatment. J Chem 2014:1

Chapter 22

Zeta Potential

Qin Zhang, Rongxue Zhou, Guo Huang, Yan Zhang, and Xiaonan Sui

Abstract

As an important characterization technique, experimentally determined zeta potential values can be used to effectively predict the surface charge of colloidal particles and then estimate the strength of mutual repulsion or attraction between particles. The dispersion mechanism of molecules or particles within colloidal systems can also be elucidated by measuring the zeta potential. The information obtained by measuring this value is crucial for the intentional control of a colloidal system's electrostatic dispersion. Moreover, the isoelectric point of protein can also be obtained by determining its zeta potential. Obtaining zeta potential values of any particle provides a vital foundation for understanding and modifying its surface electrical properties. At present, the measurement of zeta potential is widely used in food, cosmetics, health care, environmental protection, marine chemistry, and other industries. This chapter provides a comprehensive guide to instrumental methods for zeta potential determination in protein samples.

Key words Zeta potential, Protein, Surface charge

1 Introduction

As a charged particle is dispersed into a solvent, ions also in solution experience electrostatic interactions with the particle's surface charges. The association of these ions with the particle's surface form a two-layer charged particles known as the electric double layer (EDL) (Fig. 1). Due to electrostatic attraction, ions and molecules with opposite charges adsorb strongly to the surface of the charged particle to form a dense inner layer, known as the Stern layer. The electrostatic effects caused by the surface charge on the particles decrease in the outer layer of the EDL, known as the diffuse layer, according to Debye's law. This law states that the field strength of the particle's surface charge decreases by a factor of 1/e with each Debye length's distance [1]. This results in more labile association between the particle's charged surface and the ions in the diffuse layer. There is also a slipping plane where the Stern layer meets the diffuse layer. When a charged particle moves through a solvent, ions within the slipping plane are strongly attracted to the

Yonghui Li (ed.), *Plant-Based Proteins: Production, Physicochemical, Functional, and Sensory Properties*, Methods and Protocols in Food Science, https://doi.org/10.1007/978-1-0716-4272-6_22,

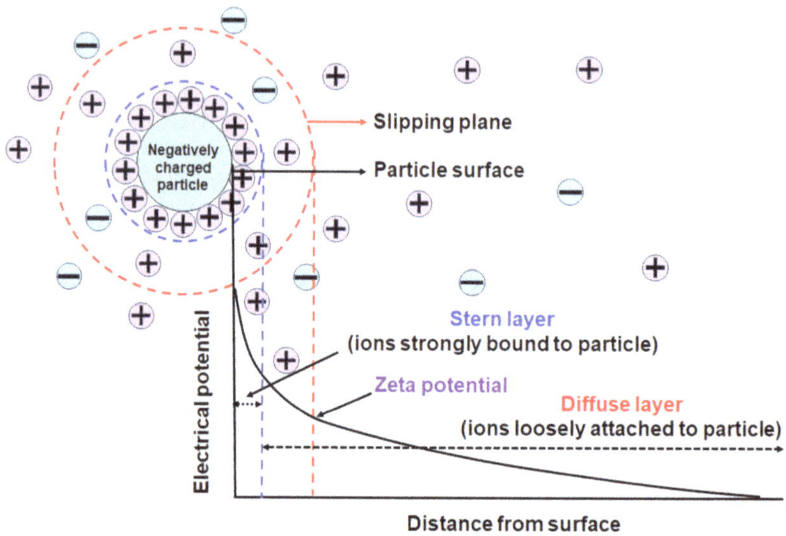

Fig. 1 The interpretation of EDL on a negatively charged particle [3]

particle's charged surface and "follow" the particle's trajectory through the solution. Ions located outside of the slipping plane experience less attraction to the particle's charged surface and therefore migrate through the solution with the particle to a lesser extent. These factors ultimately result in the surface boundary of the slipping plane. In such situations, the electrical potential of the slipping plane for a charged particle in solution with respect to distance from the particle is called the Zeta potential [2]. Thus, Zeta potential is the potential, i.e., the electrophoretic mobility, of a moving colloidal particle at the slipping plane within an electric field. It represents the difference in electrokinetic potential at the slipping plane between the EDL of moving particles within the electric field and the dispersion layer surrounding them.

Because zeta potential is related to the potential difference between the ions in the dispersion medium and the layers of attracted ions that comprise the colloidal system, it can be used to characterize the stability of colloidal protein systems [2, 4]. Molecules that possess higher absolute values of zeta potential correspond to more stable colloidal systems. Conversely, lower absolute values of zeta potential correspond to colloidal systems that are more likely to fall out of suspension or agglomerate, as the magnitude of the attractive force is exceeded by the magnitude of the repulsive force (Fig. 2) [5]. It is worth noting that Zeta potential is a systematic property, not a material property. This means that although the zeta potential of some colloidal particles is predictable, the entire system can be changed by various factors to produce virtually any desired zeta potential value [6]. One of the most important parameters used to alter the Zeta potential of a system is pH, and convention dictates that research data reported without

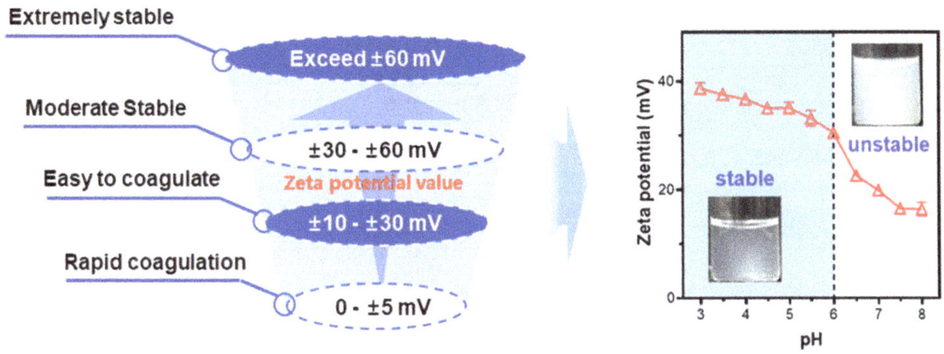

Fig. 2 The relationship between zeta potential value and stability of colloidal system [9]

including the pH value of a system is essentially meaningless [7]. The pH of a solution is dependent on the net charge of the water ions comprising a solution. It is a measurement of the ratio of hydroxide and hydronium ions present within an aqueous system. Additional charged species in a solution will interact with any solvated particles, causing a change in the zeta potential. Similar to pH, the ionic strength of the particle suspension also has a significant effect on a system's zeta potential. As ionic strength increases, the value of zeta potential decreases. When the ionic strength of the polymer dispersion becomes infinitely high, the zeta potential approaches a limit of zero [1]. Additionally, the zeta potential of a sample is also affected by special additives [8]. Characterizing the zeta potential of colloids can provide a scientific basis for the development of food products with stable formulations.

Zeta potential values are difficult to measure directly. A common method is to calculate Zeta potential from electrophoretic mobility measurements of charged particles under the applied electric field. The formula used for these calculations is based on Henry's equation [3]:

$$\frac{V}{E} = \frac{2\varepsilon\zeta f(ka)}{3\eta}$$

Where, V/E is the electrophoretic mobility ($m^2\ s^{-1}\ V^{-1}$), ε is the solvent dielectric permittivity (also referred to as the solvent's dielectric constant) ($kg\ m\ V^{-2}\ s^{-2}$), ζ is the Zeta potential, $f(ka)$ is Henry's function (a dimensionless value), and η is the viscosity ($kg\ m^{-1}\ s^{-1}$). Among these, the value of $f(ka)$ varies between 1 and 1.5. It relates the ratio of the particle size to the Debye length. Generally, the electrophoretic mobility of particles within the applied electric field can be characterized by two methods: electrophoretic light scattering and the electroacoustic phenomenon [1]. The electrophoretic light scattering method is mostly applied in food-protein colloidal systems. There are many types of

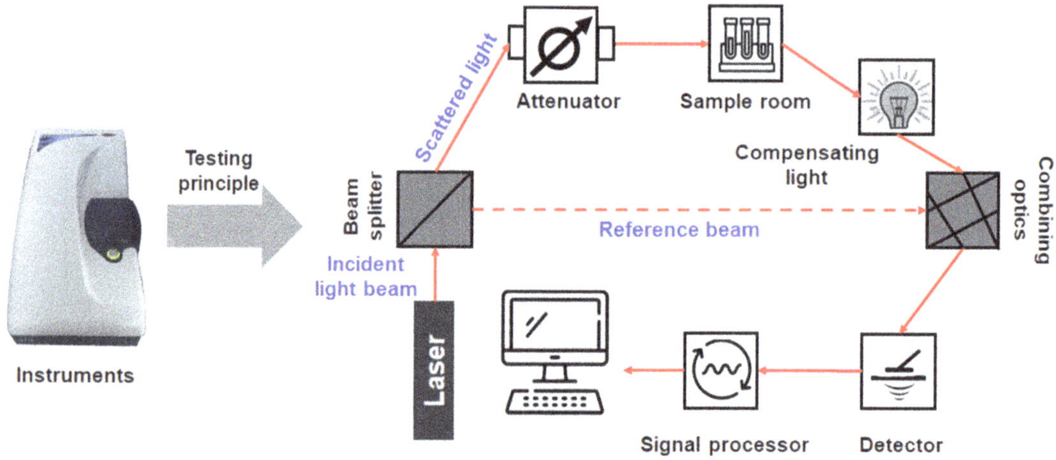

Fig. 3 Schematic diagram of the Zeta potential measuring instrumentation using electrophoretic light scattering detection

instruments used to measure zeta potential, but they all share the same basic components. These are the laser, beam splitter, attenuator, sample room, compensating light unit, combining optics, detector, signal processor, and data analysis system (Fig. 3).

The aim of this chapter is to serve as a guide for zeta potential determination, from sample preparation to experiment execution, to ensure that experiments are conducted in a standardized manner.

2 Materials

2.1 Sample Preparation for Analysis

1. 500 mL salt phosphate buffer (1 mmol/L Na_2HPO_4, 1 mmol/L NaH_2PO_4, pH 7.5).
2. Ultrapure water.
3. 0.1 μm cutoff filter.

2.2 Zeta Potential Determination

1. Zetasizer Nano ZS90 (Malvern, UK) (*see* **Note 1**).
2. Specialized cuvette (DTS1070) with a U-shaped measuring chamber and built-in electrodes.
3. 1 mL syringe.

3 Methods

3.1 Sample Preparation for Analysis

The first step is to dilute the solution to ensure that the measured particle can move freely (*see* **Notes 2** and **3**). Next, to eliminate dust particles that can affect the resulting measurement, the samples need to be filtered through a cutoff filter with 0.45 μm pores.

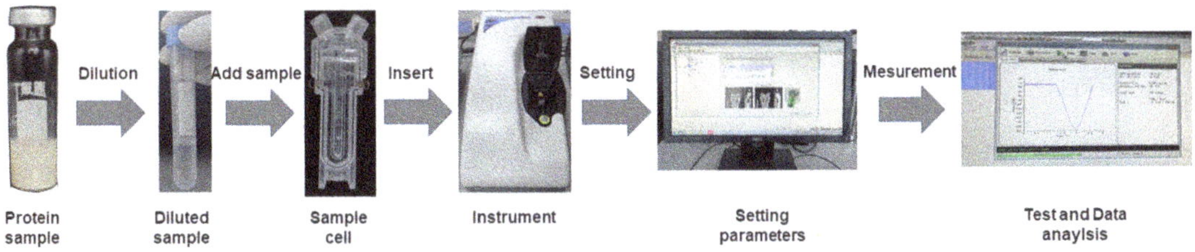

Fig. 4 Operation flow chart of protein Zeta potential test

3.2 Zeta Potential Determination

The diluted sample is placed in a cuvette and the potential of the sample is measured using a Nano Zetasizer (Fig. 4) (ZS90, Malvern, Worcestershire, UK).

1. First, turn on the device and preheat for 30 min to allow the system to balance.
2. Rinse the cuvette with ultrapure water, then condition the cuvette with an aliquot of sample solution.
3. Transfer the sample into the cuvette to reach 1/3–2/3 of the total height of cuvette (*see* **Note 4**).
4. Place cuvette with sample into the card slot of the Zetasizer (*see* **Note 5**).
5. Set analysis parameters in Zetasizer software and measure the Zeta potential. Specifically, set "measured material" as protein, "dispersion liquid" as water, Smoluchowski model, F (Ka) value is 1.5, test temperature is 25 °C, balancing time as 120 s, DTS1070 Cell, and 10–12 runs.
6. Repeat the experiment three times to avoid error.
7. Finally, export the data for analysis.

4 Notes

1. The concentration range of the samples that can be measured by the Zetasizer is from 0.1 ppm to 40% (w/v), and the particle size range is from 3.8 nm to 100 μm.
2. The best dilution solvent is phosphate buffer, rather than distilled water. The advantage of phosphate buffer is that it prevents voltage drop and ohmic heating of the electrode interface.
3. The samples to be examined should be prepared and adjusted to a consistent pH without compromising sample stability. Conventionally, 0.1 M NaOH or HCl are applied to adjust the pH of samples to a range that either leads to the most stable particle dispersion or is consistent with the intended application media. It is important to report the pH levels at which samples are adjusted with the final data.

4. Using a 1 mL syringe, inject the sample solution into the potential cell, taking care to avoid bubbles. Afterward, the surface of the potential cell is wiped clean using a Kimtech wipe.
5. The cuvette is then loaded into the instrument, with the frosted/etched side facing toward you. It is important to avoid touching the transparent face of the cuvette cell, as well as the copper sheets on either side.
6. Electrophoresis may degrade some proteins, so samples are not suitable for reuse after Zeta potential measurement [1].
7. The electrode used for Zeta potential measurement easily reacts with metal ions (such as Fe^{3+}), which will damage the electrode and reduce the quality of the data. Therefore, it is recommended to check the electrodes regularly [10].
8. It is crucial to understand that the zeta potential is a theoretical estimate of the electrostatic potential at the hypothetical surface of hydrodynamic shear between the Stern layer and the diffuse layer, rather than a direct measurement of the surface charge on a particle [11].
9. The measured zeta potential value is an average value. For monodisperse samples, this value corresponds to the zeta potential of a single component in the sample. However, the zeta potential profile of a polydisperse sample will contain multiple peaks corresponding to multiple components. In such polydispersed systems, it is important to determine and analyze the zeta potential of each peak separately to determine the zeta potential specific to each of the system's individual components [1].

References

1. Bhattacharjee S (2016) DLS and zeta potential–what they are and what they are not? J Control Release 235:337–351
2. Srinivasan S, Barbhuiya S, Charan D, Pandey S (2010) Characterising cement–superplasticiser interaction using zeta potential measurements. Constr Build Mater 24(12):2517–2521
3. Lowry GV, Hill RJ, Harper S, Rawle AF, Hendren CO, Klaessig F, Nobbmann U, Sayre P, Rumble J (2016) Guidance to improve the scientific value of zeta-potential measurements in nanoEHS. Environ Sci Nano 3(5):953–965
4. Zhang Q, Jiang L, Sui X (2023) Incorporating chitin nanocrystal yields stronger soy protein gel: insights into linear and nonlinear rheological behaviors by oscillatory shear tests. Food Hydrocoll 135:108177
5. Zhu D, Shen Y, Wei L, Xu L, Cao X, Liu H, Li J (2020) Effect of particle size on the stability and flavor of cloudy apple juice. Food Chem 328:126967
6. Ravindran S, Williams M, Ward R, Gillies G (2018) Understanding how the properties of whey protein stabilized emulsions depend on pH, ionic strength and calcium concentration, by mapping environmental conditions to zeta potential. Food Hydrocoll 79:572–578
7. Feng Y, Kilker SR, Lee Y (2020) Surface charge (zeta-potential) of nanoencapsulated food ingredients. Elsevier, Characterization of nanoencapsulated food ingredients, pp 213–241
8. Kaasalainen M, Mäkilä E, Riikonen J, Kovalainen M, Järvinen K, Herzig K-H,

Lehto V-P, Salonen J (2012) Effect of isotonic solutions and peptide adsorption on zeta potential of porous silicon nanoparticle drug delivery formulations. Int J Pharm 431(1–2): 230–236

9. Zhang Q, Sun P, Xu Z, Qu W, Zhang Y, Sui X (2024) Chitin nanocrystals as natural gel modifier for yielding stronger acid-induced soy protein isolate gel. Carbohydr Polym 323:121446

10. Hedberg Y, Wang X, Hedberg J et al (2013) Surface-protein interactions on different stainless steel grades: effects of protein adsorption, surface changes and metal release. J Mater Sci: Mater Med. https://doi.org/10.1007/s10856-013-4859-8

11. Jacobasch H, Bauböck G, Schurz J (1985) Problems and results of zeta-potential measurements on fibers. Colloid & Polymer Sci. https://doi.org/10.1007/BF01411243

Chapter 23

Thermal Analysis Using Differential Scanning Calorimetry

Qinchun Rao and Xingyi Jiang

Abstract

The usage of plant-based proteins in the food industry has been increasing over recent years. While these proteins offer promising advantages, their thermal properties have received less attention compared to animal proteins. Understanding the thermal properties such as glass transition, denaturation, and gelation will provide insights into optimal processing conditions, appropriate storage environments, and quality assessment for plant proteins. Currently, differential scanning calorimetry (DSC) is frequently used for protein thermal analysis. DSC thermogram provides thermodynamic information about plant proteins. This chapter provides a guide to using DSC for the analysis of thermal properties of plant proteins. Moreover, it provides a summary of critical control points to enhance the reliability of the results obtained through the DSC experiment.

Key words Denaturation temperature, Differential scanning calorimetry, Glass transition temperature, Heat capacity, Sample and reference cell, Soybean protein isolate, Thermogram, Water content

1 Introduction

Thermal analysis is used to determine the physicochemical properties of a substance when it is heated, cooled, or maintained at a constant temperature. Despite various methods such as isothermal scanning calorimetry (ITC), thermogravimetry analysis (TGA), and dynamic mechanical analysis (DMA) are often applied and can work for various food components [1], this chapter will only focus on differential scanning calorimetry (DSC) in protein analysis. Since the 1970s, DSC has been the most fundamental tool for studying the protein state changes in food chemistry and food protein analysis. It measures the heat energy uptake of food proteins as a function of temperature in the form of enthalpy. Briefly, in the DSC experiment, both the sample and the reference material will be heated continuously at the same rate, and the sample will experience either an endothermic or an exothermic process, and such energy will be measured [2] (Fig. 1a, b).

Yonghui Li (ed.), *Plant-Based Proteins: Production, Physicochemical, Functional, and Sensory Properties*, Methods and Protocols in Food Science, https://doi.org/10.1007/978-1-0716-4272-6_23,

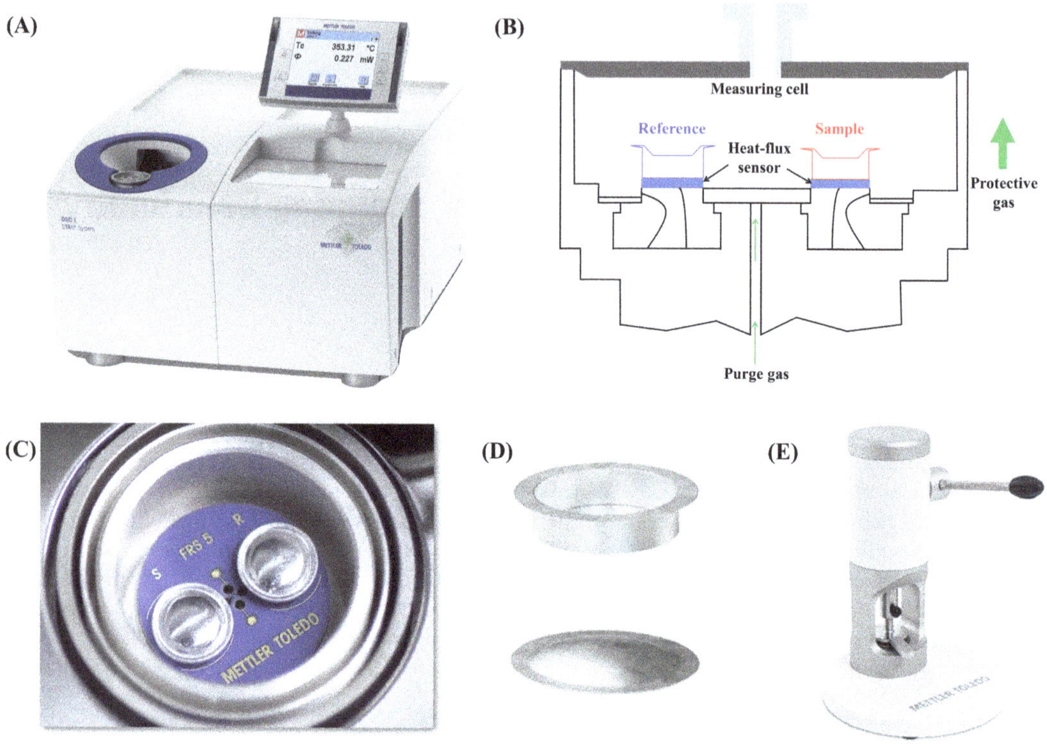

Fig. 1 (**a**) Mettler Toledo DSC 1. (**b**) Schematic representation of DSC measuring cell. (**c**) DSC sensors. (**d**) Standard 40-μL aluminum crucible with lid. (**e**) Mettler Toledo crucible sealing press

DSC enables the understanding of the phase transition, such as glass transition, melting, and crystallization, which are important for the food industry. Firstly, the glass transition temperature is a critical point that the sample changes from a rigid state to a rubbery state. It is an important parameter in determining the drying temperature during food processing, predicting the shelf life of food products, and affecting the texture changes of food products [3]. Secondly, the protein denaturation temperature is the temperature when the protein's secondary, tertiary, and quaternary structures are irreversibly changed, leaving only the primary structure remaining. Knowing about protein denaturation helps to produce food products with desired nutritional and textural properties. Thirdly, for some globular proteins, after denaturation, the formation of new bonds between protein molecules leads to protein aggregation or gelation process [4]. DSC further provides such information on protein gelation. Lastly, the crystallinity and crystallization kinetics would demonstrate the chemical purity of a protein.

The general protocol for running a DSC is straightforward: simply weighing the target sample, loading the target and reference sample to different pans, respectively, and starting the program.

However, the production of reproducible and recognized data is challenging. To start with, the sample needs to be carefully prepared based on its state. Thermal properties of proteins are dependent on various factors such as water content, pH, salt concentration, protein concentration, and food matrix. For example, for solid powder samples, the denaturation and glass transition temperature increases when water content decreases [5]. Therefore, it is always desired to measure the water content of the target sample before performing DSC. In the meanwhile, to enhance the measurement of changes in heat, if the liquid sample is used, the target protein concentration needs to be optimized and dispensed in the same solution as the reference sample. Before running DSC, deep cleaning and a pre-scan are helpful to avoid cross-contamination from previous samples. The scan rate and temperature range should be determined based on each sample. A second scan is always recommended as it ensures no further reaction takes place. Finally, the consistency of translating raw output data from DSC to protein thermal properties can be an issue. It is critical to minimize subjectivity and develop a consistent data analysis method used for all samples by different researchers. For example, the typical DSC heat flow and enthalpy diagram give three temperatures: onset, peak, and endset. It is the onset temperature that is used for the description of denaturation temperature because it is less dependent on the heating rate and sample mass.

The goal of this chapter is to be a guide for the thermal analysis of plant proteins using DSC. It describes sample preparation, setting up parameters for calibration and measurement, and data analysis.

2 Materials

Prepare all solutions and buffers using distilled water and/or analytical grade reagents. Prepare and store all reagents at room temperature (unless specified)

1. Soy protein isolate (Nuts.com).
2. Analytical balance (readability 0.0001 g).
3. Tweezers.
4. Differential scanning calorimeter 1 (DSC 1) System (Mettler Toledo, Fig. 1a–c).
5. 40 μL aluminum crucible with lid (Mettler Toledo, ref. 26,763, Fig. 1d, *see* **Note 1**).
6. Crucible sealing press (Mettler Toledo, ref. 119,410, Fig. 1e).

7. Indium pills for temperature calibration (Mettler Toledo, ref. 119442).
8. Crucible handling box (Mettler Toledo, ref. 51142765).

3 Methods

3.1 DSC Calibration

DSC instruments are calibrated on heat capacity and temperature scales. It is recommended to perform the calibration at least once a month to improve data quality. Overall, the onset temperature and the heat of fusion are evaluated automatically during the experiment. This is done by performing DSC using standard materials such as indium with known heat of fusion (*see* **Note 2**).

1. Record the combination weight of the crucible and lid for both reference and sample cell (*see* **Note 3**).
2. Use a tweezer to press soft indium pills flat, which can provide better thermal contact.
3. Put the flat indium into the sample crucible.
4. Record the weight of indium.
5. Close the crucible using the crucible press by placing the crucible in the middle of the press.
6. Put the lid in the center and rotate the lever to move the plunger down until the resistance is overcome.
7. Place the sample crucible into the sample cell.
8. Ensure the empty reference crucible is in the reference cell.
9. Select the method in STARe software, and check DSC indium exo up. Enter the sample name "indium calibration #date."
10. Enter the weight of indium and begin the measurement.
11. Choose inert nitrogen gas as the atmosphere and perform the experiment.
12. After running, the curve is automatically evaluated on the onset of indium temperature and enthalpy of fusion.
13. If the result shows that DSC is within specifications, the calibration is successful, and it can proceed with further experiments.
14. If the result is out of specification, run the second measurement using the same sample.
15. If the result is still out of specification, it is strongly recommended to contact technical support to perform the adjustment.

Table 1
Sample preparation for DSC

Sample type	Procedure
Protein powder sample (*see* **Note 4**)	Determine the moisture content of protein powder using oven drying or Karl Fischer titration method. Directly weigh 5 mg of protein powder into the aluminum crucible
Protein suspension (*see* **Note 5**)	Around 5 mg of dry protein powder will be added with different amounts of water to the desired water content. The protein suspension will be rehydrated at 4 °C overnight to reach equilibrium before analysis
Protein solution (*see* **Note 6**)	Prepare 20% (g/mL) protein solution in water or buffers such as phosphate-buffered saline, sodium/potassium acetate, sodium borate, and sodium citrate. Vortex the sample and shake to fully dissolve the proteins. The protein solution will be dialyzed against the solvent for at least 24 h with three changes of water/buffer at 4 °C. The dialyzed protein solution will be filtered through a 0.22 μm syringe filter before analysis

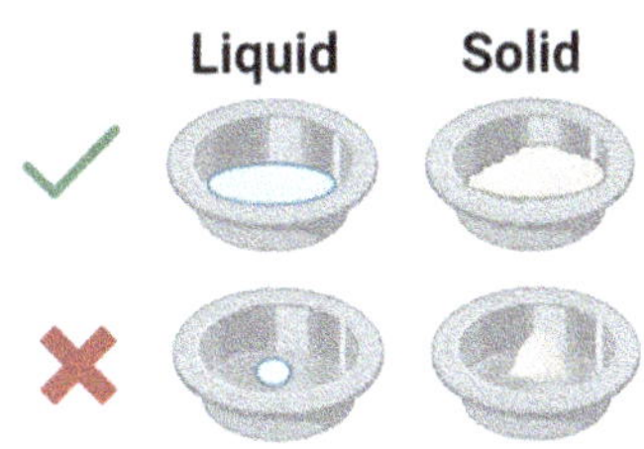

Fig. 2 Schematic representation of sample loading in DSC crucibles

3.2 Sample Preparation

1. Before sample preparation, apply the following general rules for selecting the samples:
   (a) The selected sample should be representative of the total sample.
   (b) Before analysis, the selected sample should not be changed thermally, for example, denatured or melted.
   (c) The selected sample should not go through any chemical or physical changes during sample preparation.
   (d) The selected sample should not interact with the crucible.
2. Refer to Table 1 for preparation strategies for samples in different states.
3. Spread the sample as evenly as possible to the bottom of the crucible, as shown in Fig. 2 (*see* **Note 7**).
   (a) For protein powder, use the funnel supplied in the crucible handling box.
   (b) For liquid and suspension, use a syringe needle.
4. As an example, add 5 mg of soy protein isolate and 25 μL of water to the sample crucible subsequently.

5. The lid will be closed using a crucible sealing press.
6. Record the weight of the sample and crucible (*see* **Note 8**).
7. Put the hermetically sealed crucible at 4 °C overnight to reach an equilibrium state.

3.3 DSC Operation

1. Add 25 μL of water to an aluminum DSC reference crucible (*see* **Note 9**) and close the lid using a crucible sealing press.
2. Both sample and reference crucibles will be weighed and recorded before DSC experiment.
3. Place both crucibles into DSC, and both the sample and the reference material will be held at 30 °C for 10 min in the DSC.
4. Set up the running at an initial temperature of 30 °C (*see* **Note 10**) and an end temperature of 110 °C (*see* **Note 11**). The scan rate will be 10 °C/min (*see* **Note 12**).
5. A second scan will be performed at the same condition (*see* **Note 13**).
6. After scanning, both pans will be weighed to confirm the mass lost (*see* **Note 14**).

3.4 DSC Maintenance and Cleaning

Protein denaturation and aggregation could lead to DSC system clogging and the production of inaccurate data. Proper cleaning of DSC minimizes such interference.

1. Use a brush to gently brush the cells.
2. Spray some compressed air to blow out the remaining debris (*see* **Note 15**).
3. If the contaminants are still persistent in the cell, a burn-out procedure will be adopted by using air as purge gas and following the two heating steps:
   (a) Heat from 25 to 500 °C at a rate of 20 °C/min.
   (b) Keep at 500 °C for 60 min.
4. After ending the experiment and cooling down to room temperature, repeat the brush cleaning experiment.
5. Run a water-water scan to ensure the baseline is smooth and flat without bumps or spikes.

3.5 Data Analysis

1. Select the curve generated by the DSC in STARe Evaluation Software. Results are plotted with temperature (°C) as the x-axis and heat flow (mW) as the y-axis.
2. On the typical DSC thermogram, if the enthalpy change is positive, the reaction is exothermic. Otherwise, it is endothermic (Fig. 3).
3. Select each peak from the thermogram, and go to Integration under the TA table A baseline will be automatically drawn.

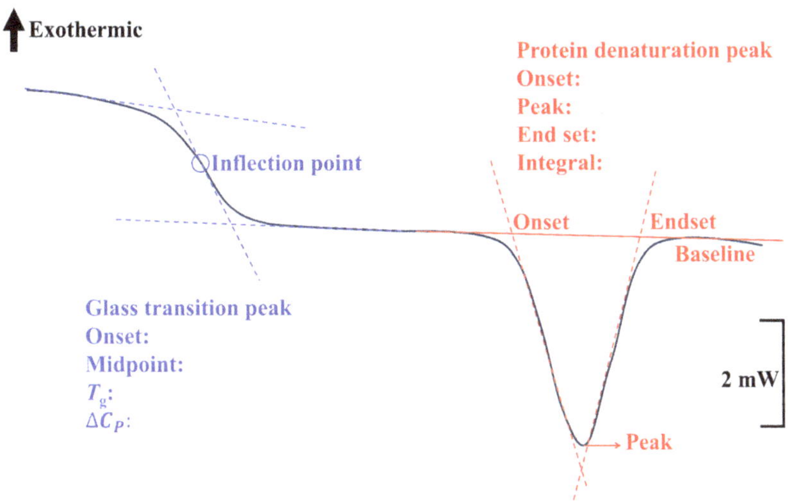

Fig. 3 Schematic representation of typical DSC thermogram

4. The integral area, onset, peak, and endset temperature will be given (Fig. 3).
5. The glass transition temperature (T_g) is reported as the inflection point of the slope [dT/dt] between the onset and endset of the slope change.
6. The protein denaturation temperature is reported as the onset temperature of the corresponding peak.

4 Notes

1. Different types of crucibles can be used for DSC analysis. They need to be inert to the samples and cannot be changed during the measured temperature range. For food protein analysis, the 40-μL standard aluminum crucible is the most frequently used one, which can stand a temperature of 640 °C.
2. Indium is selected as a calibration standard with a melting point of 156.6 °C and heat of fusion of 28.42 J/g. If the end temperature in DSC experiment goes above the melting point of indium, a second calibrant needs to be used. For example, zinc (melting point: 419.5 °C and heat of fusion: 108.26 J/g) can be selected.
3. To achieve accurate measurement results, the crucible lid combination weight on the sample and reference side should not differ by more than 0.2 mg.
4. Protein thermal parameters such as denaturation temperature are dependent on water content. Therefore, moisture determination must be performed before DSC analysis. The water

content on the dry basis or wet basis must be reported together with DSC data.

(a) Wet basis: x grams of water/100 g of food ($\frac{x\ g \text{ of water}}{100} \times 100$).

(b) Dry basis: x grams of water/100 g of solids ($\frac{x\ g \text{ of water}}{100 - x} \times 100$).

5. For those powder samples with a small amount of water content, protein denaturation may not be seen during DSC scanning. Therefore, a certain amount of water could be added to the powder. Alternatively, such protein powder could be put in a desiccator with controlled humidity using saturated salt solutions.

6. For protein solution analysis,

   (a) Dialysis of protein solution is important as it achieves an equilibration state between the protein and the solvent. Especially when a buffer is used to prepare a protein solution, the protein molecule might interact with the ionic buffer ingredient, leading to different buffer compositions, thus interfering with the heat capacity difference between the buffer and the protein sample. The dialysis procedure could be modified depending on the tubing molecular weight cut-off (MWCO), dialysis time, and the times of solvent changes.

   (b) DSC-measured parameters such as melting point were concentration-dependent. When protein concentration is too low, DSC might not be sensitive enough to generate reproducible results. When protein concentration is too high, protein-protein interaction might change the thermodynamics. If possible, perform DSC experiments using a range of four or five different protein concentrations.

7. Make sure the sample size is no more than half of the crucible, as the sample might experience thermal expansion. Also, the sample will evaporate as long as the lid is open; therefore, close the lid as soon as possible.

8. The system is considered tight if the weight does not decrease over 20 μg for 5 min.

9. The reference crucible should contain the reference material that the sample is in. If the proteins are dissolved in the buffer, such buffer will be added to the reference crucible. When the powder is used, the reference crucible will be empty.

10. The starting temperature should always be selected to be lower than the beginning of the first transition to see a segment of a flat baseline. It is usually recommended to select a temperature about 3 min before the first thermal effect occurs. For example,

if the heating rate is 10 °C/min, and the first protein denaturation peak is expected to be seen at 60 °C, the starting temperature would be 60–10 × 3 = 30 °C.

11. The ending temperature is selected to be lower than the decomposition temperature of the sample to avoid system contamination. It should be terminated about 2 min after the last thermal effect has happened. For example, if the heating rate is 10 °C/min, and the last protein denaturation peak is expected to be seen at 90 °C, the starting temperature would be 90 + 10 × 2 = 110 °C.
12. The scan rate used in DSC varies. In general, the faster the scan rate, the more sensitivity but less resolution. The slower the scan rate, the less sensitivity but the higher resolution. The most commonly used scan rate is 10 °C/min.
13. The second scan is performed to ensure that (1) all proteins have been denatured and (2) no more interactions happen.
14. If the weight is reduced by over 30 μg after DSC scanning, the measurement needs to be repeated.
15. For those contaminants that are still attached to the cell after brushing and compressed air, use a cotton swab wetted with a mild solvent such as acetone, alcohol, and hexane.

Acknowledgments

This work was supported by the National Institute of Food and Agriculture, US Department of Agriculture (2022-70001-37580).

References

1. Seenivasan A, Panda T (2017) Protein characterization by thermal property measurement. In: Glass transition and phase transitions in food and biological materials. Wiley-Blackwell
2. Gabbott P (2008) A practical introduction to differential scanning calorimetry. In: Gabbott P (ed) Principles and applications of thermal analysis. Blackwell Publishing Ltd, Oxford
3. Saavedra-Leos M, Alvarez-Salas C, Esneider-Alcalá M, Toxqui-Terán A, Pérez-García S, Ruiz-Cabrera M (2012) Towards an improved calorimetric methodology for glass transition temperature determination in amorphous sugars. CyTA – Journal of Food 10:258–267. https://doi.org/10.1080/19476337.2011.639960
4. Fitzsimons SM, Mulvihill DM, Morris ER (2007) Denaturation and aggregation processes in thermal gelation of whey proteins resolved by differential scanning calorimetry. Food Hydrocoll 21:638–644. https://doi.org/10.1016/j.foodhyd.2006.07.007
5. Rao Q, Labuza TP (2012) Effect of moisture content on selected physicochemical properties of two commercial hen egg white powders. Food Chem 132:373–384. https://doi.org/10.1016/j.foodchem.2011.10.107

Part V

Functional and Sensory Properties of Plant-Based Proteins in Food Systems

Chapter 24

Solubility

Hyukjin Kwon, Bipin Rajpurohit, and Yonghui Li

Abstract

Protein solubility stands as a critical functional attribute, particularly for numerous plant proteins which often exhibit lower solubility compared to their animal-derived counterparts. This chapter provides an overview of common techniques employed in the field of food science for soluble protein quantification. Moreover, detailed procedures for two colorimetric methods for protein solubility measurements: Bicinchoninic acid (BCA) and Bradford assays are covered. The typical experimental parameters used for the two assays are also reported. This chapter could be a valuable guideline for obtaining solubility profile of plant proteins.

Key words Protein solubility measurements, Bicinchoninic acid (BCA) assay, Bradford assay

1 Introduction

The utilization of various food proteins can impart unique functional properties to food matrices. Essential functional attributes like swelling capacity, fat binding capacity, and solubility play crucial roles in influencing the sensory and overall quality of the end food product. Among these desirable properties, solubility holds utmost importance, as it also affects other functional aspects of the protein, such as emulsifying capacity. In general, proteins used for their functional applications require high aqueous solubility [1], whereas proteins with low solubility often lead to unfavorable functionality [2].

Determining the relative solubility of a given protein solution necessitates the quantification of protein content in the supernatant fraction. Multiple methods are employed for quantifying food proteins, including the Kjeldahl method, Dumas method, UV spectrophotometry, and direct amino acid analysis. Each method has its own advantages and disadvantages, as summarized in Table 1. Kjeldahl analysis involves digesting the protein of interest with a strong acid and quantifying the nitrogen content through titration. This nitrogen content is then converted into protein concentration

Yonghui Li (ed.), *Plant-Based Proteins: Production, Physicochemical, Functional, and Sensory Properties*, Methods and Protocols in Food Science, https://doi.org/10.1007/978-1-0716-4272-6_24,

Table 1
Methods for measuring protein solubility: advantages and disadvantages

Method	Steps involved	Advantage	Disadvantage	Reference
Kjeldahl	Acid hydrolysis and nitrogen quantification with titration	Standard method, easily comparable with other research	Nitrogen conversion factor vary by protein, harsh chemical	[7, 8]
Dumas	Protein combustion	No chemical used, multiple samples can be analyzed	Equipment requirement, nitrogen conversion factor vary by protein	[7, 8]
BCA	Protein-copper chelation	More sensitive than Bradford, accurately measure low protein concentration	Interference from reducing agents	[8, 9]
Bradford	Protein-dye binding	Relatively quick, easy to measure	Interference form detergents, susceptible to protein variation	[8, 9]
Direct measurement	Acid hydrolysis and HPLC quantification	Accurate	Harsh chemical, expensive initial setup	[7, 8]

using a conversion factor. Despite being considered the standard method for protein quantification [3], Kjeldahl method requires corrosive reagents, large sample volumes, and involves laborious procedures. Another nitrogen-based method, the Dumas method, utilizes combustion to determine the total nitrogen content. Both Dumas and Kjeldahl methods are destructive and susceptible to interference from non-protein nitrogen sources. Furthermore, direct measurement of amino acid content using HPLC can provide accurate and true protein content, albeit high setup expenses. While the aforementioned methods have definite advantages (Table 1), colorimetric assays have recently gained widespread popularity as an alternative due to their ease of implementation and small sample requirements [4]. This chapter focuses on using two commonly employed colorimetric assays: BCA and Bradford assays, for protein solubility quantification.

2 Material

2.1 Protein Solubilization

Dried protein of interest, DI water, pH meter, 1 N HCl, and 1 N NaOH.

2.2 Protein Clarification

Centrifuge, microfilter (optional), and test tubes.

2.3 BCA Assay

Bovine serum albumin (BSA) standard protein, BCA analysis kit (Thermo Fisher Scientific), and 3.5 mL cuvettes.

2.4 Bradford Assay

Bovine serum albumin (BSA) standard protein, Bradford 1× Dye (Bio-Rad), and 3.5 mL cuvettes.

3 Protocol

3.1 Protein Solubilization

In preparing protein dispersions, DI water is commonly used as the solvent, unless specific buffer or ionic conditions are required. Dried protein powder should be added to DI water in an accurate amount. Remember that the relative solubility of the protein depends on its initial concentration, which should be chosen based on the intended application [2]. For plant storage proteins, the typical range is 0.1 ~ 5.0% by weight (Table 2). When adding the protein, ensure a slow and gentle dispersion, stirring carefully to avoid aeration and foaming. To achieve the desired pH, use 1 N HCl or NaOH as needed (*see* **Notes 1** and **2**). Monitoring the pH over a 30-min period will confirm if the system has reached equilibrium (Fig. 1). To prevent microbial growth, sodium azide (0.02%) can be added to the protein solution [2]. Once the pH has been adjusted, vigorously stir the solution for at least 30 min. This process ensures proper solubility and homogeneity of the protein solution.

3.2 Protein Clarification

After reaching equilibrium, centrifuge the protein solution samples. The speed and duration of centrifugation vary among researchers (Table 1). Make sure to keep the temperature similar to the temperature used in pH calibration. Avoid disturbing the formed pellets during this process. Once the centrifugation is complete,

Table 2
The range of parameters used for BCA and Bradford assays in plant protein solubility measurements

Parameters	Range	Reference
pH range	2 ~ 11	[10–14]
Initial concentration	0.1 ~ 5.0% (w/w)	[10, 11, 15–17]
Centrifugation speed	4000 ~ 20,000 g	[10, 16, 18]
Centrifugation time	15 ~ 30 min	[10, 16, 18]
Dispersion time	30 min ~ overnight	[10, 11, 17]

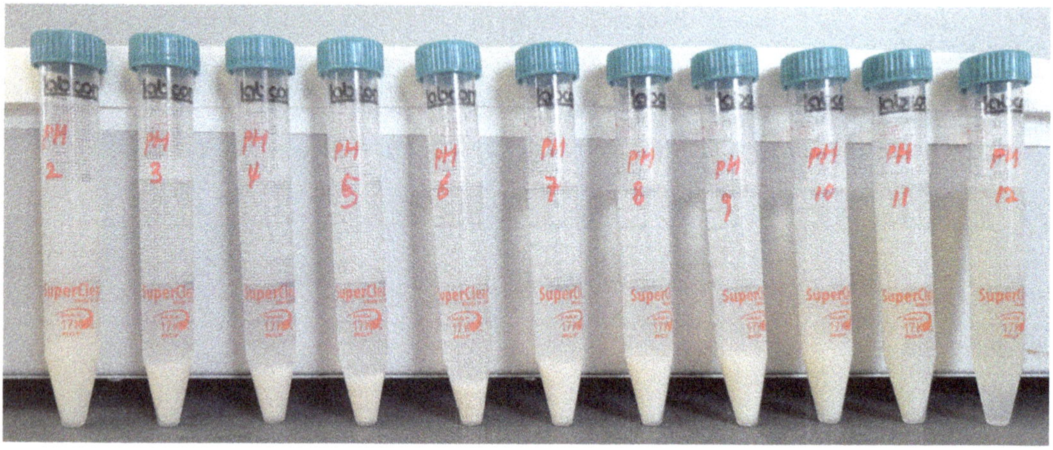

Fig. 1 Pea protein isolate solutions in water (1% w/w), ranging from pH 2 to 12

promptly transfer the supernatant to empty test tubes (*see* **Notes 3** and **4**). To reduce the likelihood of transferring any sediment, it is advisable not to attempt to remove every drop of the supernatant. If desired, the transferred supernatant can undergo an optional filtration step using microfilter tubes. This additional filtration can help further clarify the supernatant and remove any remaining particles.

3.3 BCA Assay

The procedure introduced is based on the official BCA kit manual from Thermo Fisher Scientific [5].

3.3.1 Preparation of BSA Standards

Prepare protein standards by diluting the bovine serum albumin (BSA) standard ampule in BCA kit or dissolving BSA powder in DI water. Using a clean vials or tubes, prepare 9 standards with following concentrations: 0 (blank), 25, 125, 250, 500, 750, 1000, 1500, and 2000 μg/mL (Table 3). Each measurement for standard calibration requires 0.1 mL of the prepared standards.

3.3.2 Preparation of BCA Working Reagent

Mix 50 parts BCA reagent A with 1 part BCA reagent B (50:1). The mixture should be clear and green solution. The solution can be stored for several days at room temperature. Approximate the total volume of the working solution required prior to mixing. Consider that each sample and replicates, including the standards, require 2.0 mL of the working reagent to be measured.

3.3.3 Solubility Measurement

Transfer 0.1 mL of each sample and standards into labeled test tubes. Add 2.0 mL of the working reagent to each test tube and mix well. Cover the tubes and incubate at 37 °C for 30 min to promote BCA reaction (*see* **Note 5**). For even temperature control, water bath is recommended over dry oven. After cooling the samples to room temperature, set the spectrophotometer wavelength

Table 3
Preparation of BSA standards from stock solution of 2.0 mg/mL [5]

Tube no.	Volume of DI water (μL)	Volume and source of BSA (μL)	Final BSA concentration (μg/mL)
1	0	300 of stock	2000
2	125	375 of stock	1500
3	325	325 of stock	1000
4	175	175 of tube 2	750
5	325	325 of tube 3	500
6	325	325 of tube 5	250
7	325	325 of tube 6	125
8	400	100 of tube 7	25
9	400	0	0 (Blank)

to 562 nm and blank the instrument with cuvette filled with the blank standard. Measure the absorbance of the standards and proceed with sample measurements. Prepare standard curve and use it to determine the concentration of the soluble proteins (Fig. 2a). The measurements of the samples should be finished within 10 min, as color development would slowly continue.

3.4 Bradford Assay

Similar to BCA assay, the protocol used is based on the official manual provided from Bio-Rad [6].

3.4.1 Preparation of BSA Standards

Prepare BSA standards by dissolving BSA in DI water. Using clean vials or tubes, prepare 6 standards with following concentrations: 0, 125, 250, 500, 750, and 1000 μg/mL (Table 3). Each measurement for standard calibration and unknown samples requires 40.0 μL of the solutions.

3.4.2 Solubility Measurement

Pipette 2.0 mL of 1× dye reagent into separate cuvette of test tubes. Mix them with 40.0 μL of each standard and unknown sample. Incubate them at room temperature for more than 5 min (*see* **Note 5**). Make sure not to incubate for longer than an hour. Measure their absorbances at 595 nm, and establish a standard curve (Fig. 2b)

3.5 Data Analysis

Calculate the concentration of each unknown sample using the generated standard curve. Dilution factors must be considered according to the linear range of the absorbances by the assays (*see* **Note 3**). By dividing the concentration of proteins at each pH by the total protein content, estimate the percentage solubility at each pH values (*see* **Note 6**). If no protein sediment is detected for

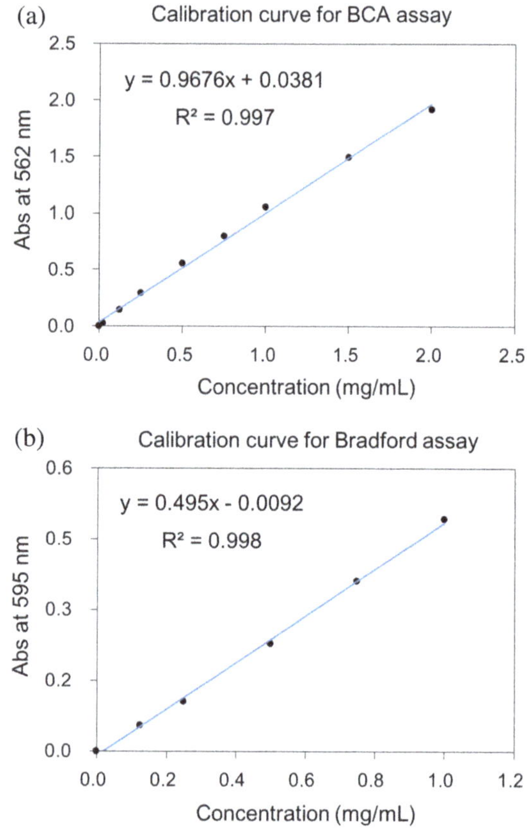

Fig. 2 Calibration curves obtained from BSA Standards: (**a**) BCA assay and (**b**) Bradford assay

pH 12 solution, the concentration at pH 12 can be used as the total protein content. Otherwise, the protein concentration in 1 M NaOH can be used as the total concentration.

4 Notes

1. As protein solubility is dictated by the pH of the solvent, maintaining stable pH is the most crucial factor in solubility determination. It is advised to monitor the pH over 30 min and adjust to the constant pH value.
2. Once pH equilibrium is reached, refrain from storing the protein solutions in refrigerator. As protein solubility differs by temperature, it is necessary to maintain stable temperature throughout all the steps listed.
3. Both BCA and Bradford assay have limited linearity ranges for BSA standard. It is necessary to fix the concentration of the unknown samples within the linearity range to ensure accurate

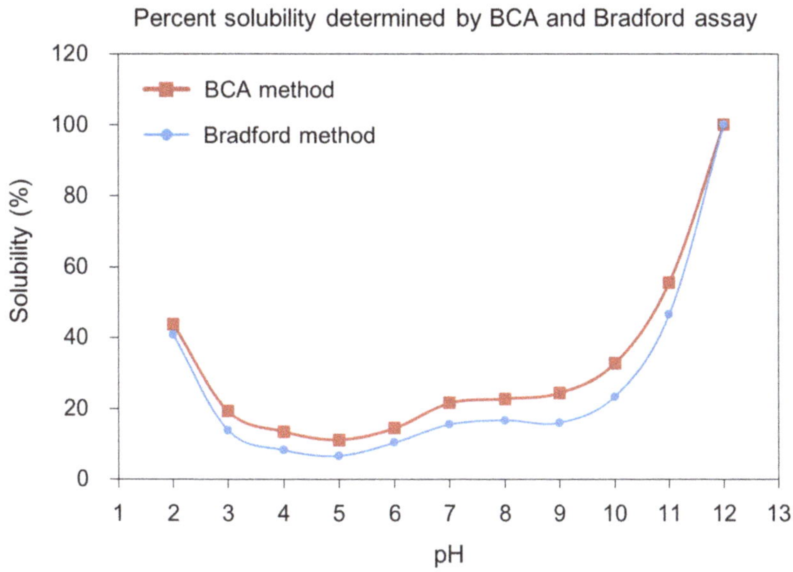

Fig. 3 Percent solubility profiles of pea protein isolate solution (1.0% w/w) determined using BCA and Bradford assays

Note: 1% pea protein solution was prepared and shaken on platform shaker for 30 min after pH adjustment. Centrifugation was done for 15 min at 10,000 g. After centrifugation the supernatant was diluted ten-fold for BSA assay and five-fold for Bradford assay

measurements. Therefore, appropriate dilution could be necessary after centrifuging the protein solutions.

4. A strongly colored sample protein could potentially disrupt the colorimetric measurements of BCA and Bradford. If the color difference is notable, consider opting for non-colorimetric techniques like Kjeldahl or Dumas analysis.
5. The optimal incubation time for colorimetric assays will be different depending on the type of proteins analyzed. Therefore, it is important to keep the same incubation time for each assay.
6. Plant protein solubility plots typically exhibit a U-shape, with the lowest solubility at the isoelectric point and increased solubility as move away from it (Fig. 3).

Acknowledgments

This is contribution No. 24-263-B from the Kansas Agricultural Experimental Station. This work was supported in part by the USDA National Institute of Food and Agriculture Hatch project (Grant Accession No. 7003330).

References

1. Nakai S, Li-Chan E (1985) Structure modification and functionality of whey proteins: quantitative structure-activity relationship approach. J Dairy Sci 68(10):2763–2772
2. Vojdani F (1996) 2 Solubility. In: Methods of testing protein functionality. St. Edmundsbury Press, London, 11–60
3. Didpinrum P, Siriangkhawut W, Ponhong K, Chantiratikul P, Grudpan K (2021) A newly designed sticker-plastic sheet platform and smartphone-based digital imaging for protein assay in food samples with downscaling Kjeldahl digestion. RSC Adv 11(58): 36494–36501
4. Hueso D, Fontecha J, Gómez-Cortés P (2022) Comparative study of the most commonly used methods for total protein determination in milk of different species and their ultrafiltration products. Front Nutr 9:925565
5. T Reagent, Compatibility, and Reconstitution Buffer. "Pierce® BCA Protein Assay Kit–Reducing Agent Compatible."
6. Manual, Instruction. "Quick Start™ Bradford protein assay." (2019)
7. Mæhre HK, Dalheim L, Edvinsen GK, Elvevoll EO, Jensen I-J (2018) Protein determination—method matters. Food Secur 7(1):5
8. Hayes M (2020) Measuring protein content in food: an overview of methods. Food Secur 9(10):1340
9. Khramtsov P, Kalashnikova T, Bochkova M, Kropaneva M, Timganova V, Zamorina S, Rayev M (2021) Measuring the concentration of protein nanoparticles synthesized by desolvation method: comparison of Bradford assay, BCA assay, hydrolysis/UV spectroscopy and gravimetric analysis. Int J Pharm 599:120422
10. Tan E-S, Ying-Yuan N, Gan C-Y (2014) A comparative study of physicochemical characteristics and functionalities of pinto bean protein isolate (PBPI) against the soybean protein isolate (SPI) after the extraction optimisation. Food Chem 152:447–455
11. Boye JI, Aksay S, Roufik S, Ribéreau S, Mondor M, Farnworth E, Rajamohamed SH (2010) Comparison of the functional properties of pea, chickpea and lentil protein concentrates processed using ultrafiltration and isoelectric precipitation techniques. Food Res Int 43(2):537–546
12. Kutzli I, Griener D, Gibis M, Schmid C, Dawid C, Baier SK, Hofmann T, Weiss J (2020) Influence of Maillard reaction conditions on the formation and solubility of pea protein isolate-maltodextrin conjugates in electrospun fibers. Food Hydrocoll 101:105535
13. Xu Y, Obielodan M, Sismour E, Arnett A, Alzahrani S, Zhang B (2017) Physicochemical, functional, thermal and structural properties of isolated Kabuli chickpea proteins as affected by processing approaches. Int J Food Sci Technol 52(5):1147–1154
14. Chang L, Lan Y, Bandillo N, Ohm J-B, Chen B, Rao J (2022) Plant proteins from green pea and chickpea: extraction, fractionation, structural characterization and functional properties. Food Hydrocoll 123:107165
15. Barac MB, Pesic MB, Stanojevic SP, Kostic AZ, Bivolarevic V (2015) Comparative study of the functional properties of three legume seed isolates: adzuki, pea and soy bean. J Food Sci Technol 52:2779–2787
16. Zheng T, Li X, Taha A, Wei Y, Tan H, Fatamorgana PB, Zhang Z et al (2019) Effect of high intensity ultrasound on the structure and physicochemical properties of soy protein isolates produced by different denaturation methods. Food Hydrocoll 97:105216
17. Jiménez-Munoz L, Brodkorb A, Gómez-Mascaraque LG, Corredig M (2021) Effect of heat treatment on the digestion behavior of pea and rice protein dispersions and their blends, studied using the semi-dynamic INFOGEST digestion method. Food Funct 12(18):8747–8759
18. Manassero CA, Vaudagna SR, Añón MC, Speroni F (2015) High hydrostatic pressure improves protein solubility and dispersion stability of mineral-added soybean protein isolate. Food Hydrocoll 43:629–635

Chapter 25

Foaming Capacity and Stability

Suchismita Roy, Gunvantsinh Rathod, and Jayendra Amamcharla

Abstract

Foaming properties are important functional properties of plant proteins. The adsorption properties of plant proteins on the air-water interface help in stabilizing foams. This property is important for product applications where a fluffy/aerated texture is required in products such as bread, cakes, ice cream, etc. This chapter discusses measuring the foaming capacity and foaming stability of protein solutions of soy, pea, and gluten proteins step by step. Foams were created by whipping the solution and its stability over time which was analyzed here.

Key words Functional properties, Foaming capacity, Foam stability, High shear homogenizer, Soy protein isolate, Gluten

1 Introduction

Foaming properties are important in food applications, such as bread, cakes, ice cream, whipped cream, and leavened products, which need a fluffy aerated texture. The plant proteins, due to their interfacial properties, can create a protective layer around the air bubbles after being adsorbed at the air-water interface [1]. The foaming properties are generally characterized by foaming capacity and foam stability. Foam formation depends on the ability of the foaming agent to reduce surface tension and its adsorption on the air-water interface [2]. The foaming capacity of a component is the measurement of how much foam can be produced by mixing protein solution vigorously. Foams are a thermodynamically unstable system [3]. The ability of the foam to retain its volume as a function of time and storage conditions is referred to as foam stability [4]. The stability of the foams is dependent on factors such as the physical and chemical properties of the adsorbed protein layer, drainage, coalescence, disproportionation, etc. [4–6].

The amino acid composition, sequence, size, shape, and charge distribution of the plant protein sources affect the molecular flexibility, solubility, surface charge, and hydrophobicity which in turn

Yonghui Li (ed.), *Plant-Based Proteins: Production, Physicochemical, Functional, and Sensory Properties*, Methods and Protocols in Food Science, https://doi.org/10.1007/978-1-0716-4272-6_25,

affect the foaming properties [7, 8]. The cultivars of the plant proteins, the extraction method, and the processing after extraction of the proteins can also play an important role [8]. For the measurement of foaming properties, there are direct methods such as the measurement of foam and drained liquid volume, foam strength, bubble size, and indirect methods as measuring the surface charge, surface tension, film thickness, elasticity, microscopic imaging, etc. [9, 10]. The selection of the method depends on the objective of the application of foam and the type of protein system. Different methods are used for foam formation, e.g., shaking, whipping, high shear mixing, gas incorporation, etc., which can give widely variable results. The characterization of foams in terms of foam height, foam volume, change of conductivity, imaging, etc., can also influence the outcome. In this chapter, the volume gain is referred to as foaming capacity.

The drainage of liquid from the foams, the movement of gas from smaller to bigger bubbles (coalescence), and the decreasing number of bubbles with increasing size reduce the foam stability [4, 11]. The time-dependent change in foam volume can be used for measuring foam stability [12, 13]. The drainage of liquid over time and the change of conductivity of the foam over time can also be used for the determination of foam stability [14, 15]. The mechanical formation of foam and time-dependent stability of foam are used for measuring foaming properties in this chapter as it is an easy and affordable method.

2 Materials

2.1 Sample Preparation for Analysis

The soy, gluten, and pea protein samples were reconstituted at 1% protein (w/w) solution at room temperature and stored overnight in refrigerated conditions to ensure complete hydration. The analysis was done the next day after adjusting them to room temperature.

3 Method

3.1 Sample Preparation

The foaming capacity and foam stability of all the samples were measured according to the method described by Shen et al. [16]. Reconstituted protein solution was diluted to a 1:10 ratio using DI water and 100 mL was taken in a 600 mL beaker. All the samples were analyzed at room temperature.

3.2 Analysis

The foaming capacity and foaming stability are analyzed in the following method.

1. Each sample was whipped for 2 min at high speed (Speed 4) using a handheld high shear homogenizer (CAT Scientific X120 Poly-Science, Niles, IL).
2. The foam was poured immediately into a 500 mL graduated cylinder and the total volume of the foam was measured immediately. There will be some residual liquid after foaming which might be transferred to the measuring cylinder while pouring foam. Note the volume.

   Foaming capacity $= \frac{(V2 - V1)}{V1} * 100$.

   V_2 = (Volume of foam immediately after stirring—volume of residual liquid)

   V_1 = Initial volume of protein solution
3. The difference in the volume was expressed as the volume of the foam. The foam stability was determined by measuring the fall in the volume of the foam after 30 min and reported as a percentage.

   Foam stability$= \frac{V30}{V2} * 100$

   V_{30} = Volume of foam after 30 min

   V_2 = Volume of foam immediately after stirring
4. All the experiments were performed in triplicate and the results are the average of three values.

4 Result

4.1 Soy Protein Isolate

Soy protein isolate was collected from a bulk supplier through Amazon. It contained 90% protein by weight. The 1% (w/w) protein solution of soy protein isolate was rehydrated overnight. The suspension was diluted to 1:10 ratio and 100 mL was taken in a 600 mL beaker for stirring. After stirring for 2 min using a handheld High shear homogenizer (CAT Scientific X120 Poly-Science, Niles, IL, US) the foam was poured in a 500 mL measuring cylinder and foaming capacity and foam stability were measured according to the method described. Foaming capacity was 60.00 ± 14.14% and foam stability was 9.21 ± 3.60% for this sample (Figs. 1 and 2).

4.2 Pea Protein Isolate

A commercial sample of pea protein isolate was collected from Anthony's Premium Pea protein. It contained 80% protein by weight. The same method was followed for the pea protein isolate. Foaming capacity was 325.00 ± 35.35% and foam stability was 65.69 ± 4.51% for this sample (Figs. 3 and 4).

4.3 Gluten

A commercial sample of gluten was collected from Anthony's Premium Vital Wheat Gluten. It contained 76.6% protein by weight. The same method was followed for the gluten sample. Foaming capacity was 275 ± 39.04% and foam stability was 27.67 ± 7.13% for this sample (Figs. 5 and 6).

Fig. 1 Foam of soy protein isolate after whipping at 0 min

Fig. 2 Foam of soy protein isolate after 30 min

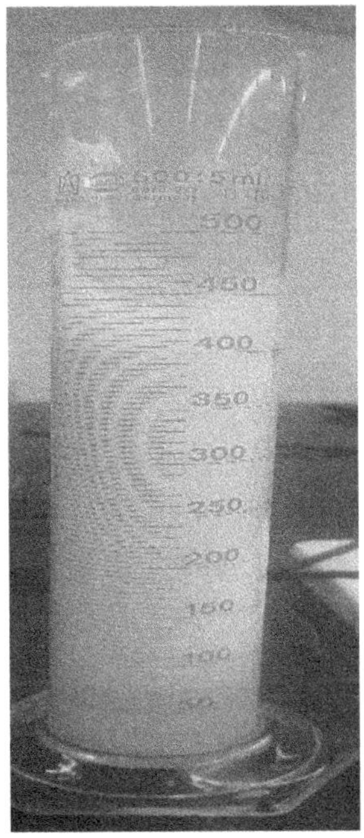

Fig. 3 Foam of pea protein isolate after whipping at 0 min

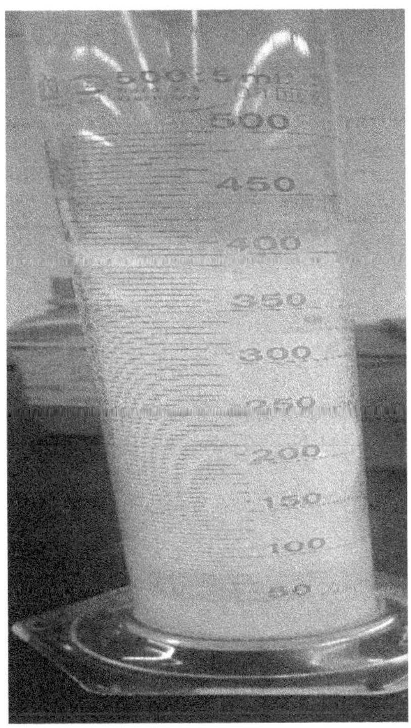

Fig. 4 Foam of pea protein isolate after 30 min

Fig. 5 Foam of gluten after whipping at 0 min

Fig. 6 Foam of gluten after 30 min

5 Notes

1. Different sources of plant proteins have different foaming capacities. Various grades of the same protein source can also provide different results. Therefore, it is important to standardize the protein content in the solution while analyzing. Sometimes if the foaming capacity is too high, the foam can overflow the beaker while whipping or a 500 mL measuring cylinder might not be sufficient for holding all the foams. The protein concentration and whipping time can be adjusted in that case.
2. Another instance may be the formation of very weak foam, making it very difficult to pour into the measuring cylinder after stirring. Protein content should be increased in that case.
3. Foaming capacity depends on the temperature solution; therefore, temperature should be maintained.
4. Beaker dimension should not change for samples to be compared, as the change in dimension changes the surface in contact with a high shear mixture, therefore it will create variation in the surface area of whipping and thereby foaming capacity and foam stability.
5. Similarly, the height of high shear homogenizer and solution volume should be maintained the same to avoid variation.
6. Avoid having any detergent or oil residue on the glassware used in the analysis.
7. If the foam is created using sparging of gas through the solutions, the cylinder diameter, the flow rate of gas, and sample quantity need to be standardized depending on the type of samples.
8. The data presented here are some examples of plant proteins collected from commercial samples. The data may differ depending on the manufacturer.

References

1. Ma KK, Greis M, Lu J, Nolden AA, McClements DJ, Kinchla AJ (2022) Functional performance of plant proteins. Foods 11(4):594. https://doi.org/10.3390/foods11040594
2. Dickinson E (1992) Introduction to food colloids. Oxford University Press
3. McClements DJ (2015) Food emulsions: principles, practices, and techniques. CRC press
4. Huppertz T (2010) Foaming properties of milk: a review of the influence of composition and processing. Int J Dairy Technol 63(4): 477–488. https://doi.org/10.1111/j.1471-0307.2010.00629.x
5. Sánchez CC, Patino JMR (2005) Interfacial, foaming and emulsifying characteristics of sodium caseinate as influenced by protein concentration in solution. Food Hydrocoll 19(3): 407–416. https://doi.org/10.1016/j.foodhyd.2004.10.007
6. Patino JMR, Conde JM, Linares HM, Jiménez JJP, Sánchez CC, Pizones V, Rodríguez FM (2007) Interfacial and foaming properties of enzyme-induced hydrolysis of sunflower protein isolate. Food Hydrocoll 21(5–6): 782–793. https://doi.org/10.1016/j.foodhyd.2006.09.002

7. Damodaran S (2005) Protein stabilization of emulsions and foams. J Food Sci 70(3):54–66. https://doi.org/10.1111/j.1365-2621.2005.tb07150.x
8. Amagliani L, Silva JV, Saffon M, Dombrowski J (2021) On the foaming properties of plant proteins: current status and future opportunities. Trends Food Sci Technol 118:261–272. https://doi.org/10.1016/j.tifs.2021.10.001
9. Grossmann L, Moll P, Reichert C, Weiss J (2020) Influence of energy density on foamability: comparison of three foaming methods. Food Res Int 129:108794. https://doi.org/10.1016/j.foodres.2019.108794
10. Zhang Y, Sharan S, Rinnan Å, Orlien V (2021) Survey on methods for investigating protein functionality and related molecular characteristics. Food Secur 10(11):2848. https://doi.org/10.3390/foods10112848
11. Sadoc JF, Rivier N (eds) (2013) Foams and emulsions, vol 354. Springer Science & Business Media
12. Raymundo A, Empis J, Sousa I (1998) Method to evaluate foaming performance. J Food Eng 36(4):445–452. https://doi.org/10.1016/S0260-8774(98)00063-6
13. Rathod G, Beckman S, Amamcharla JK (2023) Production and functional evaluation of nonfat dry milk with whey proteins as fibrils. J Dairy Sci 106(12):8479–8492. https://doi.org/10.3168/jds.2023-23599
14. Kato A, Takahashi A, Matsudomi N, Kobayashi K (1983) Determination of foaming properties of proteins by conductivity measurements. J Food Sci 48(1):62–65. https://doi.org/10.1111/j.1365-2621.1983.tb14788.x
15. Patino JMR, Delgado MDN, Fernández JL (1995) Stability and mechanical strength of aqueous foams containing food proteins. Colloids Surf A Physicochem Eng Asp 99(1):65–78. https://doi.org/10.1016/0927-7757(95)03129-2
16. Shen Y, Tang X, Li Y (2021) Drying methods affect physicochemical and functional properties of quinoa protein isolate. Food Chem 339:127823. https://doi.org/10.1016/j.foodchem.2020.127823

Chapter 26

Emulsifying Properties

Shan Hong, Hyukjin Kwon, Yixiang Xu, and Yonghui Li

Abstract

Emulsions form a major basis of various processed food products. Due to their surface-active property, food proteins, especially the plant proteins that are gaining increasing popularity recently, are important emulsifiers to form and stabilize emulsions. Proteins contribute to a stable emulsion by forming a protective layer at the oil-water interface thus to lower the interfacial tension and provide the electrostatic and structural barrier. The demand for developing more food emulsion-based products to meet customers' sensorial preferences necessitates the characterization of protein emulsification properties. This chapter delineates three commonly applied methods to characterize protein emulsification ability, namely, the centrifugation method, the turbidimetry method, and the conductivity method. These methods vary in different principles and give diverse indices that underline protein emulsification mechanisms, i.e., emulsion activity (EA), emulsion stability (ES), emulsion activity index (EAI), emulsion stability index (ESI), and emulsion capacity (EC). The detailed protocols and information overviewed in this chapter provide important insights in choosing appropriate proteins to advance the development of food emulsion-based products.

Key words Food emulsions, Plant proteins, Emulsification property, Centrifugation, Turbidity, Conductivity

1 Introduction

Emulsification is one of the most important processes in food manufacturing, as food emulsions are widely occurring in many food products, such as milk, salad dressing, ice cream, butter, cakes, etc. [1]. Emulsions are dispersions composed of two or more immiscible liquids (e.g., water and oil) in which one of them is dispersed in the other as small droplets (0.1–100 μm) [2]. They are typically formed by the process of homogenization and in the presence of an emulsifier [3]. Emulsions are inherently thermodynamically unstable systems, where over time the two phases move towards a more stable state that minimizes free energy. An emulsion may break down via several physical processes, such as creaming/sedimentation (gravitational separation), flocculation, coalescence, Ostwald ripening, and phase inversion [1, 4].

Yonghui Li (ed.), *Plant-Based Proteins: Production, Physicochemical, Functional, and Sensory Properties*,
Methods and Protocols in Food Science, https://doi.org/10.1007/978-1-0716-4272-6_26,

Owing to their amphiphilic nature (having both hydrophobic and hydrophilic groups), proteins are surface-active ingredients that are widely used as emulsifiers to form and stabilize emulsions in food products [4]. During emulsification, protein molecules rapidly migrate to the oil-water interface and adsorb on the oil droplets by positioning the hydrophobic amino acids within the oil phase and the hydrophilic amino acids within the aqueous phase, forming a viscoelastic film at the interface [2]. As a result, the adsorbed protein layer serves to lower the interfacial tension and free energy, as well as provides electrostatic repulsion and steric barrier thereby to prevent the accumulation of oil droplets and oppose the destabilization process [2, 5]. Emulsifying properties of proteins are affected by multiple factors, including the native (e.g., protein type, structure, and concentrations), environmental (e.g., pH, ionic strength, and temperature), and processing factors (e.g., emulsification energy) [4]. As plant proteins have been increasingly recognized as potential alternatives of animal proteins due to their good nutritional quality and sustainability, their techno-functionalities such as emulsification activity are also witnessed rising interests. Emulsification gives food products the characteristic appearance and sensory properties that significantly influence consumer's perception of such products. Thus, to produce better food emulsion-based products that meet customers' sensorial preferences, characterizing protein emulsification properties is important and necessary to provide information guiding product development.

Numerous methods have been proposed to measure protein emulsifying properties. Here, we overviewed the three most commonly used methods due to their ease of operation and the simple required equipment that are available in many laboratories, namely, the centrifugation method, turbidimetry method, and conductivity method.

Both the centrifugation and the turbidimetry method involve preparing an oil-in-water emulsion with a known protein concentration and a fixed oil fraction. According to a protocol proposed by Yasumatsu et al. [6], following the preparation of emulsion, the centrifugation method subsequently involves a controlled centrifugation process (time and speed) to separate the emulsion to an emulsified layer and an un-emulsified layer. The emulsion activity (EA) of the protein is then calculated as the ratio of the height of emulsion layer to the total height of the mixture in the centrifuge tube. The emulsion stability (ES), referring to the ability of emulsion droplets to remain dispersed without aggregation, is estimated after heating the emulsion at 80 °C for 30 min to accelerate the emulsion breakdown and then processed with the same centrifugation condition. The ES is thus expressed as the ratio of the height of emulsified layer after heating to the total height of the mixture.

The turbidimetry method was initially proposed by Pearce and Kinsella [7] and then modified by Cameron et al. [8]. As mentioned previously, the ability of a protein to help form an emulsion is closely related to its ability to adsorb to and stabilize the oil-water interface. Thus, the capacity of a protein to stabilize an emulsion is assumed to be associated with the interfacial area that can be coated by the available protein. According to the Mie theory for light scattering, a simple relationship exists between the turbidity and the interfacial area of an emulsion [9]. In this regard, the turbidity of an emulsion, could serve as an indication of a protein emulsifying property. Specifically, after emulsion preparation, an aliquot of the freshly prepared emulsion is diluted with 0.1% w/v SDS solution which is used to break up any flocs formed. The turbidity of the diluted emulsion is then measured at 500 nm. The emulsion activity index (EAI), presenting as the maximal interfacial area (m^2) covered by per gram of protein, is then calculated from the absorbance (Eq. 3). Moreover, as a function of particle size distribution, changes in turbidity indicate the change in particle size and concentration, serving as an implication of emulsion instability. Therefore, the emulsion stability index (ESI) is thus determined by measuring the turbidity as a function of time and calculated according to Eq. (4). Both the centrifugation and the turbidimetry methods are simple and efficient approaches to determine protein emulsification ability. Though the turbidimetry method describes the behavior of proteins at the interface, it is unable to predict the capacity of a protein that can be applied to create a stable oil-in-water emulsion. This limitation arises from its association with a fixed protein concentration that is used to coat the interfacial area in emulsions, whereas it does not reflect the maximum area that could be covered by varying protein contents.

Kato et al. (1985) proposed a conductivity-based method to characterize protein emulsifying property, by which, the emulsifying capacity (EC) of a protein (mL of oil emulsified by per gram of protein) is determined [10]. This method is based on the principle that oil is a nonconductor, whereas a protein solution can conduct electricity at pH far from its isoelectric point. Therefore, the conductivity of an emulsion is largely dependent on the amount of mobile proteins in the dispersed phase. With the increasing of the oil fraction, more mobile proteins will be consumed by adsorbing to the oil-water interphase which causes the reduction of conductivity. When the conductivity reaches zero, a phase inversion occurs at which point all the proteins have presumably partitioned at the interface. Guo et al. [11] discovered a linear regression between the conductivity and the final protein concentration in the emulsion. Therefore, the EC of a protein can be calculated from the initial protein concentration in the aqueous phase and the theoretical final protein concentration in the emulsion before phase inversion (Eq. (5)).

This chapter aims to provide a comprehensive description of the procedures and the special points to note when employing the above-mentioned methods to characterize protein emulsification ability. Such information could be valuable in choosing appropriate proteins in food development.

2 Materials

2.1 Emulsion Preparation for Analysis

1. Protein powder.
2. Deionized (DI) water.
3. Vegetable oil.
4. Analytical balance.
5. Spatula.
6. Weighing paper.
7. Beaker.
8. Pipette.
9. Pipette tips.
10. Homogenizer.
11. Kimwipes.

2.2 Centrifugation Method

1. Centrifuge.
2. Conical centrifuge tube, 50 mL.
3. Ruler.

2.3 Turbidimetry Method

1. 0.1% (w/v) Sodium dodecyl sulfate (SDS): 0.1% SDS dissolve in DI water.
2. Conical centrifuge tube, 15 mL.
3. Pipette.
4. Pipette tips.
5. Vortex.
6. 4.5 mL disposable cuvettes: path length 1 cm.
7. UV spectrophotometer.

2.4 Conductivity Method

1. Benchtop conductivity meter.
2. Washing bottle with DI water.
3. Kimwipes.

3 Methods

All procedures are carried out at room temperature unless otherwise stated.

3.1 Centrifugation Method

3.1.1 Emulsion Preparation

1. Weigh an accurate amount of protein powder (*see* **Note 1**) and place it in a beaker.
2. Add 25 mL DI water into the beaker (*see* **Note 2**).
3. Stir the mixture carefully using a spatula to break apart powder clumps and ensure an even dispersion.
4. Add 25 mL vegetable oil into the beaker (*see* **Note 3**).
5. Homogenize the mixture at a controlled condition (controlled speed and time, e.g., 16,000 rpm, 2 min) using a homogenizer (*see* **Note 4**).

3.1.2 Emulsion Activity (EA) Measurement

1. The above freshly prepared emulsion is divided equally into two 50 mL centrifuge tubes.
2. The tubes are balanced and centrifuged at 1100 g for 5 min (room temperature, RT).
3. Record the height of the emulsified layer and the whole mixture in the tube using a ruler (*see* **Note 5**) (Fig. 1).
4. EA is calculated according to Eq. (1) (Table 1).

$$\text{EA}\ (\%) = \frac{\text{Height of emulsified layer}}{\text{Height of total mixture in the tube}} \times 100 \quad (1)$$

3.1.3 Emulsion Stability (ES) Measurement

1. The above centrifuge tubes that contain emulsion are heated at 80 °C for 30 min in water bath, followed by cooling at RT for 15 min.
2. The tubes are centrifuged at 1100 g for 5 min, RT.
3. Record the height of the emulsified layer and the whole mixture in the tube after heating (Fig. 1).
4. ES is calculated according to Eq. (2) (Table 1).

$$\text{ES}\ (\%) = \frac{\text{Height of emulsified layer after heating}}{\text{Height of total mixture in the tube after heating}} \times 100 \quad (2)$$

3.2 Turbidimetry Method

3.2.1 Emulsion Preparation

Emulsion preparation is the same as in Subheading 3.1.1.

3.2.2 Emulsion Activity Index (EAI) Measurement

1. An aliquot of the above freshly prepared emulsion is immediately and carefully drawn from the bottom of the test beaker and diluted in a specific volume of 0.1% w/v SDS solution (used to calculate the dilution factor) (*see* **Note 6**).
2. The diluted emulsion is vortexed for 5 s and then transferred 2 mL to a disposable cuvette.

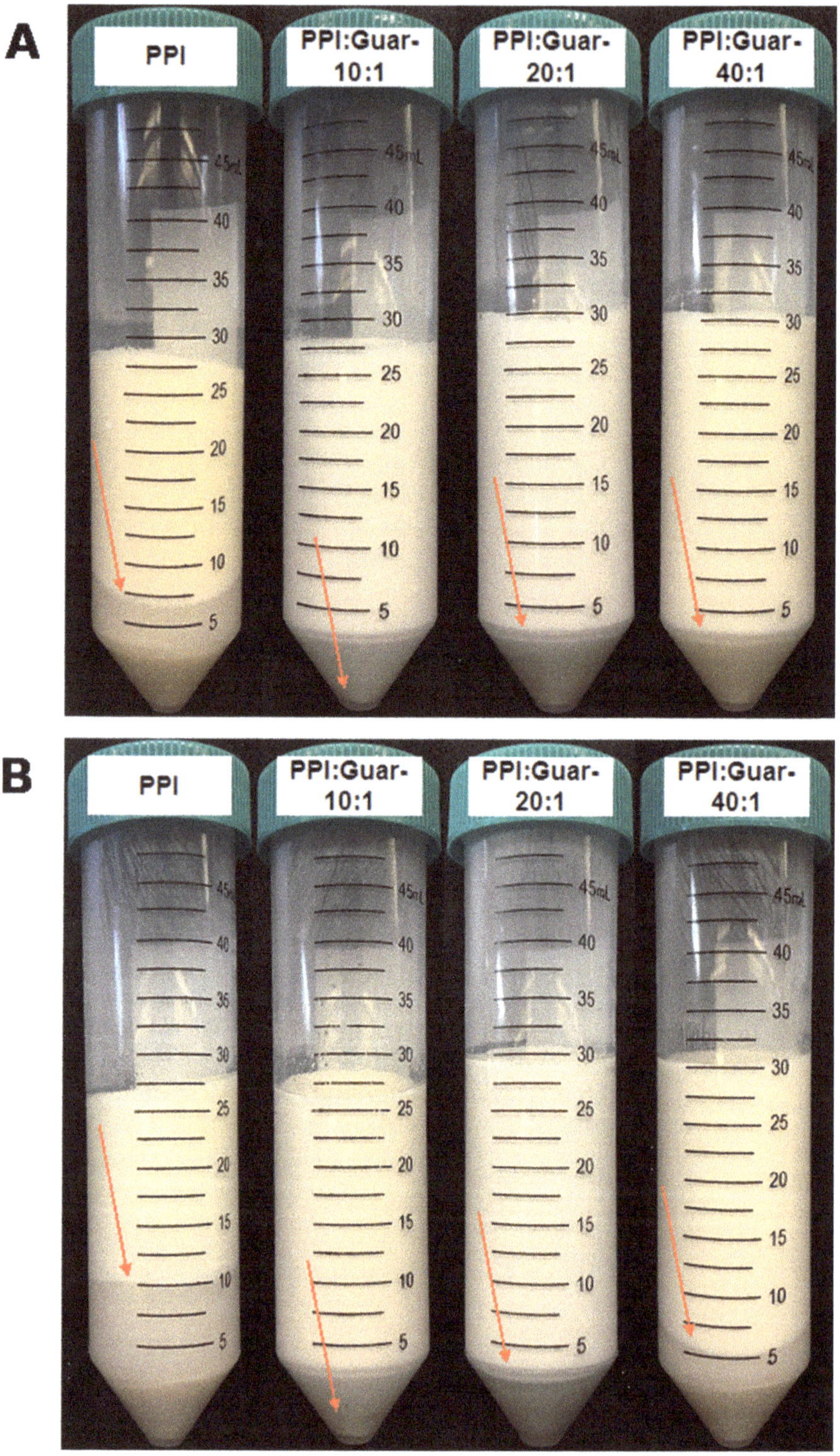

Fig. 1 Emulsions formed by different biopolymers after centrifugation at 1100 g for 5 min, before (**a**) and after (**b**) heating at 80 °C for 30 min. The top white layer is the emulsified layer while the bottom liquid layer is the un-emulsified layer. The red arrows indicate the separation phase between the emulsified layer and the un-emulsified layer. Emulsions were formed by dispersing 0.75 g sample to 25 mL DI water followed by homogenizing with 25 mL canola oil for 2 min at 16000 rpm. PPI, pea protein isolate; PPI:Guar-10:1, mixture of PPI and guar gum at mass ratio of 10:1; PPI:Guar-20:1, mixture of PPI and guar gum at mass ratio of 20: 1; PPI:Guar-40:1, mixture of PPI and guar gum at mass ratio of 40:1

Table 1
Emulsifying ability of different biopolymers characterized by centrifugation method and turbidimetry method

	Centrifugation method		Turbidimetry method	
Sample	**EA (%)**	**ES (%)**	**EAI (m^2/g)**	**ESI (min)**
PPI	71.8 ± 0.8	62.0 ± 1.7	566 ± 16	105 ± 20
PPI:Guar-10:1	99.1 ± 0.2	97.2 ± 1.3	1930 ± 23	3135 ± 46
PPI:Guar-20:1	91.3 ± 1.5	89.8 ± 0.8	1607 ± 66	1405 ± 154
PPI:Guar-40:1	89.3 ± 1.5	78.4 ± 0.6	1015 ± 36	593 ± 16

Note: Emulsions were formed by dispersing 0.75 g sample to 25 mL DI water followed by homogenizing with 25 mL canola oil for 2 min at 16000 rpm. PPI, pea protein isolate; PPI:Guar-10:1, mixture of PPI and guar gum at mass ratio of 10:1; PPI:Guar-20: 1, mixture of PPI and guar gum at mass ratio of 20:1; PPI:Guar-40:1, mixture of PPI and guar gum at mass ratio of 40:1

Abbreviations: *EA* emulsion activity, *ES* emulsion stability, *EAI* emulsion activity index, *ESI* emulsion stability index

3. The absorbance of the diluted emulsion is recorded against the 0.1% SDS solution as control at 500 nm using a UV spectrophotometer.
4. EAI is calculated according to Eq. (3) (Table 1).

$$\text{EAI (m2/g)} = \frac{2T}{C \times (1-\emptyset)} = \frac{2 \times 2.303 \times A_0 \times D}{C \times L \times (1-\emptyset) \times F} \quad (3)$$

Where T is the turbidity, A_0 is the absorbance of the diluted emulsion at 0 min, D is the dilution factor, C is the protein concentration in the aqueous phase (g/mL) (e.g., for 0.75 g protein in 25 mL water and 25 mL oil to form a 50 mL emulsion, C corresponds to 0.75/25 = 0.03 g/mL here), L is the light path length (0.01 m), ϕ is the oil volume fraction in the emulsion (e.g., for 0.75 g protein in 25 mL water and 25 mL oil to form a 50 mL emulsion, ϕ corresponds to 0.5), and F is the unit conversion factor (in this case, it's 10,000; *see* **Note 7**).

3.2.3 Emulsion Stability Index (ESI) Measurement

1. The absorbance of the above diluted emulsion is measured again at 10 min (*see* **Note 8**).
2. ESI is calculated according to Eq. (4) (Table 1).

$$\text{ESI (min)} = \frac{A_0}{A_0 - A_{10}} \times t \quad (4)$$

Where A_{10} is the absorbance of the diluted emulsion at 10 min, t is the time interval (10 min) (*see* **Note 9**).

3.3 Conductivity Method

3.3.1 Emulsion Preparation (See Note 10)

1. In several beakers, weigh an accurate amount of protein powder into each beaker.
2. Add a specific amount of water to each beaker to prepare protein suspensions with a fixed protein concentration.
3. Add different volumes of vegetable oil to each protein suspension according to different ratios of mL oil/g protein (e.g., 20, 40, 60, 80, 100 mL oil/g protein, etc.) (*see* **Note 11**).
4. Homogenize the mixtures at a controlled condition (controlled speed and time, e.g., 16,000 rpm, 2 min) using a homogenizer.

3.3.2 Emulsion Capacity (EC) Measurement

1. Immediately measure the conductivity of the above prepared emulsions (*see* **Note 12**) using a benchtop conductivity meter which is calibrated with standard calibration buffers prior usage. The conductivity of each emulsion should be recorded in duplicate.
2. Generate the linear regression between the conductivity and the final protein concentrations in emulsions, calculate the theoretical final protein concentration in emulsion (mg/mL) when conductivity becomes zero (c_2) (Fig. 2 and Table 2).
3. EC is calculated according to Eq. (5) (Table 2).

$$\text{EC (mL oil per gram protein)} = \frac{V_{\text{oil}}}{m} = \frac{V_{\text{oil}}}{c_1 \times V} \times 1000 = \frac{c_1 - c_2}{c_1 \times c_2} \times 1000 \tag{5}$$

Where V_{oil} is the amount of oil added (mL), V is the starting volume of protein solution (mL), m is the protein weight (g), c_1 is the initial protein concentration in aqueous phase (mg/mL), and c_2 is the theoretical final protein concentration in emulsion (mg/mL) derived from the regression formula when conductivity becomes zero.

4 Notes

4.1 Centrifugation Method

1. Protein emulsifying property is closely related to protein concentration. Therefore, an appropriate amount of protein should be incorporated in the formation of emulsion; usually protein concentrations in the aqueous phase of 0.1–10% w/v are prepared.
2. DI water is commonly served as the aqueous phase and neutral pH is usually studied, unless other specific buffer or ionic conditions or other pH values are specified.
3. Emulsion property is also affected by oil types and oil fractions; thus, a specific oil type and fixed oil fraction should be tested throughout the same research study.

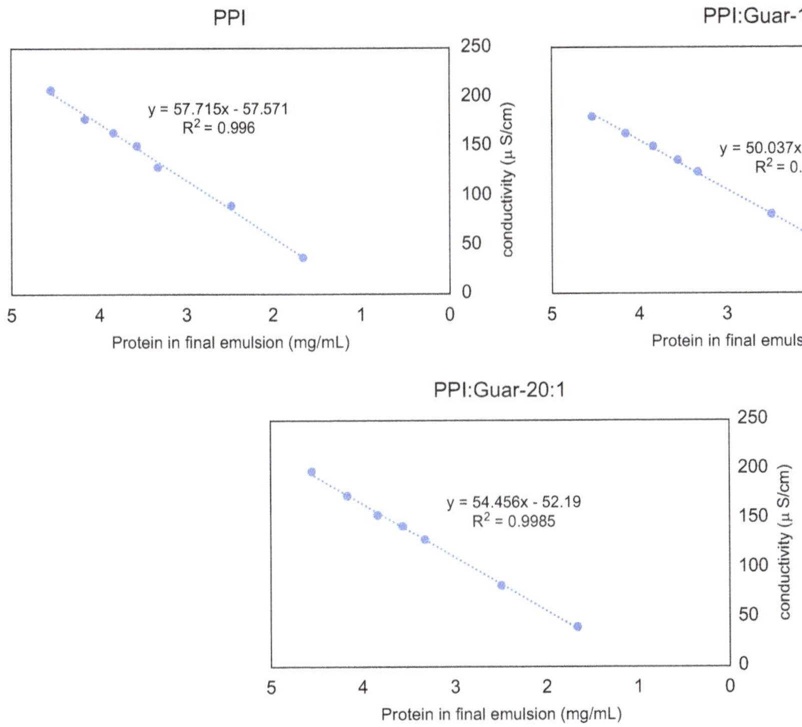

Fig. 2 Linear regression between concentrations of protein in final emulsion (mg/mL) and the conductivity of final emulsion (μS/cm). The obtained linear regression is used to calculate the theoretical final protein concentration in emulsion (c_2, mg/mL) when conductivity turns to zero. PPI, pea protein isolate; PPI:Guar-10:1, mixture of PPI and guar gum at mass ratio of 10:1; PPI:Guar-20:1, mixture of PPI and guar gum at mass ratio of 20:1

4. A fixed homogenization speed and time should be conducted throughout the same research study to avoid the influence of different homogenization conditions. Besides, in addition to homogenizer, other equipment can be used to form oil-in-water emulsions, such as high-shear blenders, colloid mills, high-pressure valve homogenizers (HPVHs), microfluidizers, sonicators, etc.
5. The sample layer is usually not horizontal after centrifugation, it is practical to record the height of the emulsified layer and the whole mixture according to the point with a similar angle of inclination. Besides, instead of recording height, the EA can also be calculated as the ratio of the volume of emulsified layer to the total volume of the whole mixture in the tube. The same as for ES calculation.

4.2 Turbidimetry Method

6. The dilution factor is dependent on the emulsifying ability of a protein. In general, the emulsion should be properly diluted to give an absorbance of 0.01–0.6 at 500 nm.

Table 2
Emulsion capacity (EC) of different biopolymers characterized by conductivity method

Sample	Oil/ protein (mL/g)	Protein conc. in final emulsion (mg/mL)	Conductivity (μS/cm)	C_1	C_2	EC (mL oil/ g of protein)
PPI	20	4.55	208.4	5	0.998 mg/mL; Calculated from $y = 57.715x - 57.571$ (Fig. 2)	802.5
	40	4.168	178.75			
	60	3.844	164.65			
	80	3.57	151.25			
	100	3.33	129.55			
	200	2.498	90.495			
	400	1.664	37.37			
PPI: Guar-10:1	20	4.542	180.3	5	0.892 mg/mL; Calculated from $y = 50.037x - 44.638$ (Fig. 2)	843.4
	40	4.164	162.9			
	60	3.846	149.5			
	80	3.568	135.15			
	100	3.334	123.8			
	200	2.504	80.595			
	400	1.67	37.555			
PPI: Guar-20:1	20	4.548	198.4	5	0.958 mg/mL Calculated from $y = 54.456x - 52.19$ (Fig. 2)	921.0
	40	4.168	173.8			
	60	3.842	153.85			
	80	3.574	143.65			
	100	3.332	129.2			
	200	2.5	82.45			
	400	1.666	40.125			

Note: Emulsions were formed by homogenizing 5 mg/mL biopolymer in aqueous with various oil to sample ratios (20, 40, 60, 80, 100, 200, 400 mL oil/g protein) with a fixed total volume of emulsion (50 mL) for 2 min at 16000 rpm. c_1, initial protein concentration in aqueous phase (5 mg/mL in this case); c_2, the theoretical final protein concentration in emulsion (mg/mL) when conductivity turns to zero (based on the linear regression). PPI, pea protein isolate; PPI:Guar-10:1, mixture of PPI and guar gum at mass ratio of 10:1; PPI:Guar-20:1, mixture of PPI and guar gum at mass ratio of 20:1

7. Readers should pay special attention to the number of the F (the unit conversion factor) in Eq. (3), as it is affected by the units of other factors in the equation. For example:
   (a) When C is in g/mL, L is 0.01 m, then F should be 10,000.
   (b) When C is in mg/mL, L is 0.01 m, then F should be 10.
8. For ESI measurement, the absorbance at t min can be measured based on the same diluted emulsion as stated in Subheading 3.2.3, or a new aliquot of emulsion can be taken from the bottom of the original emulsion at t min for dilution and absorbance measurement. However, based on the author's experience, when a new aliquot of emulsion is taken at t min, the absorbance might even increase compared to the absorbance of emulsion taken at 0 min, which might be due to the

disequilibrium of a freshly prepared emulsion at 0 min thus to cause a slightly lower absorbance. Therefore, the authors suggest measuring the absorbance at t min using the same diluted emulsion as used in the EAI measurement.

9. The time interval t min can be determined according to the stability of emulsion. For example, t min could be shorter for a coarse emulsion, whereas it can be longer for a fine emulsion as it's much more stable. Usually, 10 min is long enough to generate obvious absorbance differences of most coarse emulsions.

4.3 Conductivity Method

10. In the conductivity method, emulsions should be prepared with a fixed initial protein concentration in aqueous and varying oil amounts. For an easier implementation, readers could prepare the emulsions with a specific amount of protein, a specific volume of DI water, and an increasing volume of oil, in which case, the total volume of the emulsion is increasing. However, to minimize the effect of increasing volume on emulsion formation, readers could also prepare the emulsions with varying volumes of water and oil while with a fixed protein concentration in aqueous phase and a fixed total volume of emulsion.
11. A serial of emulsions needs to be formed with different ratios of mL oil/g protein to ensure the conductivity of the emulsion that contains the highest oil fraction is as close to zero as possible.
12. Since the conductivity of each emulsion should be measured immediately after preparation, it's better to prepare emulsions with different oil fractions one by one.

Acknowledgments

This is contribution No. 24-264-B from the Kansas Agricultural Experimental Station. This work was supported in part by the USDA National Institute of Food and Agriculture Hatch project (Grant Accession No. 7003330).

References

1. Hu Y-T, Ting Y, Hu J-Y, Hsieh S-C (2017) Techniques and methods to study functional characteristics of emulsion systems. J Food Drug Anal 25:16–26. https://doi.org/10.1016/j.jfda.2016.10.021
2. Lam RSH, Nickerson MT (2013) Food proteins: a review on their emulsifying properties using a structure–function approach. Food Chem 141:975–984. https://doi.org/10.1016/j.foodchem.2013.04.038
3. McClements DJ, Lu J, Grossmann L (2022) Proposed methods for testing and comparing the emulsifying properties of proteins from animal, plant, and alternative sources. Colloids Interfaces 6:19. https://doi.org/10.3390/colloids6020019

4. Zhang X, Wang Q, Liu Z, Zhi L, Jiao B, Hu H, Ma X, Agyei D, Shi A (2023) Plant protein-based emulsifiers: mechanisms, techniques for emulsification enhancement and applications. Food Hydrocoll 144:109008. https://doi.org/10.1016/j.foodhyd.2023.109008
5. Cabra V, Arreguín R, Farres A (2008) Emulsifying Properties of Proteins, Boletín de la Sociedad 2:80–89. https://api.semanticscholar.org/CorpusID:100003549
6. Yasumatsu K, Sawada K, Moritaka S, Misaki M, Toda J, Wada T, Ishii K (1972) Whipping and emulsifying properties of soybean products. Agric Biol Chem 36:719–727. https://doi.org/10.1080/00021369.1972.10860321
7. Pearce KN, Kinsella JE (1978) Emulsifying properties of proteins: evaluation of a Turbidimetric technique. J Agric Food Chem 26:716–723. https://doi.org/10.1021/jf60217a041
8. Cameron DR, Weber ME, Idziak ES, Neufeld RJ, Cooper DG (1991) Determination of interfacial areas in emulsions using Turbidimetric and droplet size data: correction of the formula for emulsifying activity index. J Agric Food Chem 39:655–659. https://doi.org/10.1021/jf00004a005
9. Kerker M (2016) The scattering of light and other electromagnetic radiation. Elsevier. ISBN 978-1-4832-2524-1
10. Kato A, Fujishige T, Matsudomi N, Kobayashi K (1985) Determination of emulsifying properties of some proteins by conductivity measurements. J Food Sci 50:56–58. https://doi.org/10.1111/j.1365-2621.1985.tb13276.x
11. Guo A, Xiong YL (2021) Electrical conductivity: a simple and sensitive method to determine emulsifying capacity of proteins. J Food Sci 86: 4914–4921. https://doi.org/10.1111/1750-3841.15930

Chapter 27

Water and Oil Holding Capacity

Bipin Rajpurohit, Gengjun Chen, and Yonghui Li

Abstract

In recent years, there has been an increasing demand for plant-based proteins with functional characteristics for various applications. Among these, water holding capacity (WHC) and oil holding capacity (OHC) of plant proteins and their end-use products are important functional attributes, particularly in relation to the juiciness of plant-based meat products. The measurement of the WHC or OHC includes mixing the protein samples with water or oil, respectively, to form a dispersion. This is followed by centrifugation of the dispersion to remove excess water or oil and estimating the water or oil held by samples by difference in the weights. The WHC and the OHC are expressed as g water/g protein and g oil/g protein, respectively. It is worth noting that the method variables such as mixing time and centrifugation conditions may affect the test results. Additionally, factors like protein content, source, cultivar, extraction method, and protein structure can also affect the results. Standardization of the testing method can help in the comparison of the various studies.

Key words Water holding capacity, Oil holding capacity, Plant-based protein

1 Introduction

Water holding capacity (WHC) and oil holding capacity (OHC) of protein measure the amount of water or oil held by the protein per unit mass, respectively [1]. The methods involve dispersing a known amount of protein powder in distilled water or vegetable oil followed by vigorous shaking. Following this, the resulting slurry is centrifuged to remove excess water or oil. The difference between the mass of the sample before and after centrifugation is used to determine the WHC and OHC expressed as g water/g protein or g oil/g protein [2, 3]. There is no standard method for OHC with different protein concentrations, mixing time, holding time, and centrifugation conditions used in different studies. Although, for the WHC, there are standard methods by AACC [4, 5]. Without a standard method for OHC, it is difficult to compare the results between the various studies. When protein powder is dispersed in a liquid, it takes time for the dispersion to

Yonghui Li (ed.), *Plant-Based Proteins: Production, Physicochemical, Functional, and Sensory Properties*, Methods and Protocols in Food Science, https://doi.org/10.1007/978-1-0716-4272-6_27,

form, and for the liquid to be absorbed by the protein. This means that the variation in the mixing and holding period in WHC/OHC methods may influence the results. Similarly, the water holding capacity can also be affected by the pH and temperature of the protein solution. In the previous studies, the mixing time or the holding time varied from 0–30 min and the centrifugation speed and time varied from 300× *g* to 16,000× *g* and 10 to 30 min, respectively [6]. Some researchers determine WHC by converting the protein solution into a gel by heating, cooling, or cross-linking agents. Additionally, WHC and OHC may be measured for the food products incorporated with plant protein, for example, plant-based meat, fish, egg, or cheese [6].

One of the earliest methods for determining WHC and OHC was described by Beuchat [2]. 4-gram samples were combined with 20 mL water or peanut oil. Slurries were stirred occasionally with a glass rod over a 30-min period followed by centrifugation at 15000× *g* for 15 min. The method works well for OHC but doesn't account for the solubility of proteins in the excess water (the solubilized proteins may be removed after centrifugation), which may affect the WHC results. On the contrary, the AACC methods for WHC of protein material involve the addition of small increments of water sufficient to hydrate the protein. Therefore, the proteins are not solubilized by excess water, addressing the problem that may be encountered in the earlier traditional methods [4, 5].

The juiciness of plant-based meat, a highly sought-after attribute in meat products, is closely linked to WHC and OHC, making their quantification essential [7]. Plant proteins exhibiting good WHC and OHC find application as meat extenders, texturized vegetable protein (TVP), and in plant-based meat analogs [6]. The values also depend on the type of plant protein source, cultivar, genotype, extraction method, and the structure of proteins. The range of WHC and OHC reported for legume protein is 0.9–6.8 g/g and 0.9–6.4 g/g, respectively [8–11].

2 Materials

2.1 Water Holding Capacity (Traditional Method)

1. Glass stirring rod.
2. Conical centrifuge tube, 15 mL.
3. Centrifuge.
4. Balance, accurate to 0.001 g.
5. Kimwipes.™

2.2 Water Holding Capacity (AACC Method)

1. Glass test tubes, 16 × 100 mm.
2. Plastic transfer pipette, 1.7 mL.
3. Glass stirring rods, 5 × 150 mm.

4. Plastic syringe barrel (without plunger), 20 cc slip lock.
5. Filter cloth, cut into 2 × 4-in. pieces.
6. Conical centrifuge tube, 50 mL.
7. Centrifuge.
8. Balance, accurate to 0.001 g.

2.3 Oil Holding Capacity

1. Glass stirring rod.
2. Conical centrifuge tube, 15 mL.
3. Centrifuge.
4. Balance, accurate to 0.001 g.
5. Oil.
6. Kimwipes ™.

3 Methods

3.1 Water Holding Capacity (Traditional Method) [12]

1. Weigh 1 g protein (W_0, Tables 1 and 2) in pre-weighted 15 mL conical centrifuge tubes. Record the weight of the tube and the sample (W_1, Tables 1 and 2).
2. Add 10 mL distilled water and vortex thoroughly for 5 s (*see* **Note 1**).
3. Hold the resulting slurry for 30 min at room temperature.
4. Centrifuge the tubes at 3000× *g* for 30 min.
5. Remove the supernatant carefully, wipe excessive water from the sides of the tubes using Kimwipes™, reweight the tubes (Fig. 1, *see* **Notes 2** and **3**), and record as W_2, Tables 1 and 2).

Table 1
WHC of commercial plant proteins using the traditional method and without altering the pH

Sample ID	Tube wt (g)	Sample wt (g), W_0	Tube wt + sample wt (g), W_1	Tube + sample + water (g), W_2	Water held (g)	WHC (g/g)
SPI	6.385	1.002	7.387	11.684	4.297	4.29
SPI	6.401	1.003	7.404	11.588	4.184	4.17
PPI	6.372	1.005	7.377	11.343	3.966	3.95
PPI	6.390	1.004	7.394	11.573	4.179	4.16
VWG	6.378	1.002	7.380	8.878	1.498	1.50
VWG	6.361	0.999	7.360	8.993	1.633	1.63

Note: SPI- soy protein isolate, PPI- pea protein isolate, VWG- vital wheat gluten, *wt*- weight

Table 2
WHC of commercial plant proteins using the traditional method and adjusting pH to 5.5 (*see* Note 4)

Sample ID	Tube wt (g)	Sample wt (g), W_0	Tube wt + sample wt (g), W_1	Tube + sample + water (g), W_2	Water held (g)	WHC (g/g)
SPI	6.337	1.003	7.340	10.143	2.803	2.79
SPI	6.307	0.999	7.306	9.982	2.676	2.68
PPI	6.302	1.004	7.306	10.186	2.880	2.87
PPI	6.275	0.996	7.271	10.125	2.855	2.87
VWG	6.326	0.999	7.325	8.688	1.363	1.36
VWG	6.320	1.001	7.321	8.722	1.401	1.40

Note: SPI- soy protein isolate, PPI- pea protein isolate, VWG- vital wheat gluten, *wt*- weight

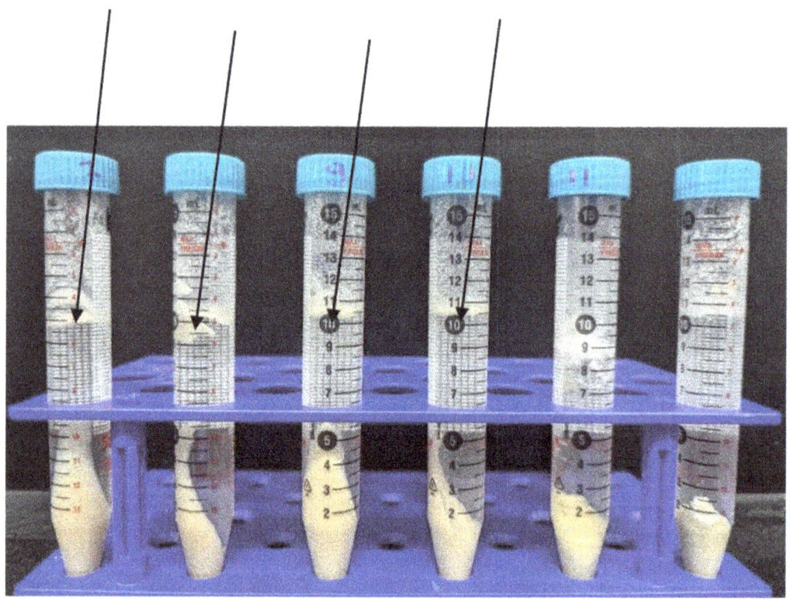

Fig. 1 Separation of excess water after centrifugation. Observe the thin layer of wet solids (indicated with black arrows) on top of the water layer in the first four tubes (*see* **Note 2**)

The WHC can be calculated according to the following formula:

$$\mathrm{WHC} = \frac{W_2 - W_1}{W_0}$$

Where W_2 is the final weight of the tube and hydrated sample, W_1 is the weight of the tube and sample, and W_0 is the weight of the sample.

Table 3
WHC of commercial plant proteins determined using the AACC method (*see* Note 7)

Sample ID	Sample wt (g), W_1	Moisture content (%), m_c	Syringe assembly wt initial (g), W_2	Water between 0.05 and 0.2 mL	Syringe assembly wt final (g), W_3	Water held (g)	WHC (g/g) dry matter basis
SPI	1.002	5.99	16.390	Yes	19.615	3.225	3.49
SPI	0.997	5.99	16.441	Yes	19.488	3.047	3.32
PPI	1.002	7.26	16.382	Yes	19.856	3.473	3.82
PPI	0.998	7.26	16.324	Yes	19.680	3.357	3.71
VWG	1.000	5.93	16.243	Yes	17.753	1.510	1.67
VWG	1.003	5.93	16.424	Yes	17.932	1.507	1.66

Note: SPI- soy protein isolate, PPI- pea protein isolate, VWG- vital wheat gluten, *wt*- weight

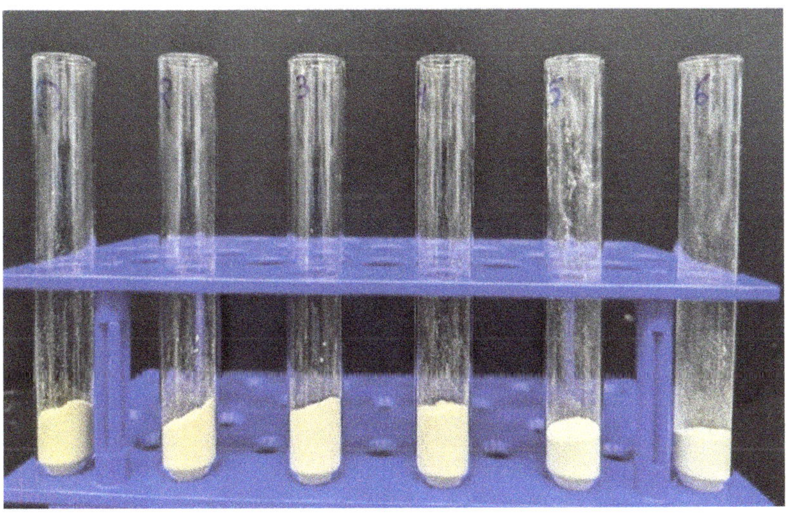

Fig. 2 Glass test tubes with 1 g sample

3.2 Water Holding Capacity (AACC Method) [4]

1. Weigh 1 g sample in the glass test tube and record the weight (W_1, Table 3) (Fig. 2).
2. Place the test tube (with sample) and filter cloth in the syringe barrel and record the weight of the syringe assembly (W_2, Table 3) (Fig. 3).
3. Remove the test tube from the syringe barrel. Add deionized water to the tube in small, unmeasured increments. Use a glass rod to mix the water and the protein powder after each water addition. Add water until the protein is thoroughly wetted/saturated with water (*see* **Note 5**). Stir the mixture for 1 min and wipe the rod on the middle of the filter cloth folded in half (2 X 2 in).

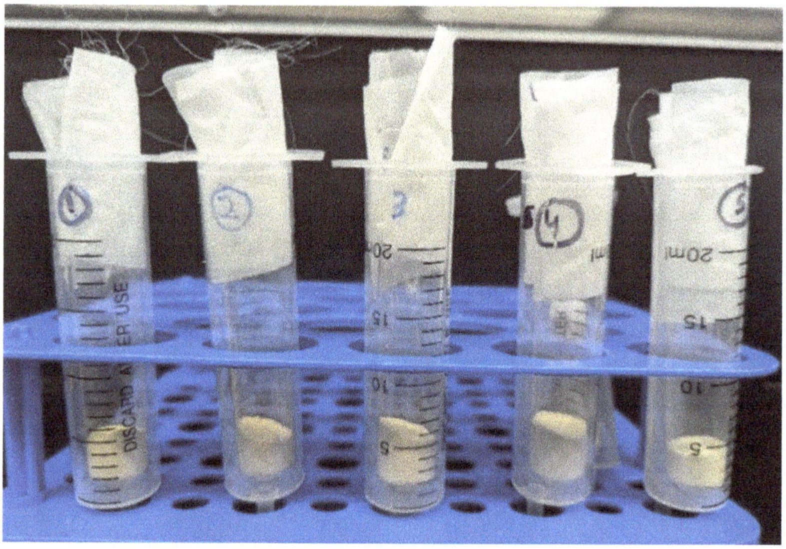

Fig. 3 The syringe assembly consists of syringe barrel, test tube with the sample, and the filter cloth

4. Place the folded filter cloth at the end of the test tube and insert the test tube in the syringe barrel inverted such that the filter cloth side is towards the nozzle of the syringe barrel.
5. Place the syringe assembly into a 50 mL conical centrifuge tube such that the nozzle of the syringe is towards the bottom of the centrifuge tube and centrifuge at 300× *g* for 10 min.
6. After centrifugation, check the amount of liquid collected at the bottom of the centrifuge tube (Fig. 4). If the amount of liquid is between 0.05–0.2 mL, proceed to the next step otherwise repeat the above steps (*see* **Note 6**).
7. Remove the syringe assembly from the centrifuge tube, weigh it, and record the weight (W_3, Table 3). WHC is estimated by the following formula:

$$\text{WHC (g water/g protein, dry matter)} = \left[\frac{(W_3 - W_2) + (W_1\ X\ m_c)}{(1 - m_c)\,W_1}\right]$$

Where W_1 is the weight in grams of the sample before water addition, W_2 is the weight in grams of the syringe assembly (syringe barrel, filter cloth, test tube, and sample) before centrifugation, W_3 is the weight in grams of syringe assembly after centrifugation, m_c is the initial moisture content % of the sample.

3.3 Oil Holding Capacity

1. Weigh 1 g protein (W_0)(Table 4) in a pre-weighted 15 mL conical centrifuge tube. Record the weight of the tube and the sample (W_1, Table 4).

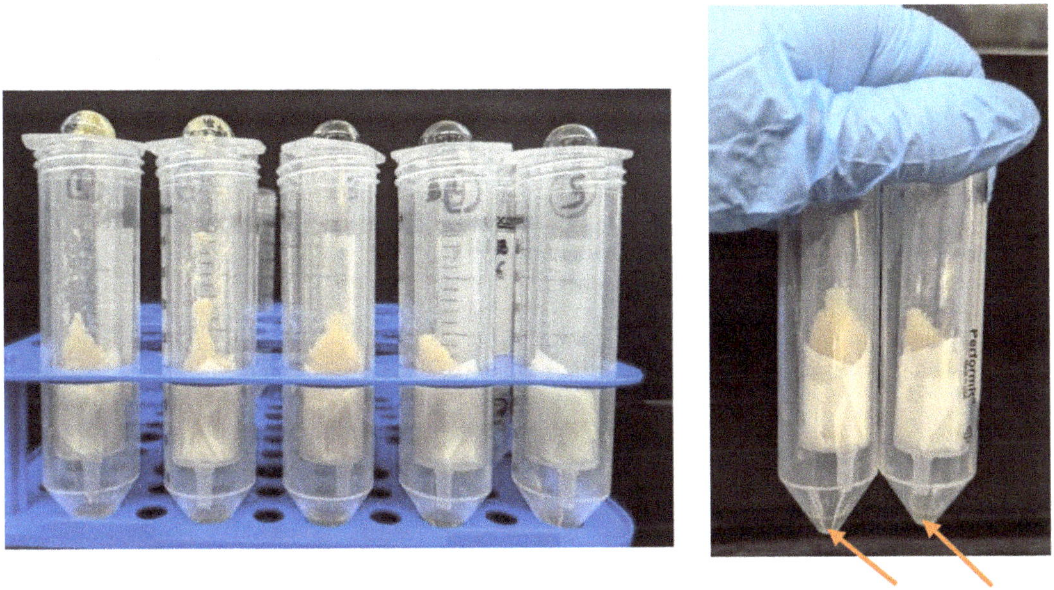

Fig. 4 After centrifugation. Observe the water collected at the bottom of the centrifuge tube. The image on the right is an enlarged image to show the water (marked with arrows) at the bottom of the centrifuge tube

Table 4
OHC of commercial plant proteins

Sample ID	Tube wt (g)	Sample wt (g), W_0	Tube wt + sample wt (g), W_1	Tube + sample + oil (g), W_2	Oil held (g)	OHC (g/g)
SPI	6.311	0.995	7.306	8.467	1.161	1.17
SPI	6.315	0.998	7.313	8.604	1.290	1.29
PPI	6.381	0.997	7.378	8.534	1.157	1.16
PPI	6.367	0.997	7.364	8.579	1.215	1.22
VWG	6.359	1.003	7.362	8.473	1.111	1.11
VWG	6.290	0.996	7.286	8.376	1.091	1.10

Note: SPI- soy protein isolate, PPI- pea protein isolate, VWG- vital wheat gluten, wt- weight

2. Add 10 mL oil and vortex thoroughly for 5 s (Fig. 5, *see* **Note 1**).
3. Hold the resulting suspension for 30 min at room temperature.
4. Centrifuge the tubes at 3000× g for 30 min.
5. Remove the supernatant carefully. Wipe off excessive oil from the side of the tubes (*see* **Notes 2** and **3**). Reweight the tubes (W_2, Table 4). The OHC is calculated according to the following formula:

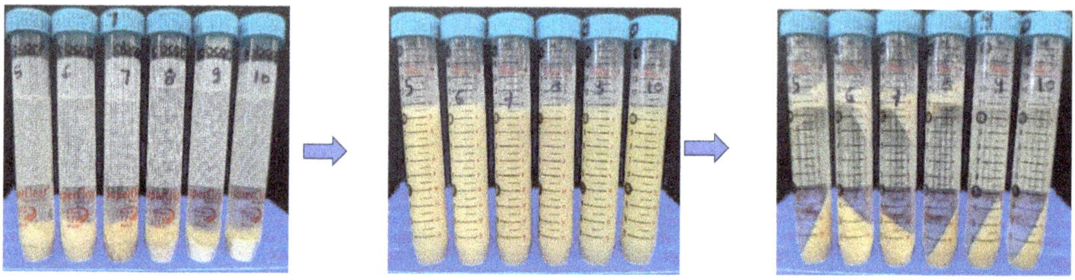

Tubes with plant protein and oil. Observe the dry layer of protein at the bottom and two separate phases.

Formation of suspension of oil and plant protein. Glass stirring rod was used to mix and disperse the protein powder followed by vigorous shaking (see note 1).

Separation of excess oil by centrifugation. The excess oil (supernatant) was carefully removed.

Fig. 5 Pictures depicting the OHC test

$$\mathrm{OHC} = \frac{W_2 - W_1}{W_0}$$

Where W_2 is the final weight of the tube and sample with oil, W_1 is the weight of the tube and sample, and W_0 is the weight of the sample.

4 Notes

1. Even after vortexing, the protein powder may not be dispersed uniformly in the water/oil (Fig. 5). Use a glass stirring rod to disperse the protein in the water/oil followed by vortexing and shaking the tubes.
2. In some samples, there may be a thin layer of soaked solids (protein) on top of the water/oil layer (Fig. 1). This might happen if the sample has very fine particles. While removing the excess water/oil from the tube, block off any solid with the help of a filter paper such that only excess water/oil is removed, and the soaked solids are not removed.
3. Be careful while wiping excessive water/oil from the sides of the tube. Make sure not to scrape the pellet. Some procedures state inverting the tubes to remove the excessive water/oil. In our experience, this should be avoided as loose parts of the pellet tend to come out of the tube with the water/oil which results in error.
4. One constraint of the traditional method to measure WHC is excess water may solubilize the protein and mislead the interpretation. To rectify this, in a parallel experiment, pH was adjusted to 5.5 (close to the isoelectric point, based on some

previous research) to render protein insoluble in water and then measure the WHC. However, results indicate that changing the pH to around the isoelectric point reduced the WHC of the proteins (Table 2). This might be due to reduced affinity between the protein and the water molecules at pH 5.5. The AACC method discussed later addresses experimental errors that may be caused by the solubility of protein by excess water. Instead of adding large amounts of water, the method includes the addition of small increments of water to wet/saturate the proteins.

5. Although a measured amount of water is not added, keep a record of the estimate of the total water added. (This would help you to adjust water in subsequent runs if the water collected in the centrifuge tube is not between 0.05 mL and 0.2 mL, *see* **Note 6**).
6. If the amount of water is less than 0.05 mL, add more water and if it is more than 0.2 mL, repeat and add less water.
7. The WHC of the commercial plant-based proteins determined using the AACC method was found to be lower than the traditional method. The reason could be the absence of any holding time in the AACC method, or the water is not completely removed by the traditional method.

Acknowledgments

This is contribution No. 24-265-B from the Kansas Agricultural Experimental Station. This work was supported in part by the USDA National Institute of Food and Agriculture Hatch project (Grant Accession No. 7003330).

References

1. Rajpurohit B, Li Y (2023) Overview on pulse proteins for future foods: ingredient development and novel applications. J Future Foods 3(4):340–356
2. Beuchat LR (1977) Functional and electrophoretic characteristics of succinylated peanut flour protein. J Agric Food Chem 25(2): 258–261
3. Lin MJY, Humbert ES, Sosulski FW (1974) Certain functional properties of sunflower meal products. J Food Sci 39(2):368–370
4. AACC approved methods of analysis, 11th Ed. Method 57–13.01. Water-holding capacity of pulse flours and protein materials. Approved July 2017. Cereals & Grains Association, St. Paul, MN, U.S.A. https://doi.org/10.1094/AACCIntMethod-56-37.01
5. AACC approved methods of analysis, 11th Ed. Method 56-30.01. Water hydration capacity of protein materials. Approved November 3, 1999. Cereals & Grains Association, St. Paul, MN, U.S.A. https://doi.org/10.1094/AACCIntMethod-56-30.01
6. Ma KK, Greis M, Lu J, Nolden AA, McClements DJ, Kinchla AJ (2022) Functional performance of plant proteins. Foods 11(4):594
7. Cornet SH, Snel SJ, Lesschen J, van der Goot AJ, van der Sman RG (2021) Enhancing the water holding capacity of model meat analogues through marinade composition. J Food Eng 290:110283

8. Lafarga T, Álvarez C, Villaró S, Bobo G, Aguiló-Aguayo I (2020) Potential of pulse-derived proteins for developing novel vegan edible foams and emulsions. Int J Food Sci Technol 55(2):475–481
9. Gundogan R, Karaca AC (2020) Physicochemical and functional properties of proteins isolated from local beans of Turkey. LWT 130:109609
10. Tang X, Shen Y, Zhang Y, Schilling MW, Li Y (2021) Parallel comparison of functional and physicochemical properties of common pulse proteins. LWT 146:111594
11. Hojilla-Evangelista MP, Sutivisedsak N, Evangelista RL, Cheng HN, Biswas A (2018) Composition and functional properties of saline-soluble protein concentrates prepared from four common dry beans (Phaseolus vulgaris L.). J Am Oil Chem Soc 95(8):1001–1012
12. Shen Y, Tang X, Li Y (2021) Drying methods affect physicochemical and functional properties of quinoa protein isolate. Food Chem 339:127823

Chapter 28

Gelation

Baochen Fang and Jiajia Rao

Abstract

The gelling capacity of plant protein is an important functional property that determines the food product quality in semi-solid and solid foods such as yogurt, tofu, bread doughs, and some meat products. Protein gelation is a process in which protein molecules can form a three-dimensional network to trap large amounts of water in a solid state. Different plant proteins can be applied to form protein gel, which exhibits various water holding capacity (WHC), appearance, texture, and rheological properties. In general, the least gelation concentration can be used to identify the gel formation condition, whereas the WHC value provides information on the gel's ability to prevent exudation. The texture and viscoelasticity properties of plant protein gel can be evaluated by a texture analyzer and a rotational rheometer, respectively. This chapter provides comprehensive methodologies for the characterization of gelling properties of plant proteins.

Key words Gelation, Least gelation concentration, Water holding capacity, Rheology, Viscoelastic, Texture

1 Introduction

Gelling property, one of the most important plant protein functionalities, reflects the structure and rheological properties of gel-based foods [1, 2]. Plant protein hydrogel is a three-dimensional network that entraps a large amount of water to maintain a solid state [3]. So far, a variety of methods have been developed to induce plant protein gels such as adding adjuvants (salt, acid, enzyme) or physical treatment (high pressure, temperature), but the most traditional way is through thermal-induced protein gel [4, 5]. Therefore, this chapter is mainly focusing on methods for evaluating the gelling properties of thermal-induced plant protein gel.

When the state transforms from "sol" to gel, the solid state along with its own viscoelasticity behavior is developed [6]. Understanding the gelling properties of plant protein gel is crucial for a number of reasons including properly designing food texture and

Yonghui Li (ed.), *Plant-Based Proteins: Production, Physicochemical, Functional, and Sensory Properties*, Methods and Protocols in Food Science, https://doi.org/10.1007/978-1-0716-4272-6_28,

structure, extending the shelf life of food, and increasing satiety in the final application [3]. In general, four major parameters are commonly applied to describe the gelling properties of plant protein gel including least gelation concentration (LGC), water holding capacity (WHC), rheological, and texture properties. The LGC is defined as the protein concentration when the gel sample from the inverted test tube did not slip immediately [7]. In general, the lower LGC of protein represents the better gelling ability in the setting conditions [7]. The LGC value can be affected by several thermal-induced gel formation conditions such as temperature and heating time. Once the self-supporting plant protein gel is formed, WHC is another important parameter to reflect the gelling properties of the protein gel. The WHC of plant protein gel is the ability to prevent water in the gel from exudation under mechanical deformation. This parameter identifies the juiciness sensation and reflects the moisture content of plant protein gel-based foods [8]. In terms of rheological and texture properties of plant protein gels, rheological measurements are not only necessary to evaluate the texture and quality of food, but also important for food processing control, equipment development, and transportation. To understand the rheological and/or texture properties of gels, the dynamic oscillatory shear test in small amplitude oscillatory shear is the most used method such as frequency sweep [9]. It provides the information on rheological properties of plant protein gel such as storage (elastic) modulus (G') and loss (viscous) modulus (G") in the linear viscoelastic region [10]. Please note that a strain sweep should be conducted initially to identify the linear viscoelastic region. Subsequently, the small amplitude oscillatory shear test is performed within this linear region. The texture properties of plant protein gel can be also characterized by a texture analyzer to obtain the hardness, rapture force, cohesiveness, resilience, and springiness properties of the gel.

The purpose of this chapter is to provide standardized or commonly used methodologies for comprehensive analysis of gelation properties of plant proteins. It provides detailed information on materials, analytical instruments, and methods required to evaluate LGC, WHC, rheological, and texture properties of plant protein gels.

2 Materials

Prepare all solutions using nano pure water (prepared by purifying deionized water, to attain a sensitivity of 18 MΩ-cm at 25 °C) and analytical grade reagents. Prepare and store all reagents at room temperature (unless indicated otherwise). Diligently follow all waste disposal regulations when disposing of waste materials. The required equipment for all experiments is listed below.

1. Water bath with the capacity to heat the samples at 85 ± 10 °C (10128-126, VWR, Chicago, IL, USA) (*see* **Note 1**).
2. Analytical balance, with a resolution of 0.0001 g (ME204E, Mettler Toledo, Columbus, OH, USA).

2.1 Least Gelation Concentration Determination

1. Glass tubes of 16 mm diameter and 125 mm high with plug-in caps (*see* **Note 2**).

2.2 Water Holding Capacity

1. Centrifuge with the capacity to reach 10,000× *g* (2400-37, VWR, Chicago, IL, USA).
2. Whatman Grade 1 Qualitative Filter Paper (1001-070, Cytiva, USA).
3. Centrifuge tubes (2 mL).

2.3 Rheological Properties of Gel

1. Discovery Hybrid Rheometer-2 rheometer (TA Instruments Ltd., New Castle, DE, USA).
2. Parallel geometry with a diameter of 40 mm (serial number 108004, TA Instruments Ltd., New Castle, DE, USA) (*see* **Note 3**).
3. Cylindrical shape of flat aluminum foil containers with a diameter slightly larger than 40 mm.

2.4 Texture Analysis of Gel

1. Texture analyzer model TAXT2i (Texture Technologies Corp., White Plains, NY, USA).
2. Cylinder probe of 25 mm and 2 mm in diameter.
3. Cylindrical shape of flat aluminum foil containers with a diameter slightly larger than 40 mm.

3 Methods

All procedures are carried out at room temperature. All experiments are conducted at least three times. Data is reported as mean ± standard deviation unless otherwise specified.

3.1 Determination of Least Gelation Concentration (LGC)

The LGC of protein is defined as the minimum protein concentration below which no semi-solid or solid gel can be formed in the inverted test tube [11]. The heating temperature and time required to form the thermal-induced protein gel are obtained from the preliminary experiment according to our previous studies [10] (*see* **Note 4**). Figure 1 shows a flowchart of the determination of LGC.

1. A wide range of protein concentration solutions or slurries are prepared, varying from 1% to 20% (w/v). For example, 0.1 g, 0.2 g, 0.4 g, 0.6 g, 0.8 g, 1 g, 1.2 g, 1.4 g, 1.6 g, 1.8 g, and 2 g

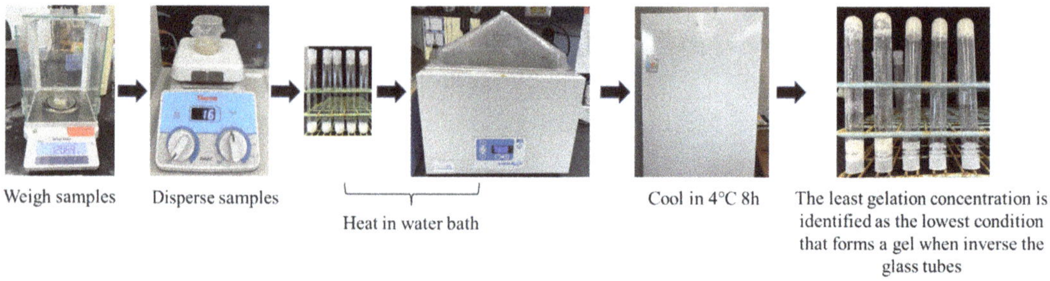

Fig. 1 Flowchart for determination of least gelation concentration

of protein powder are weighted in individual beaker, respectively. Subsequently, 10 mL of water is added to the abovementioned beakers and stirred for 30 min using magnetic stirrers to reach the protein concentration of samples, ranging from 1%, 2%, 4%, 6%, 8%, 10%, 12%, 14%, 16%, 18%, and 20% (w/v) (*see* **Note 5**). After fully dispersing of protein powder in water, 1 mL of protein solution or slurry from individual samples with varied protein concentrations is transferred into the clean glass tubes, and each glass tube is covered by a plastic cap for further experiment.

2. Preheat the water bath to 75 °C (*see* **Note 4**). Submerge the prepared glass tubes with samples from **step 1** in the preheated water bath at 75 °C for 5 mins.
3. Once the heating process is completed, samples are cooled down to room temperature immediately under tap water, and then the cooled glass test tubes are stored in a refrigerator at 4 ° C overnight (~ 8 h).
4. After that, the lass test tubes are taken from the refrigerator, inverse tubes to an upside-down direction. If the heated protein slurry doesn't fall or slip from the glass test tube, then it can be considered a gel state. The LGC is identified as the lowest protein concentration that forms a gel state when inverse the glass test tubes.

3.2 Determination of Water Holding Capacity (WHC)

The WHC of plant protein gel is carried out by measuring the weight difference of gel before and after centrifugation based on the method of Fang et al. [10]. Figure 2 shows a principle and flow chart demonstration on determination of WHC.

1. The 10% (w/v) protein slurry is prepared by weighing 1 g protein samples into the beakers with 10 mL water, and then the mixture is stirred for 30 min at room temperature to fully disperse by using magnetic stirrers (*see* **Note 6**). During the mixing time, the beaker is covered with parafilm.
2. Pre-weigh a clean, dried centrifuge tube, and record it as W_0 (g).

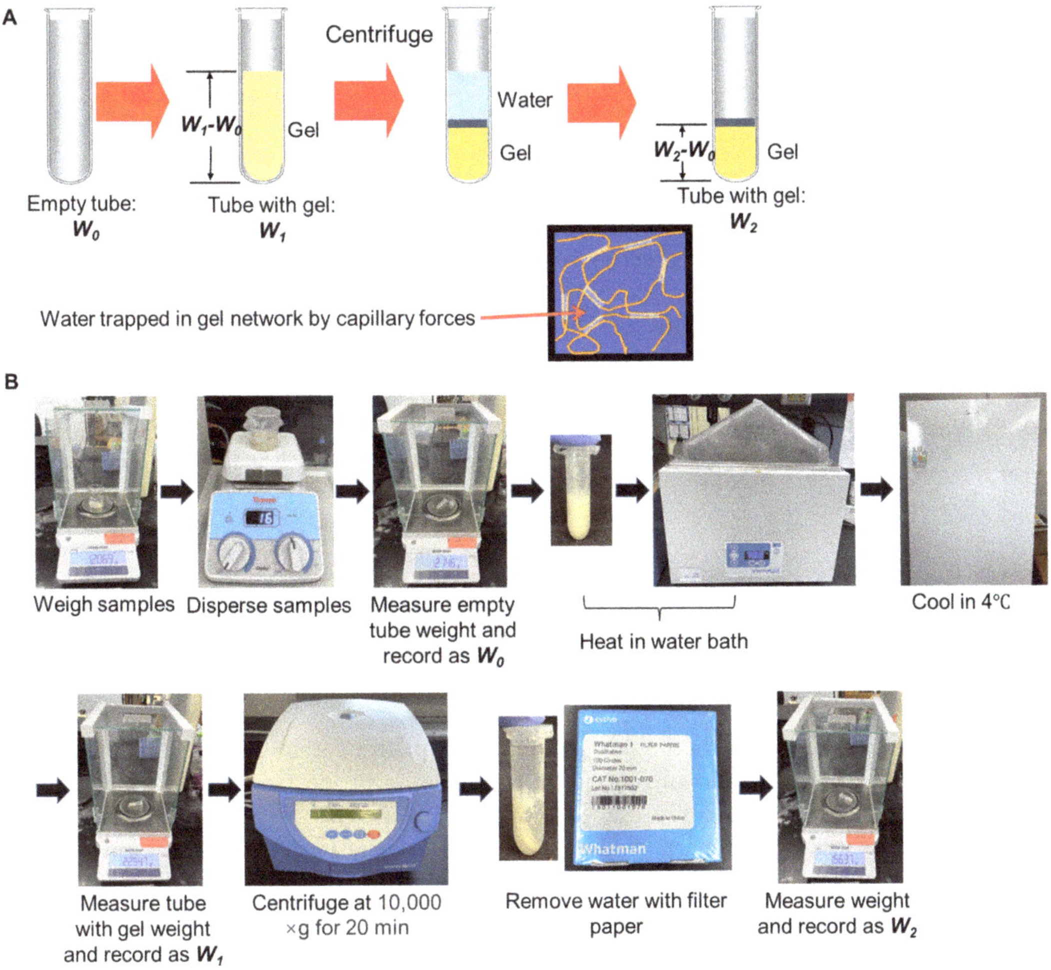

Fig. 2 Principle (**a**) and flowchart (**b**) demonstration for measuring water holding capacity of plant protein gel

3. Preheat the water bath to 85 °C (*see* **Note 6**). Transfer 2 mL of protein slurry to the pre-weighted centrifuge tubes, and then the centrifuge tube is covered by caps. Subsequently, submerge the centrifuge tubes in the preheated water bath at 85 °C for 15 min. Once the heating process is completed, samples are cooled down to room temperature immediately under tap water, and then the cooled glass test tubes are stored in a refrigerator at 4 °C.
4. Once the gel is formed, take the centrifuge tube out of the refrigerator, and then gently remove the water drops on the surface of the centrifuge tube using filter paper. Weigh the centrifuge tube again, and record it as W_1 (g).
5. Centrifuge the above mentioned tube at 10,000× *g* for 20 min. After the centrifugation, the released water in the upper layer is gently removed by filter paper. Weigh the centrifuge tube, and

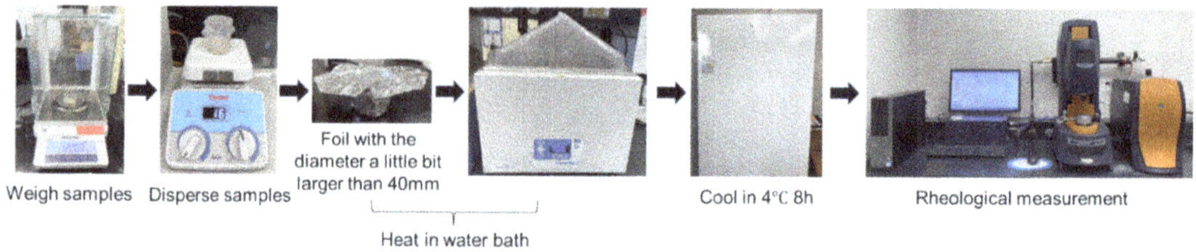

Fig. 3 Flowchart demonstration for measuring rheological properties of plant protein gel using a rheometer

record it as W_2 (g). The WHC is expressed as the percentage of weight lost before and after centrifugation.

6. Expression of Results.

The WHC is calculated using the following Eq. 1 and is expressed as a percentage (%):

$$\text{WHC}\,(\%) = \frac{W_2 - W_0}{W_1 - W_0} \times 100\% \tag{1}$$

Where: W_0 is the empty centrifuge tube weight (g), W_1 is the centrifuge tube with gel before centrifugation (g), and W_2 is the centrifuge tube with gel after centrifugation (g).

3.3 Gel Rheological Analysis

The rheological properties of plant protein gel are carried out by a rotational rheometer. Viscoelastic properties are the main properties to represent rheological properties of gel Fig. 3 elucidates a flowchart of analysis of the viscoelasticity of protein gel using a rotational rheometer according to Farooq et al. with some modifications [12].

1. The protein slurry (10%) preparation procedure is the same as in Subheading 3.2 **step 1**.
2. Prepare a cylindrical shape of flat aluminum foil containers with a diameter slightly larger than 40 mm (*see* **Note 7**).
3. Transfer 3.4 mL of 10% protein slurry to the prepared flat cylindrical shape foil covered with the flat aluminum container to obtain a ~2 mm thickness of gel (*see* **Note 8**). Submerge the bottom of the foil container in the preheated water bath at 85 °C for 15 min. Afterwards, the foil container is immediately cooled down under tap water for 20 min, and then stored in a refrigerator at 4 °C overnight (~8 h).
4. Once the foil container is taken out from the refrigerator, gently remove the gel from the container, and transfer the obtained thick gel piece to the rheometer plate (*see* **Notes 9** and **10**).
5. To determine the linear viscoelastic region, a strain sweep test needs to be run initially (*see* **Note 11**). The parameters for running a strain sweep test are shown in Table 1. The obtained data is shown in Fig. 4.

Table 1
Parameters to perform strain sweep test

Parameters	
Temperature (°C)	25
Soak time (s)	60
Frequency (rad/s)	6.28
Strain (%)	0.01–1000
Points per decade	5
Gap (mm)	2

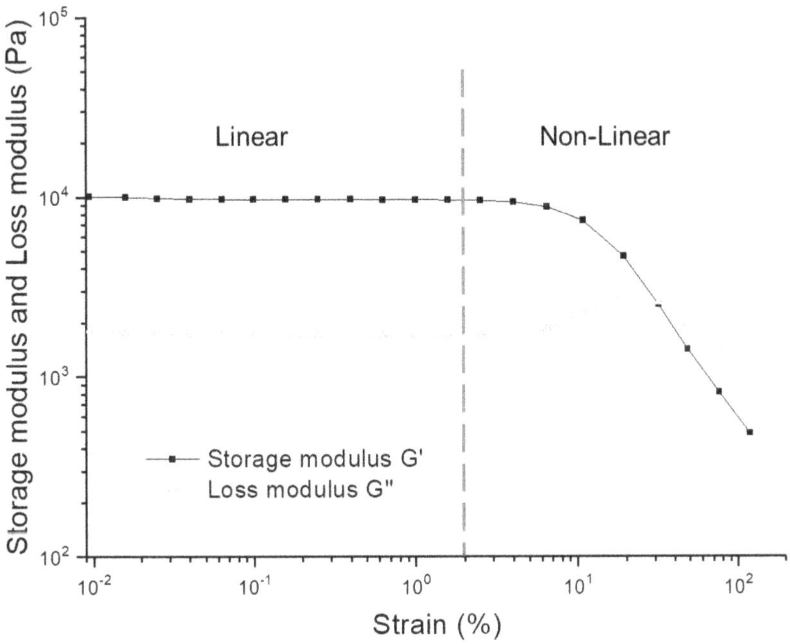

Fig. 4 Typical strain sweep curve (*see* **Note 12**)

6. Clean the plate and geometry, and then reload a new protein gel piece to the rheometer plate (*see* **Note 10**).
7. Select one strain (%) within the linear viscoelastic region obtained from **step 5**, and then a frequency sweep test is conducted. The parameters to run a frequency sweep test in the TA TRIOS software as shown in Table 2.
8. Expression of Results.

   The obtained data is shown in Fig. 5. The viscoelasticity of gel is expressed in the storage modulus (G') and loss modulus (G") as a function of angular frequency (*see* **Note 12**).

Table 2
Parameters to perform frequency sweep test

Parameters	
Temperature (°C)	25
Soak time (s)	60
Strain (%)	0.1
Frequency (rad/s)	0.1–100
Points per decade	10
Gap (mm)	2

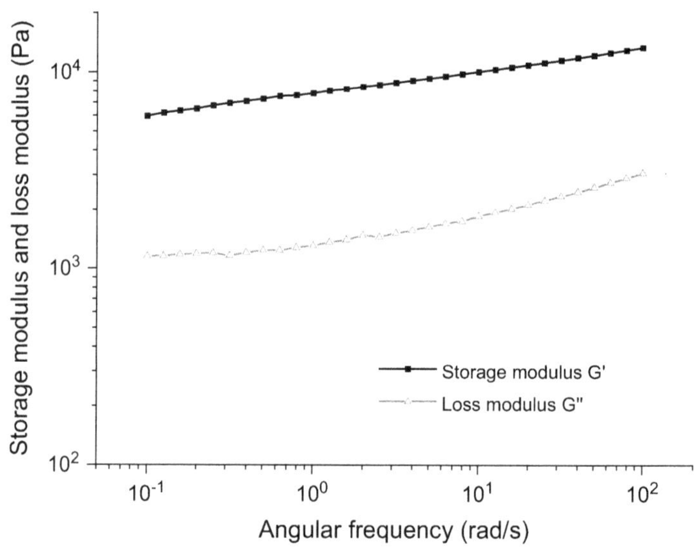

Fig. 5 Typical frequency sweep curve

3.4 Gel Texture Analysis

Gel texture analysis is performed using two types of experiments including a compression test and a rupture force test according to Qi et al. [13]. Figure 6 shows a flowchart demonstration for measuring texture properties of plant protein gel using a texture analyzer.

1. The protein slurry (10%) preparation procedure is the same as in Subheading 3.2 **step 1**. Please read the instructions on how to prepare a flat cylindrical shape foil covered with the flat aluminum container in Subheading 3.2 **step 2**.
2. Transfer 7 mL of 10% protein slurry to prepared flat cylindrical shape foil covered with the flat aluminum container to obtain a ~4 mm thickness of gel (*see* **Note 8**). Submerge the bottom of the foil container in the preheated water bath at 85 °C for

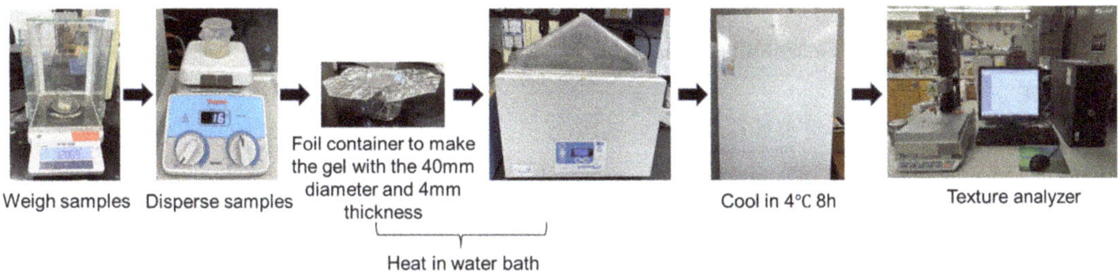

Fig. 6 Flowchart demonstration for measuring texture properties of plant protein gel using a texture analyzer

15 min (*see* **Note 6**). Afterwards, the foil container is immediately cooled down under tap water for 20 min, and then stored in a refrigerator at 4 °C overnight (~ 8 h).

3. Once the gel is formed, gently take the gel from the aluminum foil and transfer the obtained thick gel piece to the texture analyzer plate (*see* **Note 13**).
4. For the compression test, sample gel can be compressed with a cylinder probe of 25 mm. Set up the parameters in the Texture Expert Software for Windows Version 3.2 installed in the equipment as shown in Table 3.
5. Reload a new piece of gel to the plate and determine the gel rupture force with a cylinder probe of 2 mm. The parameters to run a test can be set up in the Texture Expert Software for Windows Version 3.2 that installed in the equipment as shown in Table 3.
6. Expression of Results.

   Curves are analyzed with the software to obtain the following texture profile: hardness (N), rapture force (N), cohesiveness, resilience, and springiness.

4 Notes

1. Water bath temperature may vary depending on the plant protein types, concentrations, and environmental conditions. In general, the gelation temperature is higher than the denaturation temperature of the corresponding plant protein.
2. A cap is necessary for thermal-induced protein gel formation to prevent moisture evaporation from the protein slurry during heating process in a water bath.
3. A parallel geometry with the angle of the parallel geometry 0° is selected for the experiment.
4. The LGC of plant protein is strongly influenced by the different protein materials and environmental conditions such as salts, solvents, heating temperature, and time. Before selecting

Table 3
Parameters to perform texture analysis tests

Parameters	
Compression test	
Load cell (kg)	5
Return distance (mm)	20
Return speed (mm/s)	20
Contact force (N)	0.005
Pretest speed (mm/s)	2
Test speed (mm/s)	2
Posttest speed (mm/s)	2
Rupture force test	
Load cell (kg)	5
Return distance (mm)	50
Return speed (mm/s)	20
Contact force (N)	0.005
Pretest speed (mm/s)	5
Test speed (mm/s)	5
Posttest speed (mm/s)	5

the heating temperature and time, an approximate understanding of the examined proteins is necessary to select the gel setting conditions.

5. If the examiner is not familiar with the protein, they can try the concentration range of this book chapter first since this range covers most of the plant protein. Extending the concentration range is acceptable if necessary.
6. After the LGC of the tested protein is determined, the protein concentration, heating temperature, and time for formation of thermal-induced plant protein gel can be varied based on your experiment design.
7. The diameter of geometry in the experiment is generally 40 mm. A large size of geometry is required for large gel sample preparation.
8. In general, 2 mm and 4 mm thick plant protein gel can be formed by using 3.4 mL and 7 mL of 10% protein slurry respectively, with a diameter slightly over 40 mm. Adjustment is necessary for different plant proteins and conditions.

9. Before transferring the gel pieces onto the rheometer plate, the rheometer and geometry need to be calibrated, then zero the gap.
10. After transferring the gel pieces, the operator needs to trim the gap to remove the excess gels by cutting them with a plastic card.
11. The linear viscoelastic region (LVE region) indicates the range in which a test can be carried out without destroying the microstructure of the sample. In the LVE region, G' and G'" show a constant value as a function of strain (%), the so-called plateau value. The tolerance range of deviation for G' around the plateau value at 5% deviation can be selected as a transition point from linear viscoelastic to nonlinear viscoelastic region according to the standards ISO 6721-10 and EN 14770 [14, 15].
12. Storage modulus (G') represents the gel elasticity and loss modulus (G") represents the gel viscosity.
13. Please notice that the edge of the gel may be irregular after removing the foil container. It can be trimmed to ~40 mm diameter of spherical shape piece using a knife.

Acknowledgments

This work is supported by the ND Specialty Crop Block Grant Program (FAR0034202).

References

1. Moure A, Sineiro J, Domínguez H, Parajó JC (2006) Functionality of oilseed protein products: a review. Food Res Int 39(9):945–963. https://doi.org/10.1016/j.foodres.2006.07.002
2. Sun XD, Arntfield SD (2010) Gelation properties of salt-extracted pea protein induced by heat treatment. Food Res Int 43(2):509–515. https://doi.org/10.1016/j.foodres.2009.09.039
3. Cao Y, Mezzenga R (2020) Design principles of food gels. Nat Food 1(2):106–118. https://doi.org/10.1038/s43016-019-0009-x
4. Nicolai T (2019) Gelation of food protein-protein mixtures. Adv Colloid Interf Sci 270: 147–164. https://doi.org/10.1016/j.cis.2019.06.006
5. Peyrano F, de Lamballerie M, Avanza MV, Speroni F (2021) Gelation of cowpea proteins induced by high hydrostatic pressure. Food Hydrocoll 111:106191. https://doi.org/10.1016/j.foodhyd.2020.106191
6. Banerjee S, Bhattacharya S (2012) Food gels: gelling process and new applications. Crit Rev Food Sci Nutr 52(4):334 346. https://doi.org/10.1080/10408398.2010.500234
7. Fang B, Gu Z, Ohm JB, Chen B, Rao J (2023) Reverse micelles extraction of hemp protein isolate: impact of defatting process on protein structure, functionality, and aromatic profile. Food Hydrocoll 135:108158. https://doi.org/10.1016/j.foodhyd.2022.108158
8. Nieuwland M, Bouwman WG, Pouvreau L, Martin AH, de Jongh HH (2016) Relating water holding of ovalbumin gels to aggregate structure. Food Hydrocoll 52:7–94. https://doi.org/10.1016/j.foodhyd.2015.06.018
9. Ramli H, Zainal NFA, Hess M, Chan CH (2022) Basic principle and good practices of rheology for polymers for teachers and

beginners. Chem Teach Int 4(4):307–326. https://doi.org/10.1515/cti-2022-0010

10. Fang B, Chen B, Rao J (2023) Effect of protein concentration on the structural, functional properties, linear and nonlinear rheological behaviors of thermally induced hemp protein gels. J Food Eng. https://doi.org/10.1016/j.jfoodeng.2023.111694
11. Yang J, Zamani S, Liang L, Chen L (2021) Extraction methods significantly impact pea protein composition, structure and gelling properties. Food Hydrocoll 117:106678. https://doi.org/10.1016/j.foodhyd.2021.106678
12. Farooq S, Ahmad MI, Zhang Y, Zhang H (2023) Impact of interfacial layer number and Schiff base cross-linking on the microstructure, rheological properties and digestive lipolysis of plant-derived oil bodies-based oleogels. Food Hydrocoll 2023:108473. https://doi.org/10.1016/j.foodhyd.2023.108473
13. Qi X, Simsek S, Chen B, Rao J (2020) Alginate-based double-network hydrogel improves the viability of encapsulated probiotics during simulated sequential gastrointestinal digestion: effect of biopolymer type and concentrations. Int J Biol Macromol 165:1675–1685. https://doi.org/10.1016/j.ijbiomac.2020.10.028
14. ISO 6721-10 (2015) Plastics-determination of dynamic mechanical properties-part 10: complex shear viscosity using a parallel-plate oscillatory rheometer, 3rd edn. International Organization for Standarization, Geneva
15. EN 14770 (2012) Bitumen and bituminous binders – determination of complex shear modulus and phase angle – Dynamic Shear Rheometer (DSR). Europeen Standard, Berlin

Chapter 29

Rheological Properties

Jun Yang and Prabhakar Kasturi

Abstract

Plant-based and alternative protein foods include processed foods derived from legumes, lentils, grains, fruits, vegetables, nuts and seeds, and their derived ingredients. The physical properties of the plant proteins and the final products are vital, since food processing is associated with a complex flow process. Rheological analysis is therefore an essential tool for evaluating constituent ingredients at different stages of the food system. The rheological properties of plant proteins could be fully characterized using various tools such as RVA, Mixolab, and Rheometer. These analytical tools can analyze the behavior of a broad spectrum of flour types, including white flour, whole grain, rice, corn, sorghum, soy, flax, amaranth, potato, buckwheat, quinoa, millet, teff, lentil flours, and legume proteins.

Rheological properties are crucial for the proper use of plant-based protein ingredients in food processing. Mixolab is useful in characterizing thermo-mechanical properties of novel plant protein or protein flour mixtures. The behavior of starch and protein along with their intra-molecular interactions, and with water can be captured to reflect mixing torques force, protein weakening, starch gelatinization, and starch retrogradation. In this chapter, three protein flours (semolina, chickpea, chickpea with emulsifier) have been characterized during mixing and cooking process to evaluate their thermos-mechanical profiles.

Rheological analysis can expedite innovation time for new products to market. The data obtained from rheological measurement can provide critical insights during formulation or reformulation, validating baseline ingredient performance and quality before scale-up. It is also used to screen incoming ingredients to strengthen supply chain, enabling the optimization of dry blends in baking, extrusion, frying, and compression popping platforms.

Key words Plant-based proteins, Raw ingredients, Rheological characterization, Rapid Visco Analyser (RVA), Mixolab Dough Analysis

1 Introduction

As consumers alter their dietary patterns to incorporate more plant-based protein products, for the purpose of health, animal welfare, and/or sustainability, manufacturers may face a series of new challenges. One of the biggest difficulties is the unique textures of plant

Jun Yang is an employee of PepsiCo Inc. The views expressed in this manuscript are those of the author and do not necessarily reflect the position or policy of PepsiCo Inc.

Yonghui Li (ed.), *Plant-Based Proteins: Production, Physicochemical, Functional, and Sensory Properties*, Methods and Protocols in Food Science, https://doi.org/10.1007/978-1-0716-4272-6_29,

protein derived products. It is a challenge for manufacturers to develop formulas/recipes that allow for plant-based protein nutrition without compromising on the texture of the final product. The measurement of rheological properties of raw ingredients and doughs play a critical role in evaluating the texture attributes in plant protein product, and for end product quality.

Each plant protein source differs greatly in carbohydrate, starch, oil, and protein profiles; therefore, evaluation of their performance and influence is crucial in developing and manufacturing plant-based food products. Rapid Visco® Analyzer (RVA) can be applied to characterize cooking and pasting behaviors of plant-based food systems to examine the relationship between moisture absorption, material phase change, and temperature profile change. The RVA could be used as a customizable platform to conveniently investigate ingredient performance and measure the effects of processing conditions. Other uses include ingredient screening, process monitoring, product development, and quality assurance [1, 2].

Beans, peas, and lentils are nutrient-rich, and are economical source of protein, dietary fibers, vitamins, minerals, and antioxidants. These make them ideal ingredients for novel products that satisfy consumer demands for healthy foods. Using pulse-based ingredients is challenging as pulse crops quality is hard to predict since their nutritional compositions can vary greatly with growing location, seasons, and environment. This, the addition of these ingredients into traditional formulations delivers a challenge for food manufacturers. These ingredients can greatly affect the processing characteristics and the resulting performance of traditional formulations. RVA can be served as a robust tool to monitor various batches, reformulation, and processing conditions modifications. This versatility helps ensure consistency, controllable quality, high-quality products, and adequate understanding of variables impacting new product development. RVA can determine plant protein quality by measuring the performance of samples during hydration, pasting, gelatinization, gelling, and retrogradation by continuously collecting viscosity data from a user-designed cycle of temperature and mechanic shear, which allows the user to evaluate ingredient behavior. Other uses of the RVA include testing new formulations for plant protein foods and being able to assess the effect of various plant protein ingredients on the performance of a mix for an efficient integration of plant proteins into baked, fried, extruded, and compression popped products [3].

The main challenge for using pulse crops-based ingredients in food formulations is ensuring consistent behavior during the processing; the composition of pulse ingredients can vary greatly due to cultivation conditions and variety. Further, process conditions greatly affect the quality of products using plant or alternative protein ingredients. Their different characteristics affect the ability of manufacturers to produce products with desirable texture attributes to consumers. Analytical tools such as RVA for plant proteins are constantly evolving, which provides an effective way to track the impact of changing process inputs on ingredients and its corresponding performance to fully characterize ingredients' attributes. RVA allows the flexibility of using both standard and custom-made test profile for evaluating plant protein ingredients and products. It also provides speed, precision, and automation for product testing, which makes it a unique tool for product development, process control, and quality assurance. Pursuing improvements in processing regiments, maintaining quality control, and characterizing plant protein performance in doughs and finished products is still a major focus for the food industry. Advanced analytical tools including RVA, Mixolab, and Rheometers offer the technological edge for suppliers and manufacturers to quantify ingredient quality and measure the crucial effect of changing process inputs on finished products [4, 5].

Mixolab is a dough testing instrument utilized to evaluate the quality and performance of hydrated ingredients under shear and thermal treatments. It is used to characterize the rheological behavior and pasting properties of doughs as a result of the interactions of its protein and starch components with water under simultaneous applications of mixing and heating. A dough characterizer such as Mixolab provides a comprehensive analysis of dough behavior during the mixing, heating, and cooling cycle. It also assesses dough quality criteria through the entire dough production cycle containing water absorption, effects of mixing (mixolab—development and stability), gluten resistance to heat, maximum viscosity, and starch retrogradation. The instrument is utilized to assess the quality and process performance of flours, and understand dough attributes. The test profiles can be modified to simulate the constraints that the actual dough undergoes during the production process. This allows mimicking the actual process conditions to assess the dough consistency, and quality of protein and starch components. Ingredient manufacturers can use the data generated from Mixolab testing to examine the quality of their materials, evaluate enzymatic effects, and manipulate new formulations (i.e., high protein and fiber formulations, and gluten-free) [10, 11].

The Mixolab, which uses the same principles of the farinograph and viscoamylograph, generates a single graph that can help illustrate the individual graphs generated from RVA, farinograph, mixograph, and viscoamylograph testing. It can use a customized testing protocol to evaluate dough behavior as temperature increases and decreases, to classify dough based on comprehensive quality parameters using profiler system, to exhibit a reconstituted Farinograph® curve and provide fully comparable results using simulator protocol, and to feature intuitive automated testing for a complete assessment within 45 min. The potential benefits of using a Mixolab testing includes: (1) provides a holistic picture of the entire flour quality among water, protein, and starch as it uses actual doughs during testing; (2) can be used to design specifications based on specific processes; and (3) analyzes dough behaviors with various raw materials through formulation adjustment.

The Mixolab test measures the dough consistency (in terms of torque, Nm) using two mixing blades, while subjecting the dough to the dual stress of mixing and temperature changes. The test is based on the preparation of a constant dough sample weight hydrated to obtain a target consistency during the 1st test phase. In the "Chopin+" as a standard protocol, the dough weight is set at 75 grams, while the target consistency (at C1) is set at 1.1 ± 0.05 Nm [12].

The "Chopin+" protocol is typically applied for characterizing ingredient quality, especially for plant proteins. MixoLab determines the change in torque forces as a function of temperature and time during dough development. It tracks down the changes in viscoelastic attributes among protein, starch, water, and other compositions upon heating and continuous mixing.

Using Mixolab testing benefits ingredient suppliers and manufacturers through understanding water amount adjustments and dough mixing parameters, detecting enzyme activity (e.g., alpha-amylase) in ingredients, assessing starch damage, predicting ingredient behavior during thermal processing, examining hot/cold functionality of starches and proteins, and evaluating the effects of different additives (e.g., dough conditioners) on dough mixing attributes [13–15].

2 Materials

2.1. All tests will be performed with a Rapid Visco® Analyzer (Perten RVA 4500, PerkinElmer Inc.) and its corresponding accessories. Chickpea flour is used to run RVA testing.

2.2. Semolina flour (SF), Chickpea flour (CPF), Chickpea flour with varying levels of emulsifier (CPF + emulsifier) were selected as ingredients for dough characterization using Mixolab 2.

3 Rapid Visco® Analyzer (RVA)

The Rapid Visco® Analyzer (RVA) is a thermal viscometer that combines stirring, ramped temperature, and variable shear profiles. It is optimized for testing viscous properties of starch and similar temperature-dependent hydrocolloids. Typically, RVA 4500 is applied under 100 °C; while RVA 4800 can perform both normal tests below 100 °C and high-temperature tests up to 140 °C in a specially designed pressure vessel. This method can be used for characterizing ingredient quality. RVA measures the change in viscosity of ingredients (i.e., flours, starches, plant proteins) as a function of temperature and time under simultaneous heating and continuous mixing. Evaluating the pasting properties of plant-based ingredients gives useful information in measuring hydration, swelling, thermal history, and interaction of ingredients components [6, 7].

3.1 Safety Precautions

Personal protective equipment (PPE) must be worn at all times. For this procedure, PPE includes, but is not limited to:

- Safety goggles to prevent chemicals from burning the eyes.
- Any other equipment the analyst thinks is necessary.

3.2 Method Performance

According to manufacturer's specifications: Accuracy: ± 2% for S2000 oil nom. 5000 cP

3.3 Equipment/Instrumentation/Apparatus

- Analytical Balance (three decimal place minimum).
- RVA - Perten Series 4500 (see illustration) with an associated computer.
- Thermocline for Windows software (version 3.15.3.347).
- Aluminum canisters and paddles.
- Weigh boat.
- Spatula/spoon.

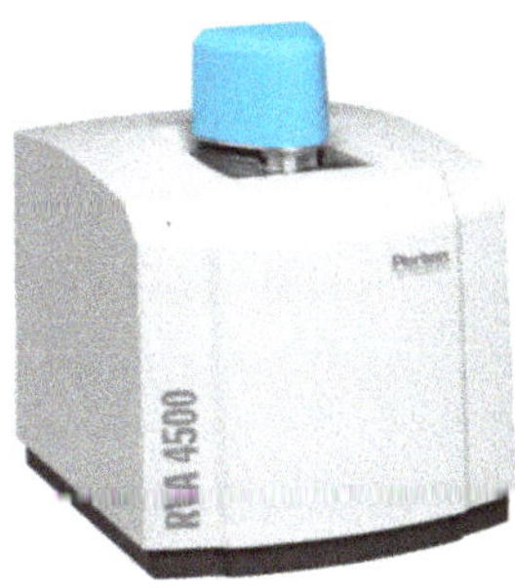

3.4 Reagents and Solutions

Reagents

- Deionized (DI) Water.

Solutions

- Not Applicable.

Calibration

Calibration of the RVA instrument is performed annually by an approved vendor.

3.5 Procedure

3.5.1 Sample Preparation

1. Grind the sample to pass through 1 mm sieve (below) to control the particle size.

2. Measure the moisture content of the sample using TGA/TGM thermobalance, HX-204, etc.
3. Calculate the sample weight to use. This weight is based on moisture content.
4. Weigh sample into a weigh boat.
5. In an aluminum canister, weigh enough water to equal a total weight of 28 ± 0.005 g when the sample and water are combined. For example, if the weight of the sample is 4.50 g, the amount of water required would be 23.50 g.
6. Load sample when the equipment preparation is ready.

3.5.2 Equipment Preparation

1. Make sure that cooling water circulation system is "on."
2. Connect RVA equipment with software in an associated computer. Allow the RVA to warm up for 30 min prior to use.
3. Start the RVA Thermocline for Windows by selecting the icon [TCW] from the window screen.
4. Select [Run] from the dropdown menu at the top of the screen and select [Zero Instrument].
5. Insert paddle into the RVA motor. The paddle is in place when the tabs are aligned, and it "clicks" into place.
6. Select [Start] on the computer screen.
7. When the "zero" section is stable, select [Zero], [Stop], and [Close]. See illustration below.

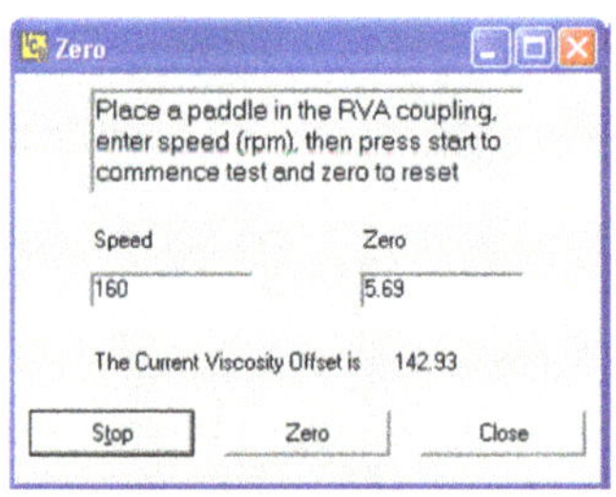

8. Under the "Run" dropdown at the top of the screen, click on [Select Configuration to Run], then select the appropriate test configuration file [rvc]. See illustration below.

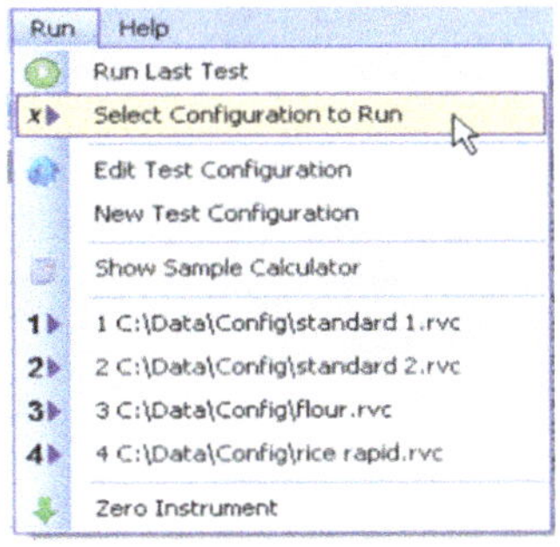

9. Select appropriate folder for the file to be saved and enter a unique filename. Select [Save].
   - A prompt to lower the sample into the viscometer will appear in the screen. Now the system is ready to run the test.

3.5.3 Analysis

1. Add the weighed sample into the aluminum canister containing water. The testing of the sample must be done immediately after the water addition.
2. Make sure that the paddle rotates freely without touching the inside of the sample container when attached. The paddle is in place when the tabs are aligned, and it "clicks" into place. See illustration below.

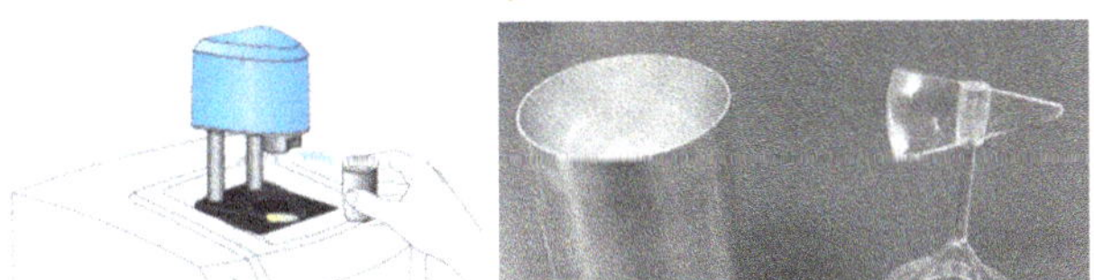

3. Press the tower into the RVA by using a smooth motion and stopping when the sample "clicks" into place. This automatically starts the instrument cycle. See illustration below.

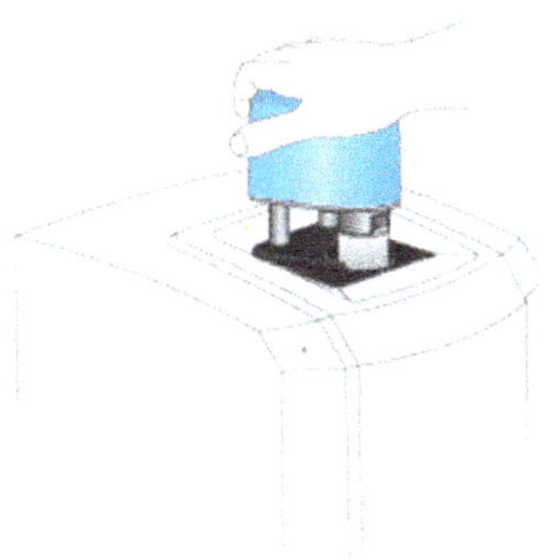

4. When the analysis is completed, the RVA tower will automatically rise from the RVA. The RVA curve will display the result of *Peak Viscosity*, *Trough Viscosity*, and *Final Viscosity*. Record these

results on the Viscosity of Ingredients by RVA Data Sheet (Appendix A).

5. If the temperature profile does not automatically appear on the display, double click on the "viscosity" axis on the left side and a menu will pop up. Select [Temperature] under the available "y-axis" dropdown (see illustration below). Select [Close].

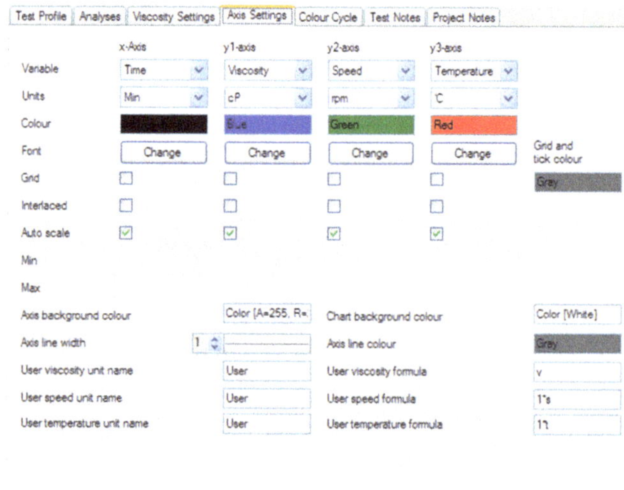

3.5.4 Chickpea Flour RVA Testing

For chickpea flour, use 5.54 g sample at AS IS (8.95%) moisture in 28.0 g of total suspension to obtain 18% solid content. Use distilled or deionized water as the solvent. The sample size was controlled using sieve size of 1 mm. The time and temperature profile for chickpea flour is in the Table 1.

The chickpea weight was calculated in Table 2. The solid content was set 18% in total 28 grams of slurry.

Manual Calculation

$$S_1 = \frac{S_0(100 - M_0)}{(100 - M_1)}$$

$$W_1 = W_0 + S_0 - S_1$$

Where:

S_1 = corrected sample weight (g)

M_0 = reference moisture content of sample (%)

M_1 = actual moisture content of sample (%)

S_0 = standard sample weight (g)

W_1 = corrected water weight (g)

W_0 = standard water weight (g)

Table 1
The customized temperature profile for chickpea flours

Time (HH:MM:SS)	Type (temp speed)	Value (C) or (RPM)
00:00:00	Temp	50
00:00:00	Speed	960
00:00:15	Speed	160
00:02:00	Temp	50
00:07:00	Temp	95
00:12:00	Temp	95
00:16:00	Temp	50
00:17:00	Temp	50
00:17:00	End	

End of test time (HH:MM:SS): 00:17:00
Initial ideal temperature (Celsius): 50 ± 1 °C
Time between readings (sec.) 2

Example Using 0% Moisture Basis (Dry Basis)
A chickpea flour sample has been submitted for RVA analysis. Using a moisture balance, it is determined to have a moisture content of 8.95%. Using the guide, 18% solids will be used in the RVA slurry; total slurry (sample + DI water) weight will be 28 g.

$$M_0 = 0\%$$

$$M_1 = 8.95\%$$

$$S_0 = 18\% \text{of } 28 \text{ g} = 5.04 \text{ g}$$

$$W_0 = (100 - 18)\% \text{of } 28 \text{ g} = 22.96 \text{ g}$$

$$S_1 = \frac{5.04 * (100 - 0)}{(100 - 8.95)} = \frac{504}{91.05} = 5.54 \text{ g}$$

$$W_1 = 22.96 + 5.04 - 5.54 = 22.46 \text{ g}$$

The RVA curve is shown in Fig. 1, and results were recorded in Table 3.

In starch RVA analysis, the starch is heated in an aqueous system after the addition of water. Starch granules absorb water and swell; furthermore, the internal crystalline structure becomes gelatinized, in which the granule itself has broken down and formed gel. The viscosity changes generated by heating and cooling starch in water generally produce a characteristic pasting curve.

Table 2
The chickpea flour weight calculator

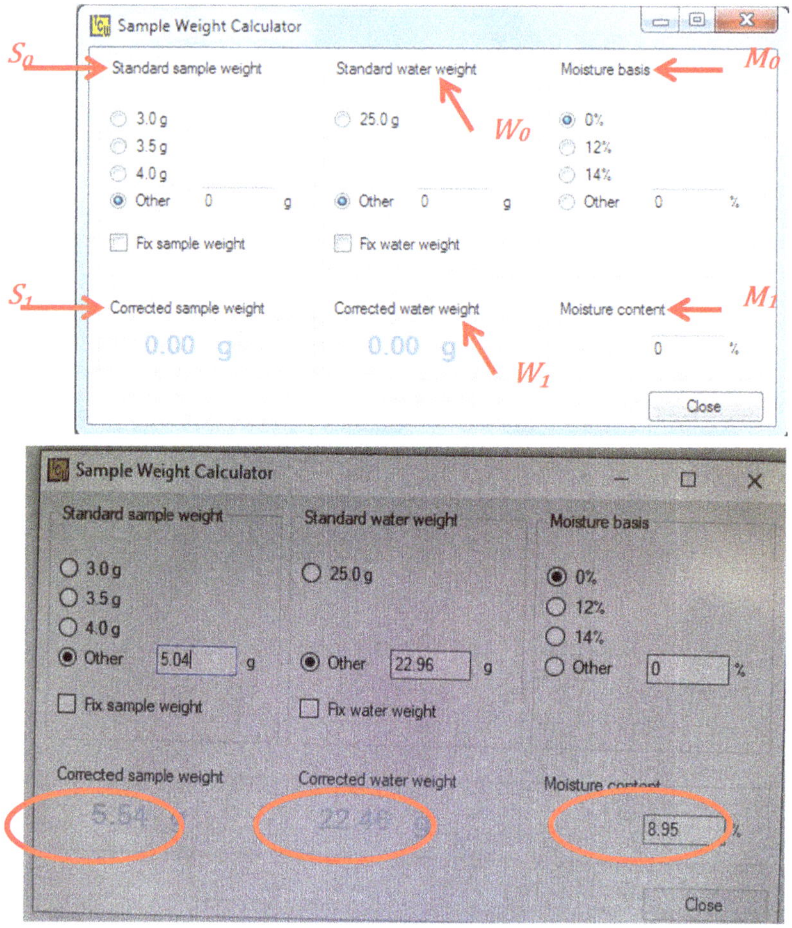

Pasting temperature is an indication of the minimum temperature required to cook, stability of other components in the system, and potential energy costs [8].

The Peak Viscosity (PV), which is determined at various maximum hold temperatures provides an indication of the sample's ability to paste under those conditions. The PV value indicates the water-holding capacity of the starch or mixture. It is influenced by factors such as sample purity and the presence of matrix proteins and dietary fibers, which may restrict viscosity development of the pulse starch granules. The PV is well correlated with final product quality and provides an indication of the viscous load likely to be encountered by a mixing cooker.

The trough viscosity (TV) reflects the rate of breakdown in viscosity in relation to holding strength, and hot paste viscosity. The TV value is associated with the nature of material, temperature,

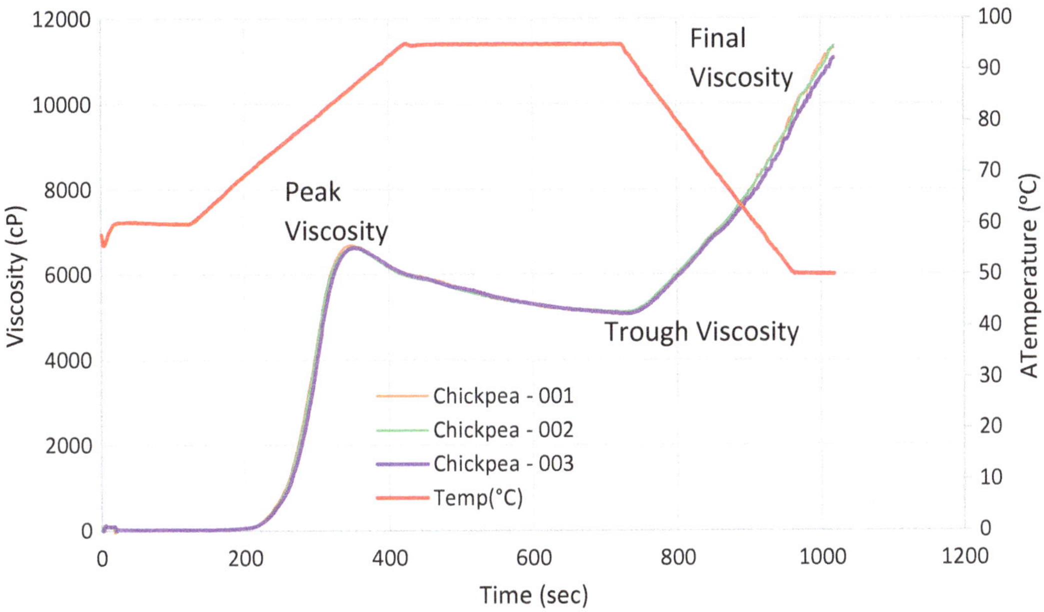

Fig. 1 RVA curve for chickpea flours

Table 3
RVA data sheet

Sample ID	Moisture content (%)	Peak viscosity (cP)	Trough viscosity (cP)	Final viscosity (cP)
Chickpea flour—001	8.95	6695.00	5064.00	11291.00
Chickpea flour—002		6623.00	5127.00	11357.00
Chickpea flour—003		6640.00	5078.00	11065.00

degree of mixing/shear stress. The ability of a material to withstand this heating and shear stress is an important attribute for many processes.

The Final Viscosity (FV) at the end of the test is an indicator of the sample's ability to set after cooling. The re-association among starch molecules during cooling is referred to as the setback. It involves retrogradation, which is the re-ordering, of the starch molecules, and has been correlated with texture of various finished products. The FV value is influenced by factors such as the degree of starch pasting and amylose content in the sample. For instance, FV is typically higher with more amylose leaching. This is commonly used parameter to measure the product's quality [9].

4 Mixolab Dough Analysis

4.1 Principle

The Mixolab generates a graph that reflects mixing, absorption, gluten, gelatinization, amylase activity, and retrogradation characteristics of the dough during heating/cooling and mixing conditions. These characteristics are measured through the five phases of dough behavior testing done in the standard "Chopin +" Mixolab test. In addition, the dough development temperatures and times corresponding to the torque values measured are also taken [12].

Phase I is defined as the initial mixing step wherein a constant temperature of 30 °C (86 °F) is maintained for 8 min to assess dough development and weakening. The parameters consist of a maximum consistency (C1 point in curve), time to reach C1, and dough stability. The dough stability is measured as the time (mins) elapsed in which torque is higher than maximum consistency. The water absorption (%) allowed for the sample is also determined through the C1 value as the water absorption used should lead to doughs with a max consistency of 1.1 ± 0.05 Nm during the phase 1 (8 mins).

Phase II measures dough weakening during heating from 30 °C to 60 °C (86–140 °F). It evaluates gluten weakening as a function of torque reduction by a decrease in dough consistency. The relevant variables from this phase are minimum consistency (C2 point in curve) and dough temperature at minimum consistency. The C2 value, which is taken at the lowest point of the curve, is used to determine protein weakening as a function of mechanical work and temperature. This value is often used to determine the quality of the proteins present in the flour.

Phase III represents starch gelatinization during the heating step at 60–90 °C (140–194 °F). The heat applied causes the starch granules to swell and leach out after absorbing water. The starch gelatinization parameters evaluated here are minimum consistency (C3 point in curve), and dough temperature at minimum consistency. The C3 value shows the maximum torque observed during the heating phase, which is a measure of gelatinization performance.

Phase IV displays dough stability during heating at 90 °C (194 °F). The important parameter is the hot gel stability minimum consistency during cooking (difference between minimum consistency of phase III at point C3 and phase IV at point C4).

Phase V assesses starch retrogradation in the dough during the cooling phase from 90 °C to 60 °C (194–140 °F). The parameter is the increase in torque between minimum consistency of phase IV and maximum consistency of phase V (C5 point in curve).

Aside from the diverse information obtained from the standard "Chopin +" protocol, the Mixolab test profiles can also be customized by varying test parameters such as mixing speed, temperature, water absorption, etc., according to the intended product design. This allows the Mixolab test method to be flexible and adaptable to the user's needs [13, 16, 17].

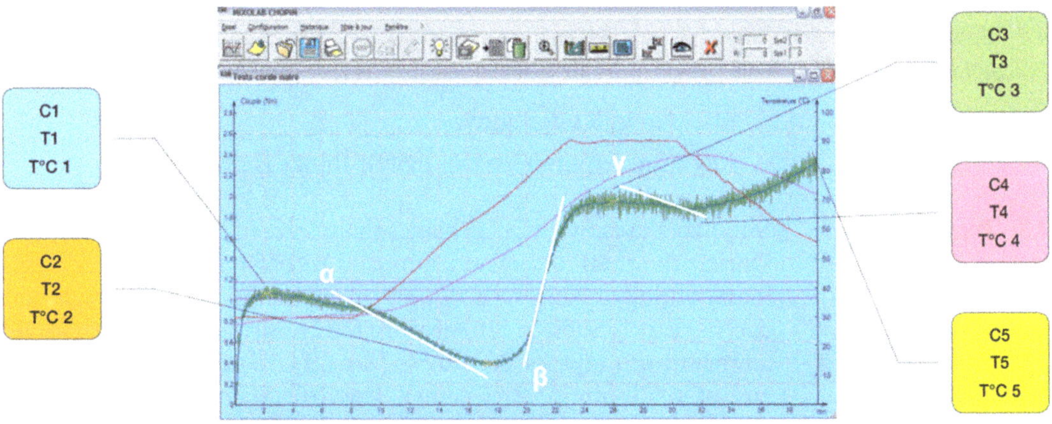

Source: Mixolab Applications Handbook [3]

Aside from these major values, (C1, C2, C3, C4, and C5), Mixolab tests also generates additional important quality measurements for the dough. Water absorption (%) indicates the amount of water needed to obtain C1 = 1.1 ± 0.05 Nm. A sample can absorb amount of water to obtain a given consistency during the phase at constant temperature.

Time (min) required to obtain C1 is the dough development time; hence, the stronger the test dough material, the longer the development time needed.

Dough stability (min) is a time during which the upper frame is greater than C1—11% (phase at constant temp). It reflects mixing resistance of dough: The longer this time is, the more the sample to be strong. Amplitude (Nm) is the width of curve to C1, which represents dough elasticity, the higher the value, the greater the elasticity of the sample.

Other parameters contain slope α, β, and γ. Slope α is a slope of curve between end of period at 30 °C and C2, suggesting the rate of protein weakening speed under the effect of heat and mixing. Slope β is a slope of curve between C2 and C3, indicating starch gelatinization speed. Slope γ is a slope of curve between C3 and C4, implying enzyme degradation speed. Compared to C1 ~ C5, the slopes are very tricky and susceptible small changes.

Gluten serves as a dividing line between a functional and a non-functional flour. Gluten is extensible and retains gas to form a proper shape and texture. Wheat flour protein levels correlate to gluten content. Hard wheat consists of a higher content of protein and gluten than soft wheat. Typically, baked products such as bread, bagels, and even pretzels depend on gluten and a higher protein level (12–14%) for structure in comparison with pastries produced from pastry flour (8–9% protein) and cookies or cakes (5–8% protein). Therefore, when combining alternative protein ingredients with wheat flour, the protein level and gluten content should be taken into consideration for proper dough development.

4.2 Safety Precautions

Personal protective equipment (PPE) must be worn at all times. For this procedure, PPE includes, but is not limited to:

- Safety goggles and gloves.
- Any other equipment the analyst thinks is necessary.

4.3 Performance

Mixolab 2 uses two mixing blades that rotate in opposite directions. During the rotation, it is tracking and recording the dough consistency during mixing in a dough mixer that also has heating and cooling capabilities. It can measure the consistency of dough in the temperature range of 30–90 °C (86–194 °F) during heating at a rate of 4 °C/min (7.2 °F/min). The instrument determines characterization of the gluten-forming proteins and starch functionality in a single assay, which dough consistency over the temperature range has been assessed. These functions can be performed by measuring ingredient resistance to mixing action of the kneading arms (torque in Nm), recorded by the computer software in a graph as changes in viscosity with time.

In terms of application of Mixolab, according to AACC International Official Method 54–60.01 (Cereals & Grains Association), the Mixolab test includes the following steps:

1. Sample moisture should be determined prior to testing. Instrument software can adjust sample weight according to the moisture value on a Dry matter/base 14%/base 15%/as-is moisture basis. The sample moisture content and desired water absorption is then input. The software can then calculate sample weight and water-added weight, respectively, based on the protocol applied.
2. The test sample is weighed out and place it into the Mixolab dough mixer. The water injector is positioned. Usually, a preliminary test is conducted to determine if the selected water absorption is adequate for the run.
3. The test is started, and verified that, during the first 8 min, the torque measured by sensor is within 1.10 ± 0.05 Nm (C1 point in curve). If in the range of torque force, testing is allowed to continue. Otherwise, stop the test and repeat procedure with an adjusted new water absorption. Once the target C1 value is achieved, the instrument continues and finish the test.

4.3.1 Equipment/ Instrumentation/Apparatus

- Analytical Balance (three decimal place minimum).
- Mixolab 2—CHOPIN Technologies (see illustration) with an associated computer.
- Mixolab software (MIXOLAB CHOPIN [4.1.2.5]).

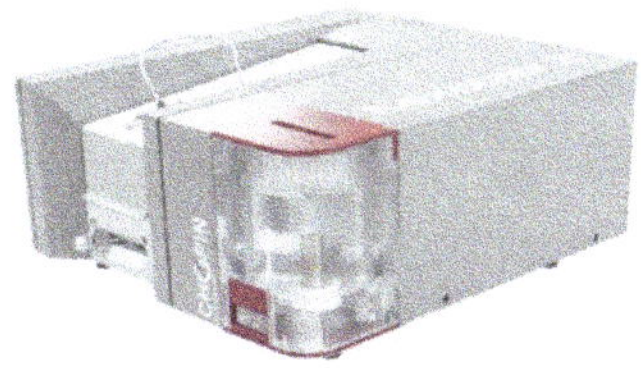

- Special funnel.

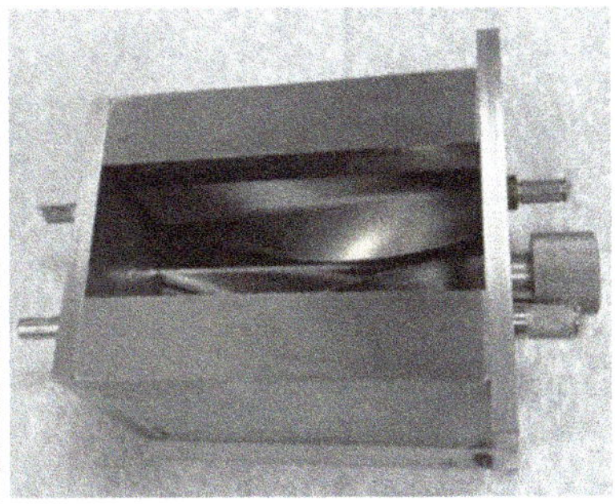

- Designed dough mixer.
- Weigh boat.
- Spatula/spoon.

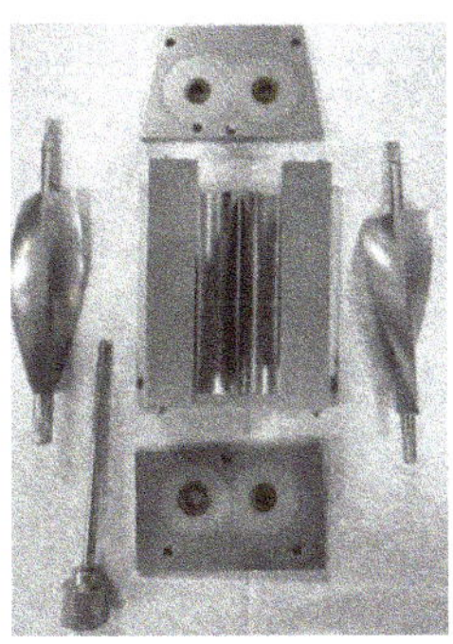

Table 4
The Mixolab target values for control flour

	C1	C2	Temp (°C) at 20 min	C3	C4	C5
Min	1.05	0.40	67.7	1.62	1.22	1.58
Target	1.10	0.48	68.7	1.82	1.46	2.1
Max	1.15	0.56	69.7	2.02	1.70	2.64

4.3.2 Reagents and Solutions

Reagents

- Deionized (DI) Water.

Solutions

- Not Applicable.

Calibration

Calibration of the Mixolab 2 instrument is performed annually by an approved vendor. The material used for calibration is "Control Flour." It is recommended to be stored in dry conditions away from heat. The Mixolab target values (13.9% moisture content, and hydration/water absorption (b14%) is 59.4%) are in Table 4. The target values could be varied batch by batch. A proper way to obtain fresh samples is to contact the equipment manufacturer (CHOPIN Technologies).

4.4 Procedure

4.4.1. Sample Preparation.

4.4.1.1. Measure the moisture content of the sample using TGA/TGM thermobalance, HX-204, etc.

4.4.1.2. Calculate the sample weight to use (see illustration below), which is based on internal moisture content and water absorption percentage.

4.4.1.3. Weigh sample into a weigh boat.

4.4.2. Sample Testing.

4.4.2.1. Connect Mixolab 2 equipment with software in an associated computer. Allow the Mixolab 2 to warm up for 30 min prior to use.

4.4.2.2. Make sure that cooling water circulation system is "on."

4.4.2.3. Select the desired protocol for plant protein flours (CP Chopin+90 g, see the screen below; chickpea flour, lentil flours).

4.4.2.4. Choose the sample moisture and approximated hydration (e.g., 60% for Semolina flour and 65% for chickpea flours).

4.4.2.5. Specify a hydration base (As-is, 14% base, 15% base, or "dry base"). It is recommended to use a 14% base for plant protein.

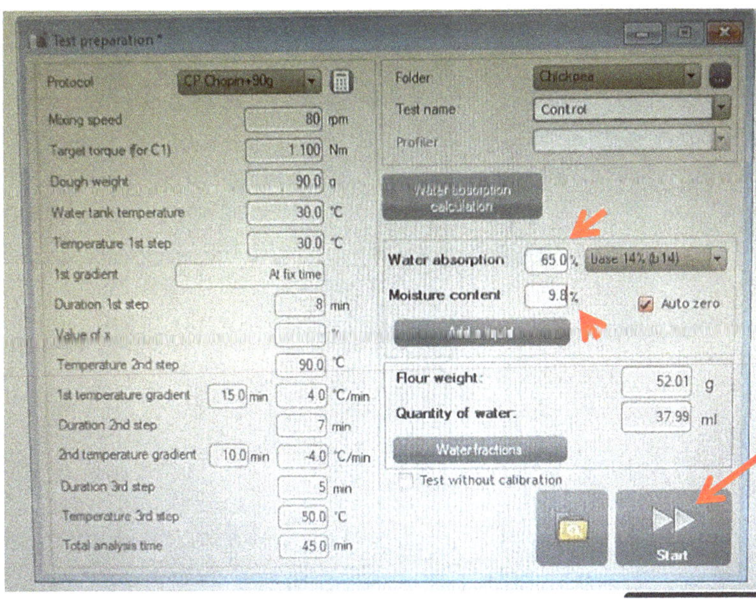

4.4.2.6. Weigh out the amount of sample calculated by the Mixolab software.

4.4.2.7. Place the dough mixer in its housing, and then close the lid.

4.4.2.8. Click the "Start," the testing begins. The dough mixer will rotate, and the samples will be added though the special funnel.

4.4.2.9. Place the water injection nozzle to add the fixed amount of water.

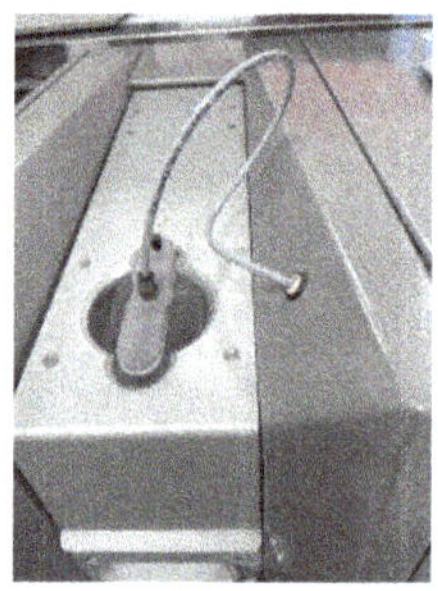

4.4.2.10. If the consistency C1 is outside the tolerances, stop the test, clean the dough mixer with running water and brush, and assemble it and place it back in the Mixolab. This has to be done immediately, as elongated running will increase the temperature, locking the machine.

4.4.2.11. Use the integrated calculation tool to predict the actual hydration of the sample, accounting for the results of the previous test (hydration used, C1 torque obtained and moisture).

4.4.2.12. Restart the test and let it run until the end after 45 min.

4.4.2.13. Upon completion of the testing, dismantle and clean the dough mixer, reset it in the instrument, and then it is ready for the next test.

4.4.3. Testing Protocol for Plant Protein Flour.

The protocol used for plant protein flours is shown in Table 5.

The water absorption based on 14% moisture content is calculated in Table 6.

Table 5
Mixolab2 protocol for plant protein flour

Plant protein flour protocol	
Mixing speed	**80 rpm**
Target torque (for C1)	1.100 nm
Dough weight	90.0 g
Tank temperature	30.0 °C
Temperature 1st step	30.0 °C
1st gradient	At fix time
Duration 1st step	8 min
Temperature 2nd step	90 °C
1st temperature gradient	15 min; 4.0 °C/min
Duration 2nd step	7 min
2nd temperature gradient	10 min, −4.0 °C/min
Temperature 3rd step	50.0 °C
Duration 3rd step	5 min
Total analysis time	45.0 min

Table 6
Sample and water calculation based on different water absorption

Water absorption (%)	14% moisture base	Theoretically calculated	Theoretically calculated water	Actual sample weight based on	Actual water weight (Based on Total 90 g)
50	14	100	50	60.00	30.00
60	14	100	60	56.25	33.75
70	14	100	70	52.94	37.06
55	9.8	95.34	59.66	55.36	34.64
60	9.8	95.34	64.66	53.63	36.37
65	9.8	95.34	69.66	52.01	37.99
70	9.8	95.34	74.66	50.48	39.52

4.5 Result Analysis

4.5.1. The hydration percentage of the 3 samples tested is listed in Table 7.

The MixoLab2 curves of three samples tested are presented in Fig. 2. As indicated, Semolina flours (SF) have a different mixing and cooking profiles in comparison with Chickpea flour (CPF) and Chickpea flour in addition of emulsifier. Semolina sample needs less

Table 7
Hydration (%) and water basis of 3 samples

Folder tests	Hydration (%)	Work basis
Semolina flour	55.0	B14
Chickpea flour	65.0	B14
Chickpea flour + emulsifier	65.0	B14

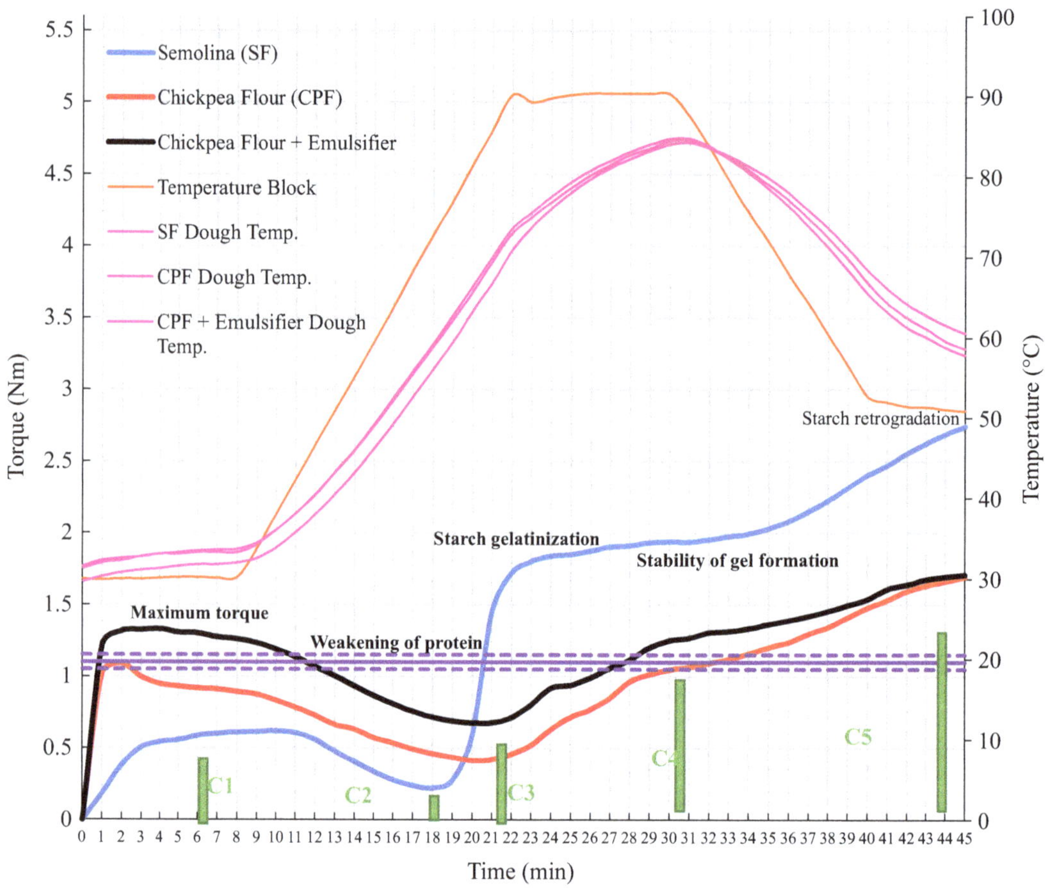

Fig. 2 The Mixolab curves of three flours. Time for C1: the time evolution of the dough. The stronger the flour is, the longer the time evolution (time to reach C1). The elasticity of the dough can be described by an amplitude. The resistance against kneaded dough can be depicted by a stability (The longer the duration, the stronger the flour). C2: attenuation of protein due to mechanical work and temperature; C3: starch gelling; C4: the stability of the hot gel formation; C5: starch retrogradation measured in the cooling phase

water to reach the targeted torque force between 1.05–1.15 Nm. On the other hand, CPF plus emulsifier needs to add more water to reach the targeted torque force. With 65% of water absorption, CPF reached the targeted torque force. The two samples of CPF and CPF plus emulsifier showed a similar pattern/trend in mixing and cooking behavior, although CPK plus emulsifier blend exhibited a stronger dough performance. In addition, the stability of gel formation (C4) of the three samples were noteworthy, suggesting a weak gel strength due to less gelatinized starches.

5 Notes

5.1. Since RVA is typically designed for starch-based ingredients or products, the solid content for those type of materials is not high with less than 13% in RVA testing. For plant-based proteins, the materials level will be increased to 15% or even 18%, to enhance substrate sensitivity.

References

1. Un-Pea-lievable: How innovative analytical technologies are helping to drive innovation in the pulse- and pea-based ingredients market. newfoodmagazine.com. Accessed 20 Aug 2023
2. Webb D, Dogan H, Li YH, Alavi S (2023) Physico-chemical properties and texturization of pea, wheat and soy proteins using extrusion and their application in plant-based meat. Food Secur 12:1586
3. Pagani MA, Marti A, Bottega G (2014) Heat milling and flour quality evaluation. In: Bakery products science and technology, 2nd edn. Wiley, pp 43–48
4. Wittek P, Walther G, Karbstein HP, Emin MA (2021) Comparison of the rheological properties of plant proteins from various sources for extrusion applications. Food Secur 10:1700
5. Serna-Saldivar SO (2010) Quality control of cereal grains and their products. In: Cereal grains: properties, processing, and nutritional attributes. CRC Press, Taylor & Francis Group, LLC, pp 470–497
6. Klosta M, Drusch S (2019) Structure formation and rheological properties of pea protein-based gels. Food Hydrocoll 94:622–630
7. Park R et al (2020) High temperature rotational rheology of the seed flour to predict the texture of canned red kidney beans (Phaseolus vulgaris). Food Secur 9:1002
8. Measure Performance of Ingredients and Finished Products by RVA 4500. Perten Instruments. www.perten.com. Accessed 20 Aug 2023
9. AACC International. Approved Methods of Analysis, 11th Ed. Method 54-40.02 (2010) Determination of rheological behavior as a function of mixing and temperature increase in wheat flour and whole wheat meal by Mixolab. Cereals & Grains Association, St. Paul
10. Tran Dang D, Konvalina P, Capouchova I, Janovska D, Lacko-Bartosova M, Kopecky M et al (2020) Comparative study on protein quality and rheological behavior of different wheat species. Agronomy 10(11):17
11. Konvalina P, Grausgruber H, Tran Dang K, Vlasek O, Capouchova I, Sterba Z et al (2017) Rheological and technological quality of minor wheat species and common wheat. In: Wanyera R, Owuoche J (eds) Wheat improvement, management and utilization. BoD–Books on Demand, Norderstedt; ISBN 978-953-51-3151-9
12. Mixolab Applications Handbook. The CHOPIN Technologies. https://www.kpmanalytics.com/products/mixolab-2
13. Accessed in September 2023. https://bakerpedia.com/processes/mixolab/
14. Pejcz E, Burešová I (2022) Rheological characteristics of model gluten-free dough with Plantago seeds and husk incorporation. Food Secur 11:536. https://doi.org/10.3390/foods11040536
15. Belc N, Duta DE, Culetu A, Stamatie GD (2021) Type and amount of legume protein concentrate influencing the technological, nutritional, and sensorial properties of wheat bread. Appl Sci 11:436. https://doi.org/10.3390/app11010436
16. Schmiele M, Ferrari Felisberto MH, Pedrosa Silva Clerici MT, Chang YK (2017) MixolabTMfor rheologicalevaluation of wheat flour partially replaced by soy protein hydrolysate and fructooligosaccharides for bread production. LWT Food Sci Technol 76:259–269
17. Dhaka V, Gulia N (2012) Application of Mixolab to assess the bread making quality of wheat varieties. J Food Process Technol 1:183

Chapter 30

Viscosity

Suchismita Roy, Gunvantsinh Rathod, and Jayendra Amamcharla

Abstract

Viscosity is an important physical property of protein dispersions, which is defined as the internal resistance to flow. This is a crucial property of protein dispersions as it influences the overall acceptability and mouthfeel of the finished product. The understanding of viscosity helps in the appropriate selection of pumps and other process equipment. In addition, it is also critical to understand how various process parameters affect the flow behavior of protein dispersions. The viscosity of plant protein dispersions is influenced by pH, protein concentration, temperature, ionic strength and thermal history, etc. This chapter discusses the measurement of viscosity as a function of shear. Pea protein and soy protein isolates were used as examples of plant proteins, and a detailed analysis of viscosity and flow behavior is elaborated in this chapter.

Key words Viscosity, Flow behavior index, Shear stress, Cone and plate geometry

1 Introduction

Viscosity is a property of liquid to resist flow. For the measurement of viscosity, two terms are important: shear stress, which is the tangential force applied on an area, and shear rate, which is the velocity gradient formed by shear stress. Some samples exhibit constant viscosity and their shear rate and shear stress have a linear relationship, which is classified as Newtonian liquid. Sugar solution, honey, tea, and olive oil are examples of Newtonian fluid [1]. When the increase of shear stress is not proportional to the shear rate, the samples are called non-Newtonian liquids. Most of the food products are non-Newtonian. When the viscosity decreases with an increase in shear, it is called shear thinning behavior. Apple sauce, banana puree, and tomato ketchup show shear thinning behavior. Viscosity is reported at a particular share rate for non-Newtonian liquids, and it is referred to as apparent viscosity at that share rate (must be specified).

Yonghui Li (ed.), *Plant-Based Proteins: Production, Physicochemical, Functional, and Sensory Properties*, Methods and Protocols in Food Science, https://doi.org/10.1007/978-1-0716-4272-6_30,

Protein sources provide different functionalities in food products which are important for product structure. They provide thickening and consistency to the food protein solution due to their water absorption and holding ability and interaction with other components [3]. All the proteins have different behavior in the matrix, giving different viscosity and consistency [2]. Depending on the environment of proteins in terms of their temperature, pH, and ionic strength, the viscosity and flow behavior of the protein solution changes. Measurement of the viscosity of the protein solution gives us an idea to optimize conditions for the protein solution to achieve the best thickening or consistency in the final products. Also, viscosity measurement enables measuring and monitoring of the product consistency and changes that take place during product production and storage. Since food protein solutions range from thin (e.g., water/milk) to thick (e.g., pudding/condensed milk) in consistency, their viscosity has a wide variation.

Traditionally, this viscosity measurement was started with visual observation of flow or spread on the surface. Then, different viscometers such as the Ostwald viscometer, falling ball viscometer, capillary viscometer, reverse flow viscometer, vibrational viscometer, Brookfield viscometer, etc. were introduced [1]. With the advanced rheometer (stress-strain measurement), viscosity measurement becomes accurate. In this chapter, we will describe viscosity measurement using a stress-strain-controlled rheometer (Anton-Paar).

2 Materials

2.1 Sample Preparation for Analysis

The soy and pea protein samples were reconstituted at 10% total solids (TS) at room temperature and kept refrigerated overnight to ensure complete hydration of proteins. The rheological analysis was done the next day after adjusting the samples' temperature to room temperature. All the samples were analyzed by a stress-strain controlled rheometer MCR 92 (Anton Paar, US) and its corresponding accessories such as cone and plate geometry and cup and bob geometry along with a temperature controller unit. The rheometer was connected to a rheology analyzing software Anton Paar RheoCompass™.

The rheometer MCR 92 from Anton Paar was used for measuring the viscosity of protein dispersions. It has an active temperature controller to maintain temperature during the experiment (Fig. 1). Two kinds of geometry are usually used for measuring apparent viscosity. They are Cone and Plate Geometry with CP-50-mm diameter stainless steel cone with 1° angle (Fig. 2) and Cup and Bob geometry with CC- 60 Length-60 mm, Bob diameter-38.7 mm, and Cup diameter-42 mm (Fig. 3). These probes

Fig. 1 Anton Paar rheometer with cone and plate geometry

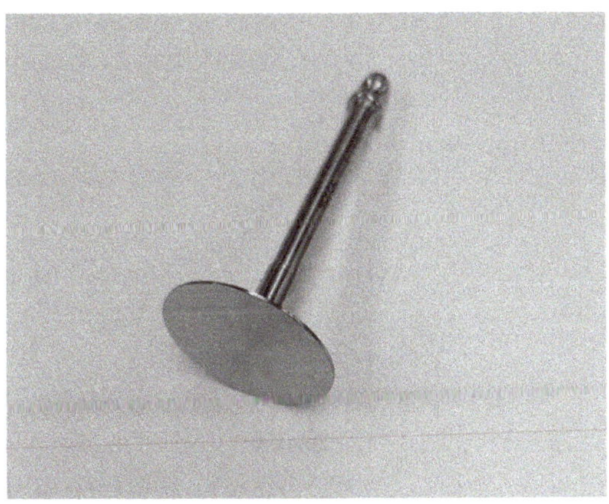

Fig. 2 CP-50 accessory for cone and plate geometry

are available in other dimensions. The selection of appropriate geometry and their dimension depends on the type of sample and their viscosity. The cup and bob geometry can be used for the sample with lower viscosity, where a larger surface area is required for measurement. Cone and plate can be used for single phased samples with no aggregates or particulates.

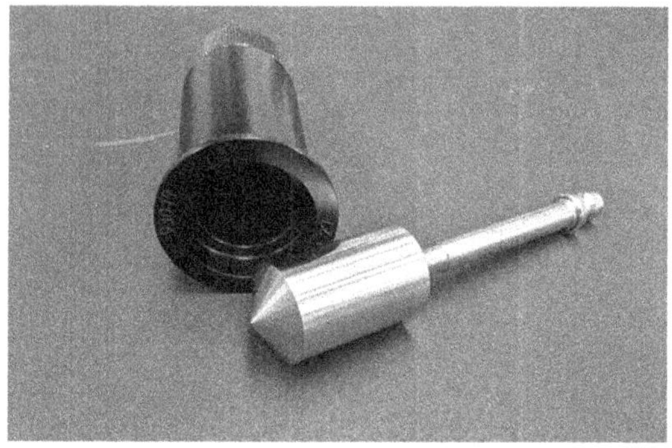

Fig. 3 Cup and bob accessories

3 Method

The value of viscosity is highly impacted by the temperature of the protein solution to be measured. Therefore, it is important to mention the temperature while mentioning viscosity. All procedures here are carried out at 25 °C unless otherwise specified. It is important to maintain a constant temperature during the measurement of viscosity. The selection of temperature during the viscosity measurement depends on the objective of the research.

3.1 Sample Preparation

All the samples are prepared at a 10% TS (w/w) level. The percentage of total solids can be decided according to the desired product application.

3.2 Analysis

The apparent viscosity was measured in the following method.

1. A 10% (w/w) solution of each sample was prepared, and the solution was allowed to rehydrate overnight in refrigerated condition (4 °C). The next day the solution was allowed to equilibrate at 25 °C for 30 min before analysis [5].
2. After starting, the MCR 92 rheometer Rheo Compass™ software was started. The viscosity profile was selected in the software and the instrument was initialized. Then the cone and plate geometry were attached to the rheometer. A zero gap was set between the probe and the plate followed by moving the probe to the sample loading position.
3. For measurement of viscosity, 1 mL solution was placed in the center of the plate (*see* **Note 1**). Next, the probe was moved to the measurement position and the excess sample was trimmed using a pipette or through a trimming spatula if it was spread outside the plate.
4. The temperature control device was lowered covering the whole geometry to maintain temperature during

measurement. (Note: Peltier-based temperature controller maintains the temperature for the plate accessory and the active temperature controller controls the surrounding temperature).

5. For measurement, a flow diagram/macro was written in Rheo Compass™ software. The set of the program includes the temperature of measurement and time to equilibrate the temperature. After temperature, it has the option to set a constant shear rate or shear ramp, logarithmic or linear. After the selection of the shear rate range, another analysis step can calculate parameters derived from stress-strain data.
6. Finally, the measurement can be carried out at the applied shear rate (10–300 s^{-1}) linear. This range of shear rate can be adjusted according to the sample consistency. If the range is too low, the instrument may not detect the shear stress. If the range is too high a shear rate might cause instability [4].
7. Viscosity was calculated using shear stress and strain values in the software. Viscosity can be reported at a particular shear rate. Using shear stress and shear rate data, consistency co-efficient and flow behavior index can be calculated using the power law model as given below [6].

   Power-law model:

$$\tau = K\ddot{\Upsilon}^{\eta}$$

   Where τ = shear stress, $\ddot{\Upsilon}$ = shear strain, K = consistency coefficient, η = flow behavior index

8. Similar to cone and plate, other geometry plate and plate (Fig. 1) for thick paste-like products, and cup and bob geometry (Fig. 4) can be used for low viscosity solution.
9. In the case of volatile samples, a solvent trap or oil should be used on the surface during analysis. Also, it is recommended to add a few drops of vegetable oil on the protein dispersion to avoid any water evaporation when measuring viscosity at high temperatures.

4 Result

4.1 Pea Protein Isolate

A commercial sample of pea protein isolate was collected from Anthony's Premium Pea protein. It contained 80% protein by weight. A 10% TS solution was prepared using pea protein isolate and was analyzed as the method described. The viscosity was obtained from the software. The pea protein solution showed a viscosity of 118.53 ± 15.29 mPa.s at 50 s^{-1} shear rate. The viscosity versus shear rate is plotted in Fig. 5. The shear rate and shear stress were plotted to fit the power law model (Fig. 6). The flow behavior index and the consistency coefficient were 0.402 ± 0.02 and 1.17 ± 0.22 mPa.s, respectively.

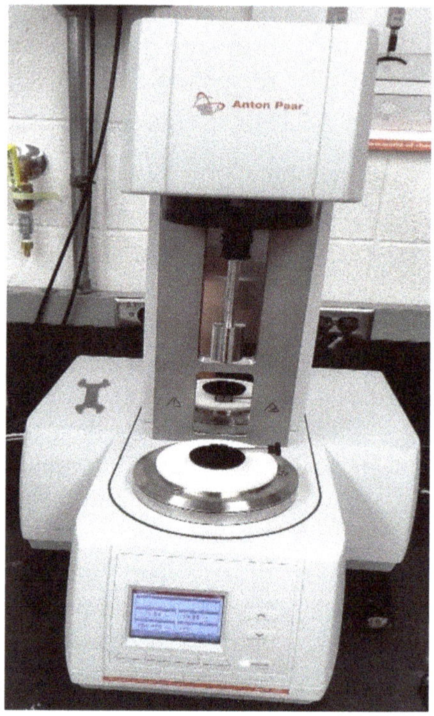

Fig. 4 Rheometer with cup and bob geometry

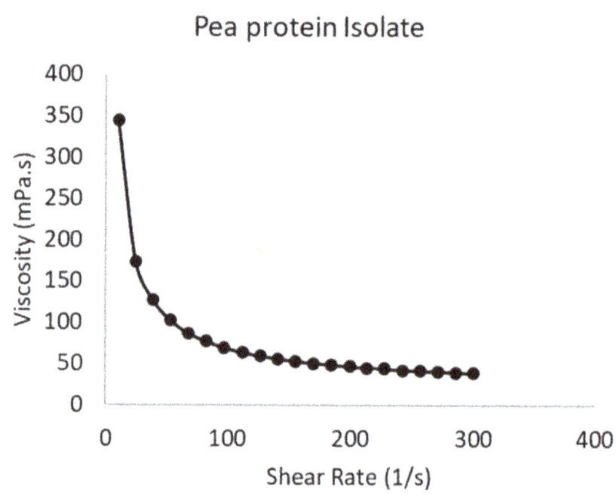

Fig. 5 Representative plot of viscosity of pea protein solution (mPa.s) as a function of applied shear rate (s^{-1})

4.2 Soy Protein Isolate

Soy protein isolate was collected from a bulk supplier through Amazon. It contained 90% protein by weight. A 10% TS solution was prepared from soy protein isolate and the viscosity was determined according to the method discussed before. The viscosity was obtained from the software. The pea protein solution showed a

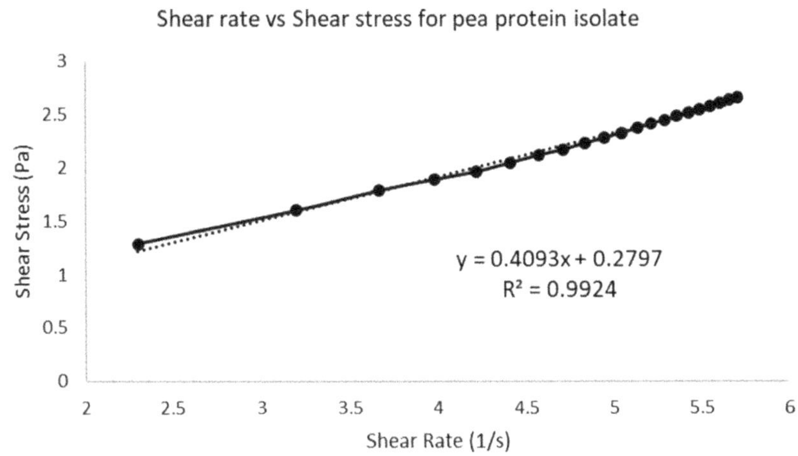

Fig. 6 Shear stress (Pa) plotted against Shear rate (s^{-1}) for representing the Power Law model

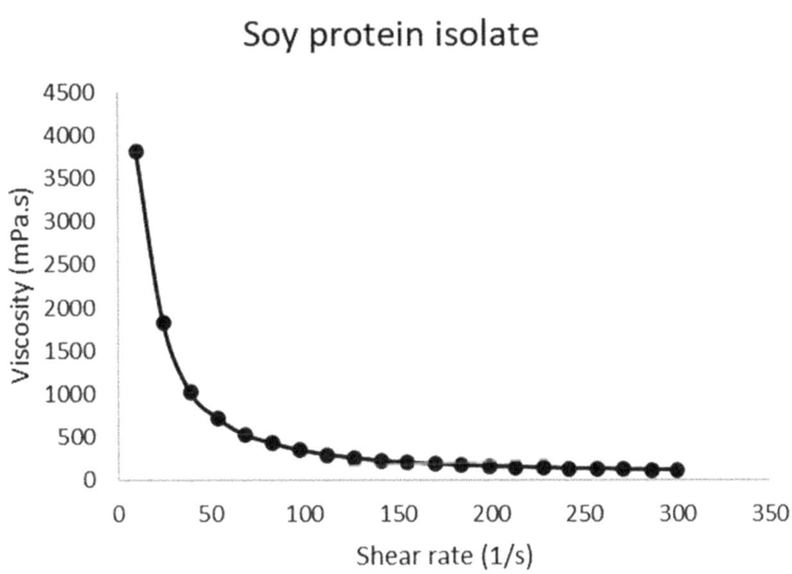

Fig. 7 Representative plot of viscosity of soy protein solution (mPa.s) as a function of applied shear rate (s^{-1})

viscosity of 915.08 ± 90.37 mPa.s at 50 s^{-1} shear rate (Fig. 7). The shear rate and shear stress were plotted to fit the power law model (Fig. 8). The rheological behavior of soy protein isolate could not be properly characterized by the Power Law equation. Hence, the same concentration solution (10% TS) cannot be used for all plant protein samples. The power law model was used in the pea protein sample of 10% TS but it was not adequate for the soy protein sample. We may need to fit another suitable model to obtain consistency co-efficient and flow behavior index.

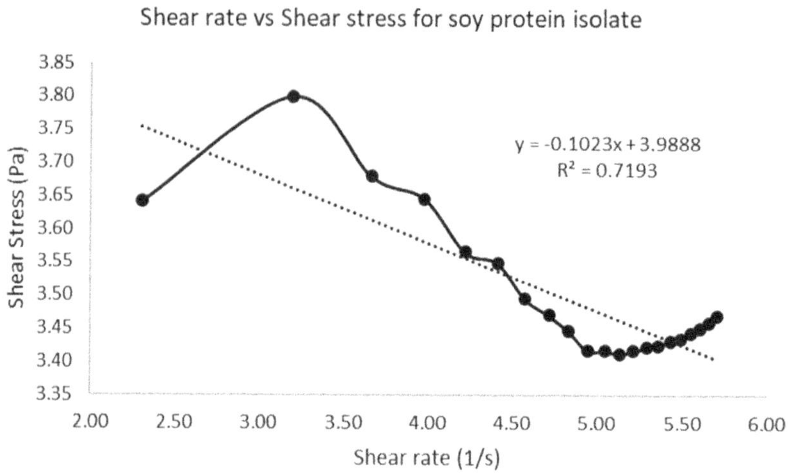

Fig. 8 Shear stress (Pa) plotted against Shear rate (s^{-1}) for representing the Power Law model

5 Notes

1. Protein solution viscosity and flow behavior vary with changes in temperature and pH; therefore, both should be controlled while measuring.
2. The temperature of measurement should be always mentioned while reporting viscosity.
3. The samples with big aggregates are not suitable for the cone and plate geometry as it has a very small gap (101 μm) between the plate and cone. For a solution with very small aggregates or particulates and lower viscosity, cup, and bob geometry should be used instead of cone and plate.
4. The probes and the instrument surface should be cleaned after use to avoid sample residue. Care should be taken not to have any scratches on the geometry.
5. The adjustment in the motor can be done through the Rheo-Compass software when needed or as recommended by the service engineer.
6. Rheometer measurement is sensitive to vibration. So, the instrument should be mounted on a strong foundation.
7. Various ingredients have different ranges of viscosities, and they show different shear stress when a shear rate is applied. Therefore, standardization of the method and application of appropriate models, e.g., Power law model, Herschel Bulkley model, Casson model, etc. are required. The adjustment of the shear rate and stress and fitting of the models can be done by the software or manually.

References

1. Bourne M (2002) Food texture and viscosity: concept and measurement. Elsevier
2. Chou DH, Morr CV (1979) Protein-water interactions and functional properties. J Am Oil Chem Soc 56(1):A53–A62. https://doi.org/10.1007/BF02671785
3. Foegeding EA (2015) Food protein functionality—a new model. J Food Sci 80(12):C2670–C2677. https://doi.org/10.1111/1750-3841.13116
4. Grossmann L, Kinchla AJ, Nolden A, McClements DJ (2021) Standardized methods for testing the quality attributes of plant-based foods: Milk and cream alternatives. Compr Rev Food Sci Food Saf 20(2):2206–2233. https://doi.org/10.1111/1541-4337.12718
5. Lamsal BP, Jung S, Johnson LA (2007) Rheological properties of soy protein hydrolysates obtained from limited enzymatic hydrolysis. LWT-Food Sci Technol 40(7):1215–1223. https://doi.org/10.1016/j.lwt.2006.08.021
6. Rathod G, Amamcharla J (2021) Process development for a novel milk protein concentrate with whey proteins as fibrils. J Dairy Sci 104(4):4094–4107. https://doi.org/10.3168/jds.2020-19409

Chapter 31

Volatile Compounds

Ali Raza, Changqi Liu, and Jing Zhao

Abstract

Protein ingredients are generally considered bland in flavor. However, during storage, various chemical processes may occur, such as oxidation of residual lipids and amino acids, degradation of amino acids, and the Maillard reaction between proteins and reducing sugars. These reactions can significantly alter the volatile profile of plant-based foods by generating desirable or undesirable odors. To better understand the flavor chemistry of plant proteins and identify the key aroma-active compounds and their formation pathways, it is necessary to analyze their volatile profile using advanced analytical techniques. One such method is the combination of headspace solid-phase microextraction (HS-SPME) and gas chromatography-mass spectrometry-olfactometry (GC-MS-O). This technique offers a rapid, non-invasive way to extract and detect volatile compounds and key odorants from complex proteinaceous foods. This chapter describes detailed protocols for the analysis of volatile compounds in plant proteins through HS-SPME GC-MS-O.

Key words Volatile compounds, Gas chromatography-mass spectrometry-olfactometry (GC-MS-O), Headspace solid-phase microextraction (HS-SPME), Flavors, Plant proteins

1 Introduction

As the world population continues to rise [1], there is growing demand for high-quality proteins. Due to the limited capacity of animal protein production, the demand for alternative proteins and plant-based food products has dramatically increased in recent times [2]. Plant protein ingredients have found applications in various products such as bakery goods, dietary supplements, and meat substitutes. Even though the market size and variety of plant protein ingredients are growing, the widespread applications of certain plant proteins are still limited by their poor functionalities and off-flavors. The undesirable characteristics often associated with alternative proteins and plant-based foods include bitter taste, astringency, aftertaste, and beany flavor [3–6]. To enhance consumer acceptance and the marketability of these products, it is

Yonghui Li (ed.), *Plant-Based Proteins: Production, Physicochemical, Functional, and Sensory Properties*, Methods and Protocols in Food Science, https://doi.org/10.1007/978-1-0716-4272-6_31,

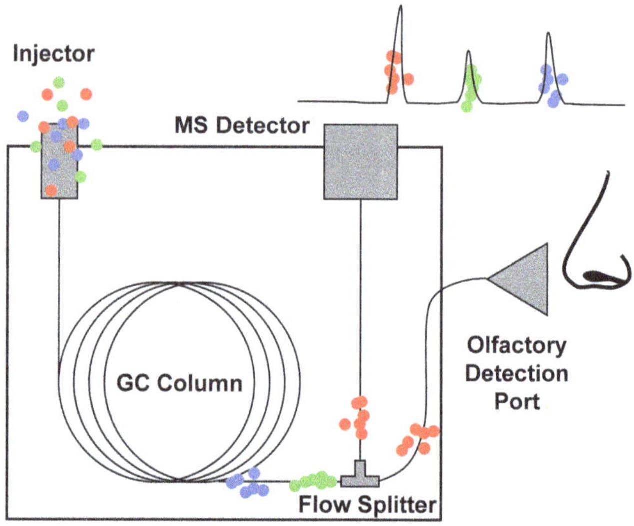

Fig. 1 Illustration of a gas chromatography-mass spectrometry-olfactometry (GC-MS-O) system

imperative to gain a better understanding of the flavor chemistry and potential flavor precursors.

To identify the key aroma compounds in plant proteins and to monitor the development of undesirable flavors during the processing and storage of these products, an effective volatile isolation and identification technique is required. Gas chromatography-mass spectrometry-olfactometry (GC-MS-O) is a powerful tool to analyze volatile and aroma-active compounds. It allows the simultaneous identification of the chemical nature and the odor characteristics of the separated volatiles (Fig. 1). The samples for GC-MS-O analysis can be prepared using different techniques, such as static headspace (S-HS), dynamic headspace sampling (DHS), purge and trap (P&T) [7], and solvent extraction techniques like simple solvent extraction (SE), simultaneous distillation-extraction (SDE), and solvent-assisted flavor evaporation (SAFE) [8–10]. Among these methods, solid phase microextraction (SPME) is a technique that concentrates the headspace volatiles on the surface of fiber coatings through analyte diffusion, absorption, and adsorption phenomenon [11, 12]. Headspace solid-phase micro-extraction (HS-SPME) is a rapid, simple, non-invasive, and reliable volatile extraction technique widely used for the identification of volatile compounds in foods [13–16] in combination with GC-MS-O. Using HS-SPME also reduces degradation to the chromatography column and the overall GC-MS instrument [17]. HS-SPME has been combined with other volatile extraction techniques (S-HS, DHS, P&T, etc.) for the isolation and identification of potential volatile compounds in plant-based foods [15, 18–21]. This chapter includes a detailed protocol for HS-SPME GC-

MS-O analysis of plant protein ingredients. This protocol can also be adopted for protein-based foods with appropriate modifications. The analysis will enable the identification of key odorants and off-flavors in plant protein ingredients.

2 Materials

2.1 Solid Phase Microextraction Arrangement (Figs. 2 and 3)

1. SPME sampling stand and holder to hold the SPME fiber (MilliporeSigma™, Burlington, MA, USA).
2. Headspace sampling vials, 40 mL, with Teflon-rubber disc mounted on screw-top (MilliporeSigma™, Burlington, MA, USA).

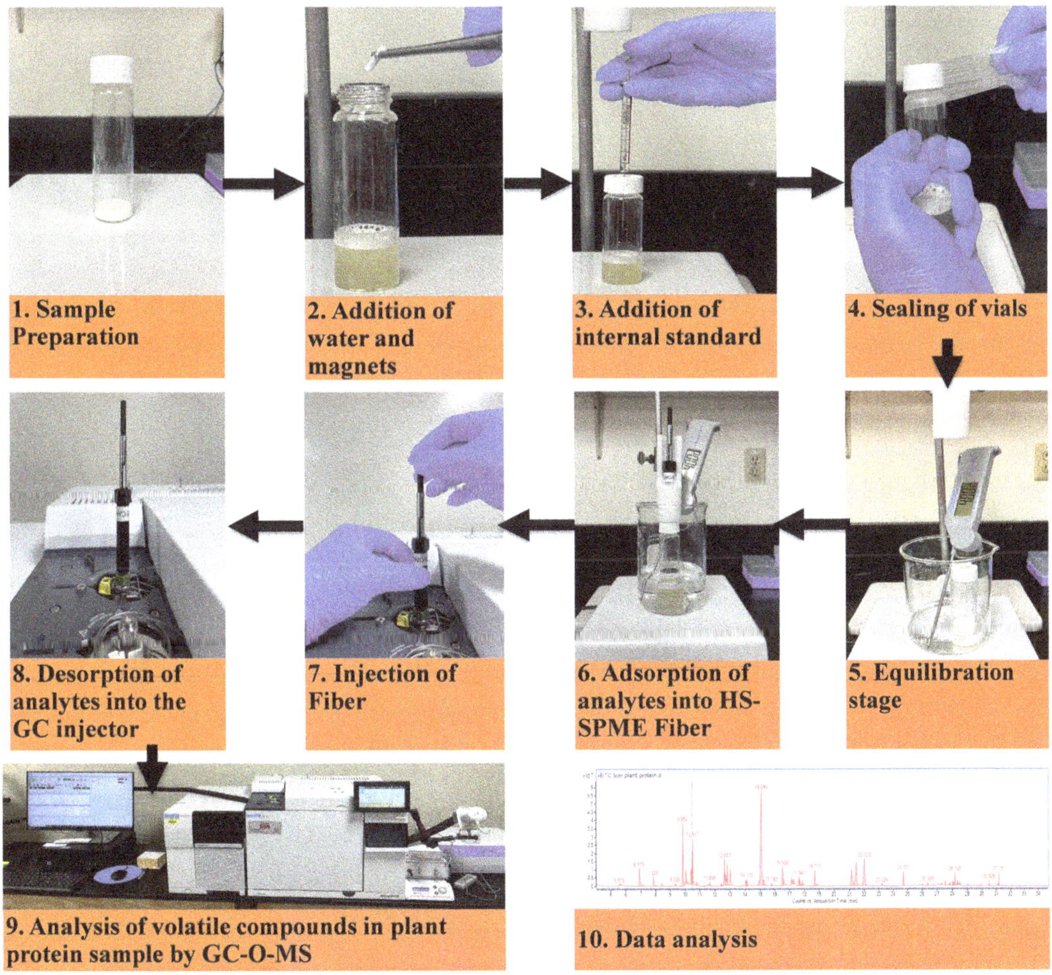

Fig. 2 Schematic arrangement of headspace solid phase microextraction (HS-SPME) and gas chromatography-mass spectrometry-olfactometry (GC-MS-O) analysis of volatile compounds for plant protein ingredients

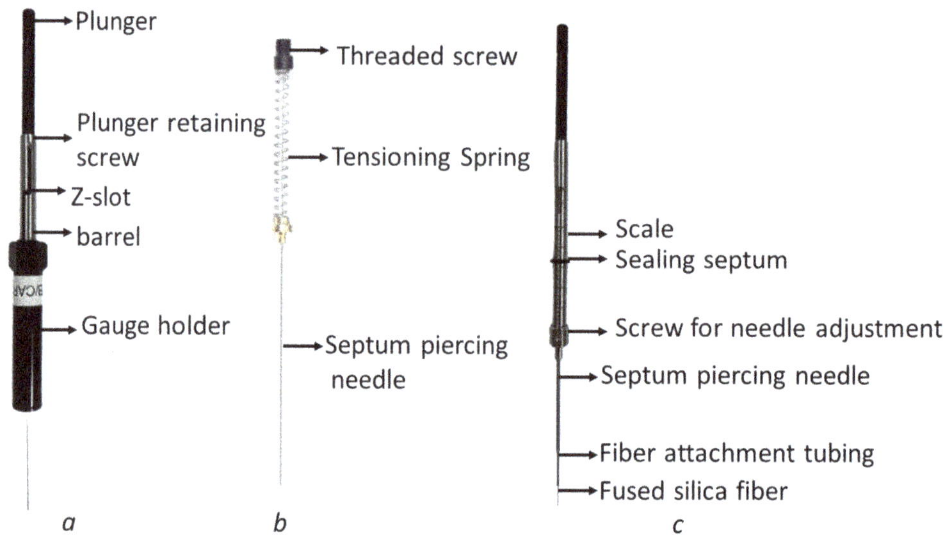

Fig. 3 Anatomy of a solid phase microextraction holder and fiber assembly

3. SPME stable flex fiber Divinylbenzene/Carboxen™/Polydimethylsiloxane (DVB/CAR/PDMS), df 50/30 μm (Supelco, Bellefonte, PA, USA).
4. Magnetic stirrer heating plate and a beaker.

2.2 Sample and Chemicals

1. Plant protein powder (*see* **Note 1**).
2. Ultrapure water purified using a Millipore water purifier system (Waltham, MA, USA).
3. 2-Methyl-3-heptanone (internal standard).
4. *n*-Hexane, chromatography grade (> 99% purity).

2.3 Gas Chromatography-Mass Spectrometry-Olfactometry (GC-MS-O)

1. Agilent 8890 gas chromatography (GC, Agilent Technologies, Santa Clara, CA, USA).
2. Agilent 5977 C mass-selective detector (MSD, Agilent Technologies, Santa Clara, CA, USA).
3. Gerstel olfactory detection port ODP-4 (Linthicum, MD, USA).
4. Agilent J&W HP-5 column (30 m, 0.25 mm inner diameter, and 0.25 μm film thickness, Agilent Technologies, Santa Clara, CA, USA).
5. C7–C30 saturated alkanes, chromatography grade (Millipore-Sigma™, Burlington, MA, USA).
6. Carrier gas: ultra-high purity helium (99.999% purity) with a flow rate of 2 mL/min.
7. Makeup gas: nitrogen gas (99.999% purity) should be installed with a flow rate of 0.59 mL/min (*see* **Note 2**).

2.4 Data Analysis

1. Qualitative analysis: Agilent MassHunter Workstation Qualitative Analysis Version 10.0 (Agilent Technologies) software.
2. Quantitative identification of volatile compounds: Agilent MassHunter Workstation Quantitative Analysis Version 10.0 (Agilent Technologies) software.

3 Methods

3.1 Sample Preparation

1. Weigh 1 g of protein powder into a 40 mL SPME headspace vial and add 9 mL ultrapure water to hydrate it (*see* **Note 3** and Fig. 2, and **steps 1** and **2**).
2. Add 5 μL of internal standard (2-methyl-3-heptanone) in 995 μL of solvent (*n*-hexane) to make a 1 mL diluted standard solution. Add 1 μL of the diluted internal standard solution into the sample from the top screw septum of the SPME vial using a 10-μL liquid injection syringe (*see* Fig. 2, **step 3**).
3. Close the SPME sample vial with a screw cap and seal it using a silicone-lined/polytetrafluoroethylene (PTFE) septum. Cover the top of the vial with an adequate amount of parafilm tape to prevent any volatile loss during the extraction process (*see* Fig. 2, **step 4**).
4. Pre-heat the SPME vial containing the protein slurry at 40 °C for 30 min using a water bath (*see* Fig. 2, **step 5**) (*see* **Notes 4** and **5**).

3.2 Volatile Extraction

1. Condition the StableFlex fiber (DVB/CAR/PDMS) prior to the analysis (*see* **Note 6**).
2. With the fiber retracted, insert the SPME piercing needle through the septum of the headspace vial (*see* Fig. 2).
3. Expose the fiber by pushing down the holder plunger (*see* Fig. 2, **step 6**) (*see* **Note 7**).
4. Extract the headspace volatiles at 40 °C for 60 min (*see* **Note 8**).
5. After the extraction process, retract the fiber into the SPME piercing needle and pull out the SPME holder from the SPME stand.

3.3 Setting Up the Parameters

3.3.1 Inlet and Carrier Gas Parameters

1. Use the online data acquisition software (MassHunter Acquisition version 10.0, Agilent Technologies) to set the GC-MS operational parameters (*see* Fig. 4).
2. To change the chromatography conditions, select **Method** and then **Edit entire method**. Alternatively, navigate to **Instrument, GC edit parameters** (*see* Fig. 5) to adjust the GC parameters.

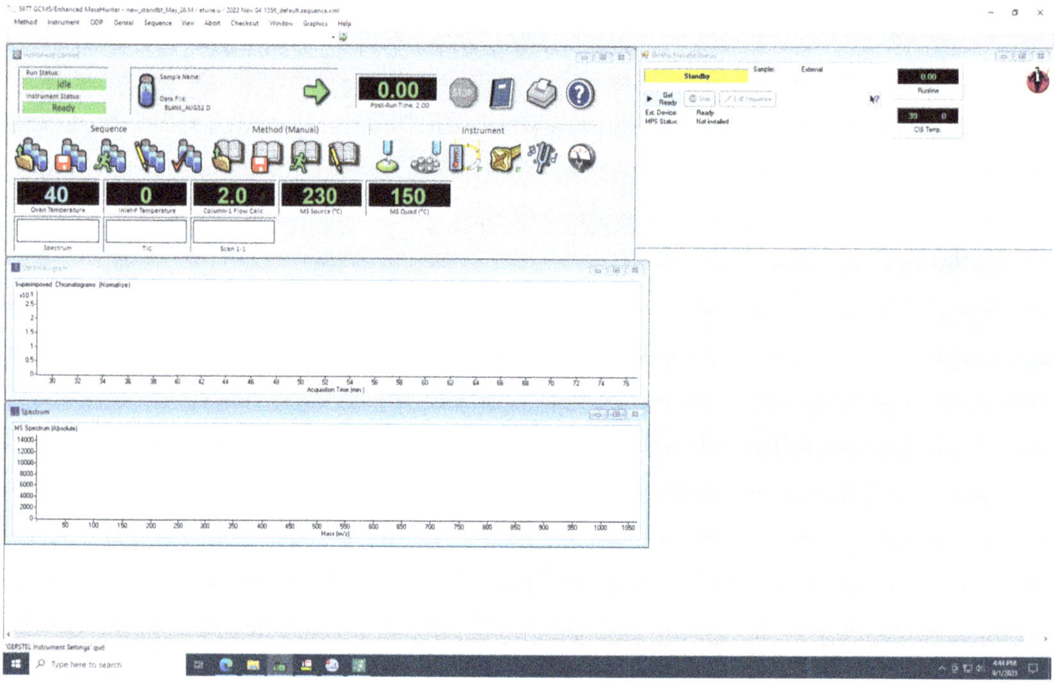

Fig. 4 Agilent MassHunter Acquisition software main page

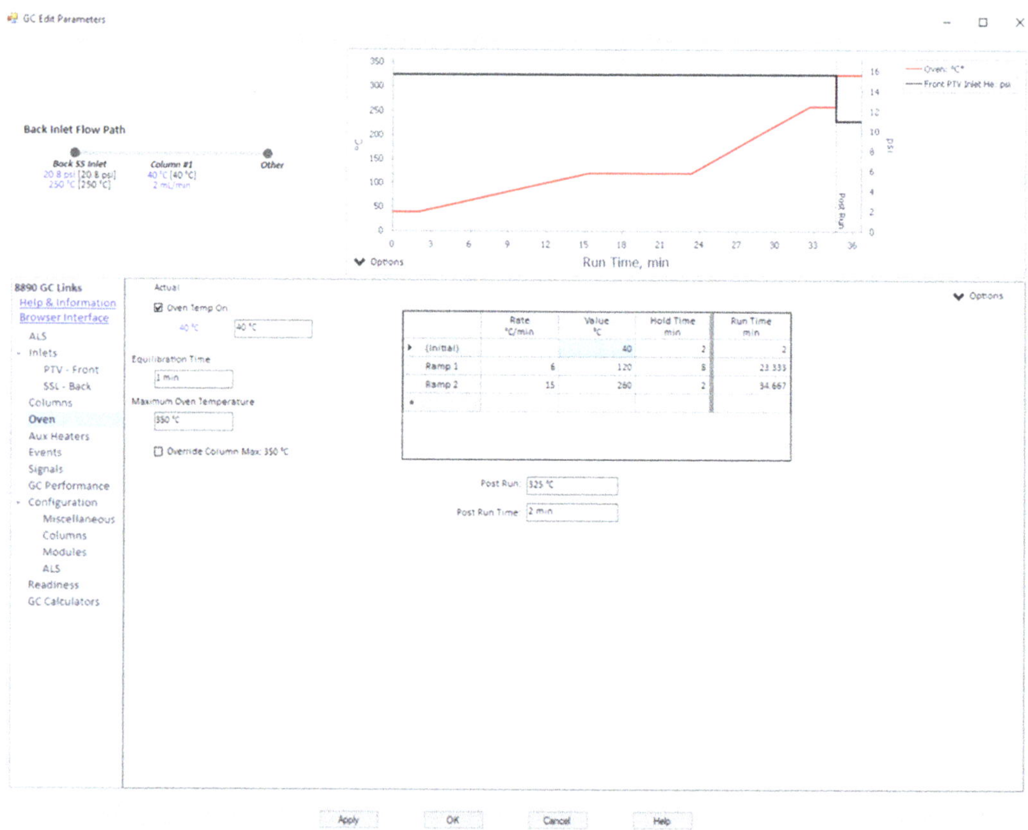

	Rate °C/min	Value °C	Hold Time min	Run Time min
(Initial)		40	2	2
Ramp 1	6	120	5	23.333
Ramp 2	15	260	2	34.667

Fig. 5 Gas chromatography (GC) method editor dialog box

3. Configure the inlet and carrier gas settings.
   (a) Heater: Set the temperature of the heater to 250 °C.
   (b) Pressure: Adjust the helium pressure to 20.81 psi.
   (c) Inlet mode: Use a splitless mode for the injection of the samples.
   (d) Purge flow: Adjust the septum purge flow to 3 mL/min and program the purge flow to split the vent at 60 mL/min.
   (e) Flow rate: Set a constant flow rate of 2 mL/min for the carrier gas (helium).

3.3.2 GC Oven Parameters

1. Initial oven temperature: 40 °C with a hold time of 2 min.
2. Ramp 1: raise temperature to 120 °C at a heating rate of 6 °C/min and hold for 8 min.
3. Ramp 2: increase the temperature to 260 °C at a rate of 15 °C/min and hold for 2 min.
4. Post run: 325 °C for 2 min using the same chromatography conditions.
5. The total chromatography runtime will be 34.66 min. GC-MS operational parameters are presented in Table 1.

3.3.3 MSD Parameters

1. In MassHunter Acquisition software, go to **Single Quadrupole MS Method Editor** to adjust the MSD parameters (*see* Fig. 6).
2. MSD parameters (*see* Fig. 6).
   (a) Tune type: EI.
   (b) Solvent delay: 4 min (*see* **Note 9**).
   (c) Gain factor: 10.00.
   (d) Acquisition type: scan.
   (e) Mass scanning range m/z: start mass: 40.00, end mass: 550.00 (*see* **Note 10**).
   (f) MSD Temperatures: mass source temperature: 230 °C; MS Quad temperature: 150 °C.
3. Other MSD conditions.
   (a) Use the default settings for the following parameters: electron impact ionization energy: 70 eV, electron multiplier voltage: 900 V.

3.3.4 Olfactory Detection Port (ODP) Parameters

1. A split ratio of 2:1 for ODP and MSD needs to be set during instrument setup (*see* **Note 11** and Fig. 7).
2. Set the ODP transfer line and mixing chamber temperatures to 280 °C and 150 °C, respectively using Gerstel C200 controller.

Table 1
Summary of gas chromatography-mass spectrometry-olfactometry (GC-MS-O) operational parameters

GC parameters	
Mobile phase	He >99.999%, flow rate 1.5 mL/min
Inlet mode	Spitless mode
Heater temperature	250 °C
Pressure	20.81 psi
Total flow	65 mL/min
Septum purge flow	3 mL/min
Maximum oven temp	350 °C
Column	Agilent HP-5; 30 m × 0.25 μm × 0.25 μm
Initial temperature	40 °C (hold for 2 min)
Ramp 1	120 °C, 6 °C/min (hold for 8 min)
Ramp 2	260 °C, 15 °C/min (hold for 2 min)
Total run time	34.667 min
Post run time/temperature	2 min /325 °C
MSD parameters	
Mode	Full scan, 40–550 amu
MSD transfer line temperature	280 °C
Ion source temperature	230 °C
Ionization condition	70 ev
Quadrupole temperature	150 °C
ODP parameters	
Olfactory unit	ODP 4
ODP transfer line temperature	280 °C
Column effluent split ratio	2:1
Mixing chamber temperature	150 °C

MSD mass selective detector, *ODP* olfactory detection port

3.3.5 Desorption of Volatile Compounds into the GC Injector

1. Open the method file using MassHunter Acquisition software.
2. Click the run icon on the software window (*see* Fig. 4), select a data path, and name the data file.
3. Click **ok** to run the method.
4. With the fiber retracted, inject the SPME piercing needle into the GC injector port (*see* Fig. 2, **step 7**).

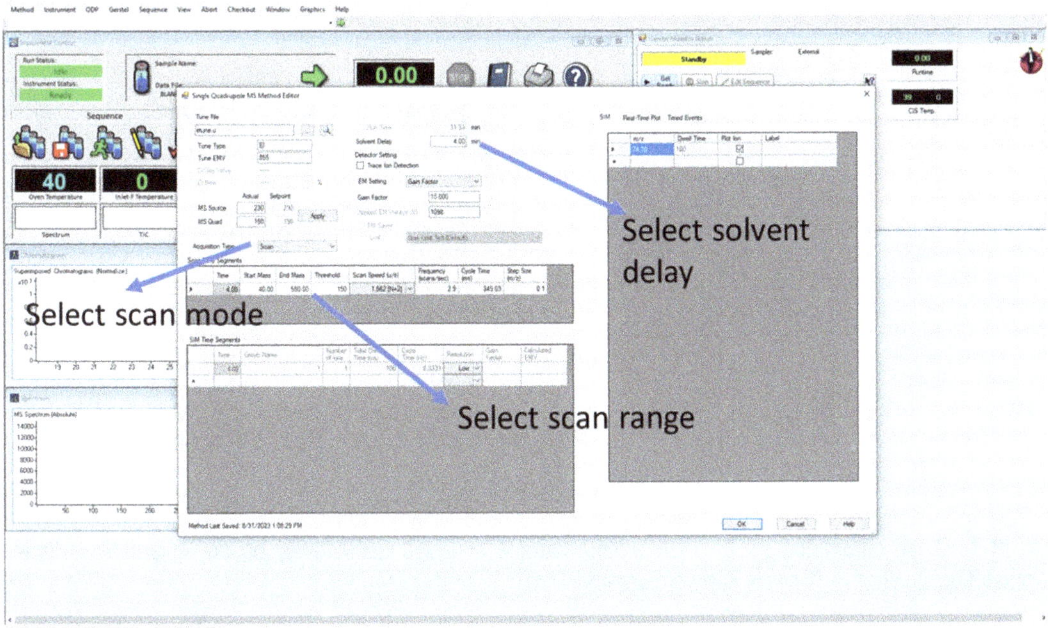

Fig. 6 Mass selective detector (MSD) method editor dialog box

Fig. 7 Gerstel olfactometry detection port (ODP) calculator

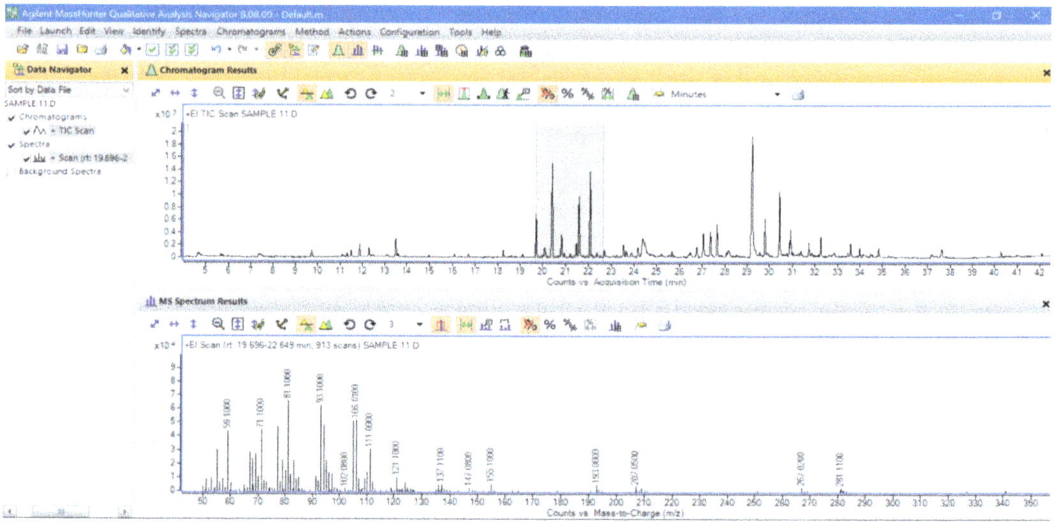

Fig. 8 Total ion chromatogram of plant protein sample analyzed qualitatively on an Agilent MassHunter Workstation Qualitative Analysis software version 10.0

5. Expose the fiber into the injector liner by pushing down the plunger forward via a Z-slot (*see* **Note 12**).
6. Desorb at 250 °C for 8 min in a splitless injection mode (*see* Fig. 2, **step 8**).
7. Retract the fiber and remove the needle (*see* **Note 13**).

3.4 Data Analysis

3.4.1 Qualitative Data Analysis

1. Open the data file in the MassHunter Workstation Qualitative Analysis software (*see* Fig. 8).
2. Integrate and extract the peak spectra of a selected peak range or the entire highlighted chromatogram (*see* **Note 14**).
3. Create a data sheet to list the identified volatile compounds from the sample.
4. Tentative identification:
   (a) Match the spectra of unknown compounds with those available in the National Institute of Standards and Technology (NIST-14) library offline or online database (Fig. 9).
   (b) Calculate the retention indices (RIs) of the unknown volatile components based on the retention times of the alkane standards. Run a mixture of C_7-C_{30} alkane standards under the same chromatography conditions (Table 1). Use the following formula to calculate the RIs of unknown compounds.

$$RI = 100N + 100n \frac{t_{Ra} - t_{RN}}{\left(t_{R(N+1)} - t_{RN}\right)}$$

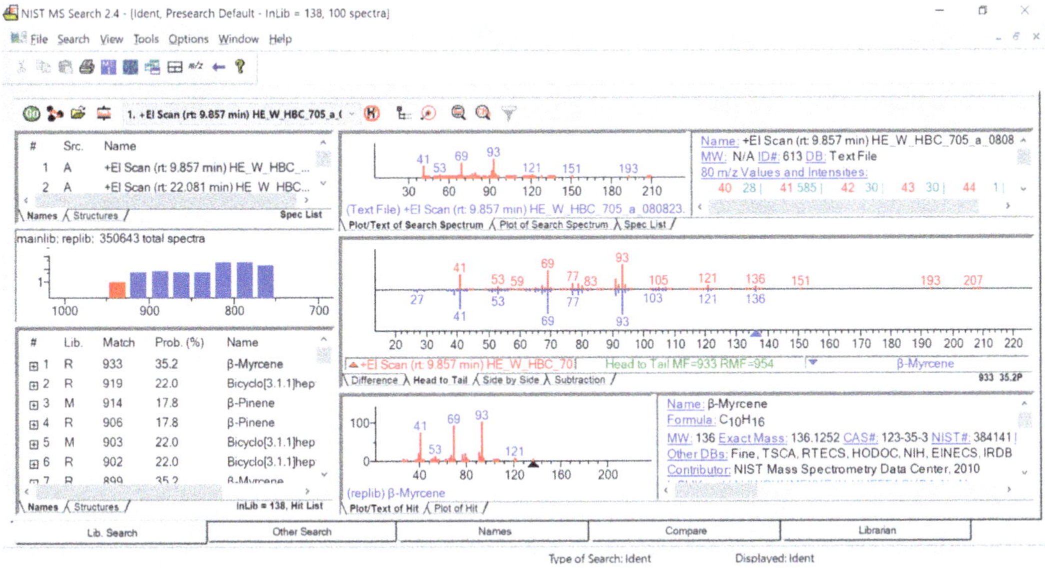

Fig. 9 Matching of unknown compounds in plant proteins through NIST offline database

where RI is the retention index of the unknown volatile compound, *N* is the carbon number of the lower alkane, *n* is the difference in the carbon number between two alkanes, t_{Ra} is the retention time of the unknown volatile component, t_{RN} and $t_{R(N+1)}$ are the retention times of alkanes eluted before (lower carbon number) and after (higher carbon numbers) the unknown volatile compound, respectively.

(c) Compare the retention indices (RIs) of the unknown compounds with those available in the National Institute of Standards and Technology (NIST-14) library, using a library match factor identification criterion of over 70%. Identify the odor characteristics of the compounds through GC-O sniffing results by matching the recordings with the corresponding retention times. Verify the odor characteristics of the compounds by searching the online flavor databases (*see* Fig. 10).

(d) For confirmation, match the retention times, retention indices (RIs), and mass spectral information of the unknown volatile compounds with those of authentic standard compounds after injecting them into the GC-MS system using the same chromatography conditions.

Flavornet Home | Kovats RI | Ethyl Ester RI | Odorants | Odors | Odor Classes

OV101	DB5	OV17	C20M	Odorant	Odor
[312]	[329]	[420]	677	trimethylamine	fish
418	427	[518]	714	ethanal	pungent, ether
[483]	500	[591]	696	methanethiol	sulfur, gasoline, garlic
[489]	506	[597]	571	propanal	solvent, pungent
500	500	500	500	pentane	alkane
[519]	536	[627]	1037	propanol	alcohol, pungent
527	505	[596]	716	dimethyl sulfide	cabbage, sulfur, gasoline
530	[547]	[591]	848	ethyl formate	pungent
[580]	597	[688]	[945]	methyl ethyl ketone	ether
[592]	609	[700]	[957]	methyl ethyl sulfide	sulfur, garlic
600	600	600	600	hexane	alkane
602	628	[719]	907	ethyl acetate	pineapple
603	[620]	[711]	[968]	methylbutenol	herb
[619]	[636]	[727]	984	pentanone	ether
[630]	647	[738]	1099	isobutanol	wine, solvent, bitter
[631]	648	[739]	1047	trans-crotonaldehyde	flower
[638]	[655]	[746]	1003	methylpentanone	mint
639	641	729	912	methylbutanal	cocoa, almond

Fig. 10 Online database of odorants: www.flavornet.org

3.4.2 Semi-Quantification of the Volatile Compounds

1. The quantities of unknown volatile compounds can be quantified by adding an internal standard into the sample at a specific concentration. The concentration of individual volatile compound (C) can be calculated using the following formula [13].

$$C_i = C_{\mathrm{IS}} \frac{A_i}{A_{\mathrm{IS}}}$$

where C_i is the concentration of the unknown volatile compound, C_{IS} is the concentration of the internal standard, A_i is the peak area of the unknown volatile compound, and A_{IS} is the peak area of the internal standard recorded on the GC-MS chromatogram.

3.4.3 Quantification of the Volatile Compounds

1. Use the selective ion monitoring (SIM) mode in the GC-MS workstation software to quantify the volatile compounds.
2. Run each of the standard compounds or a mixture of the standard compounds at five concentrations over a selected linear range.
3. Make calibration curves for each target compound.
4. Calculate the concentration of a volatile compound based on its standard curve.

4 Notes

1. When measuring samples that are not in the powder form, grind the samples into fine particles using a food grinder to enhance the volatile extraction process from the headspace of the sample.

2. Flow rates are adjustable and can be optimized prior to analysis.
3. A protein dispersion of 2–15% (w/v) is typically used for volatile analysis [22, 23]. Make sure that the final sample volume remains below 1/3 of the vial, leaving sufficient headspace to prevent the fiber from coming into contact with the sample or being contaminated by it.
4. This step is used to establish a uniform extraction temperature inside the SPME vial, which must be kept constant during the analysis. A stirring heating plate can also be used in place of a water bath. When using a heating plate, ensure a uniform and consistent extraction temperature and stirring speed.
5. SPME extraction is typically done in a temperature range of 0–60 °C [24]. The extraction conditions (temperature and time) need to be optimized. For protein samples, a temperature of 40 °C or below should be used to prevent protein denaturation and heat-induced gelation [25–27].
6. For initial conditioning, the fiber should be thermally baked at 270 °C for 30 min as per the guidelines of the manufacturer. The fiber should also be thermally baked in a conditioning chamber or GC injection port between the samples at 250 °C for 1–5 min to clean up any former residuals. The three-coated fiber (DVB/CAR/PDMS) is especially sensitive to important volatile compounds in plant proteins. However, the selection of the appropriate SPME fiber is based on sample type and literature reports. Previous studies have successfully used a three-coated StableFlex SPME fiber (DVB/CAR/PDMS) to extract volatile compounds from the headspace of plant-protein samples [3, 15].
7. Establish a predefined fiber depth setting. Ensure that the fiber remains at a specific distance from the sample to prevent any contact with the sample or potential contamination from sample droplets that may rise during stirring.
8. Prior to analysis, it is recommended to carefully optimize the SPME extraction conditions (temperature, time, and height of SPME fiber exposed to the headspace of the sample).
9. Solvent delay is used to safeguard the filaments installed in the ion source against contamination. The solvent delay of 4 min is selected specifically for *n*-hexane under the current chromatography conditions.
10. The m/z range can be adjusted based on the molecular weight range of the target compounds.
11. The split ratio of 2:1 for ODP: MSD is achieved by setting up the capillaries based on the Gerstel ODP column calculator (*see* Fig. 7).

12. To change the needle depth, adjust the sealing septum (needle guide) on the SPME holder so that the top of the ring is between scales 3 and 5 (*see* Fig. 3). Ensure consistent needle depth for each run within the same project.
13. After each sample injection, clean or thermally bake the SPME fiber in a conditioning chamber or GC injector at 250 °C for 2 min to prevent any carry-over substances between the samples and to ensure the accuracy of the volatile isolation for the next injection.
14. By selecting the option "integrate and extract the entire chromatogram," you will automatically extract an average spectrum from each of the individual peaks. Even if you opt to select a specific range of chromatograms, you still have the option to integrate the entire chromatogram or just the highlighted region.

References

1. Nations U (2019) World population prospects 2019. Department of Economic and Social Affairs, p 141
2. Zhao J, Liu C, Raza A (2024) Overview of protein flavours. In: Zhao J, Liu C (eds) Flavour and consumer perceptions of food proteins. The Royal Society of Chemistry, United Kingdom, pp 1–19
3. Yu S, Huang X, Wang L et al (2022) Characterization of selected Chinese soybean paste based on flavor profiles using HS-SPME-GC/MS, E-nose and E-tongue combined with chemometrics. Food Chem 375:131840
4. Mittermeier-Kleßinger VK, Hofmann T, Dawid C (2021) Mitigating off-flavors of plant-based proteins. J Agric Food Chem 69: 9202–9207
5. Sha L, Xiong YL (2020) Plant protein-based alternatives of reconstructed meat: science, technology, and challenges. Trends Food Sci Technol 102:51–61
6. Zhang B, Zhang J, Yu X et al (2022) Evaluation of the adsorption capacity and mechanism of soy protein isolate for volatile flavor compounds: role of different oxygen-containing functional groups. Food Chem 386:132745
7. Kremser A, Jochmann MA, Schmidt TC (2016) Systematic comparison of static and dynamic headspace sampling techniques for gas chromatography. Anal Bioanal Chem 408: 6567–6579
8. Elmore JS (2015) Extraction techniques for analysis of aroma compounds. In: Flavour development, analysis and perception in food and beverages. Elsevier, pp 31–46
9. De Jong C, Van Kekem C, Azario M et al (2022) Comparison and optimisation of headspace methods for the analyses of oxidation related off-flavour compounds in plant protein concentrates and infant formula. In: Proceedings of the 16th Weurman flavour research symposium
10. Moreira C, Ferreira-Santos P, Teixeira JA et al (2022) Active aroma compounds assessment of processed and non-processed micro-and macroalgae by solid-phase microextraction and gas chromatography/mass spectrometry targeting seafood analogs. Front Sustain Food Syst 6:1011020
11. Jeleń HH (2006) Solid-phase microextraction in the analysis of food taints and off-flavors. J Chromatogr Sci 44:399–415
12. Spietelun A, Kloskowski A, Chrzanowski W et al (2013) Understanding solid-phase microextraction: key factors influencing the extraction process and trends in improving the technique. Chem Rev 113:1667–1685
13. Song H, Liu J (2018) GC-O-MS technique and its applications in food flavor analysis. Food Res Int 114:187–198
14. Shen D-Y, Li M-K, Song H-L et al (2021) Characterization of aroma in response surface optimized no-salt bovine bone protein extract by switchable GC/GC× GC-olfactometry-mass spectrometry, electronic nose, and sensory evaluation. LWT 147:111559
15. Singh A, Shi Y, Magreault P et al (2021) A rapid gas-chromatography/mass-spectrometry technique for determining odour activity values of volatile compounds in plant proteins:

soy, and allergen-free pea and brown rice protein. Molecules 26:4104

16. Nolvachai Y, Amaral MS, Herron R et al (2023) Solid phase microextraction for quantitative analysis–expectations beyond design? Green Anal Chem 4:100048
17. Johnson CB, Kazantzis A, Skoula M et al (2004) Seasonal, populational and ontogenic variation in the volatile oil content and composition of individuals of Origanum vulgare subsp. Hirtum, assessed by GC headspace analysis and by SPME sampling of individual oil glands. Phytochem Anal 15:286–292
18. Deng H, He R, Huang R et al (2022) Optimization of a static headspace GC-MS method and its application in metabolic fingerprinting of the leaf volatiles of 42 citrus cultivars. Front Plant Sci 13:1050289
19. Cai J, Lin P, Zhu X et al (2006) Comparative analysis of clary sage (S. sclarea L.) oil volatiles by GC–FTIR and GC–MS. Food Chem 99: 401–407
20. Vichi S, Riu-Aumatell M, Mora-Pons M et al (2007) HS-SPME coupled to GC/MS for quality control of Juniperus communis L. berries used for gin aromatization. Food Chem 105:1748–1754
21. Azimi F, Fatemi M (2016) Multivariate curve resolution-assisted GC-MS analysis of the volatile chemical constituents in Iranian Citrus aurantium L. peel. RSC Adv 6:111197–111209
22. Zhogoleva A, Alas M, Rosenvald S (2023) Characterization of odor-active compounds of various pea preparations by GC-MS, GC-O, and their correlation with sensory attributes. Fut Foods 8:100243
23. Ebert S, Michel W, Nedele AK et al (2022) Influence of protein extraction and texturization on odor-active compounds of pea proteins. J Sci Food Agric 102:1021–1029
24. Bianchin JN, Nardini G, Merib J et al (2014) Screening of volatile compounds in honey using a new sampling strategy combining multiple extraction temperatures in a single assay by HS-SPME–GC–MS. Food Chem 145: 1061–1065
25. Fischer E, Cachon R, Cayot N (2021) Effects of extraction pH on the volatile compounds from pea protein isolate: semi-quantification method using HS-SPME-GC-MS. Food Res Int 150:110760
26. Tunick MH, Iandola SK, Van Hekken DL (2013) Comparison of SPME methods for determining volatile compounds in milk, cheese, and whey powder. Food Secur 2:534–543
27. Solina M, Baumgartner P, Johnson RL et al (2005) Volatile aroma components of soy protein isolate and acid-hydrolysed vegetable protein. Food Chem 90:861–873

Chapter 32

Twin-Screw Extrusion Processing for Textured Plant Protein Meat Analogs

Qiong Chen and Minwei Xu

Abstract

Texturized plant proteins (TPP) are increasingly used as meat analogs due to concerns over sustainability, health, and ethics. Twin-screw extrusion allows for efficient, continuous production of TPP meat analogs with meat-like texture. This chapter provides a comprehensive guide on using twin-screw extrusion to produce TPP meat analogs at a laboratory scale. Detailed protocols are presented on extruder setup, ingredient preparation, feed calibration, extrusion processing, extrudate analysis, and machine cleaning for both low-moisture and high-moisture extrusion of soy protein and wheat gluten mixtures. Key extrusion parameters including temperature profile, screw speed, moisture content, and screw configuration are discussed in relation to texturization and functional properties of TPP extrudates. Methods for evaluating the degree of protein texturization, hardness, and tensile properties are provided. By following the procedures outlined, this chapter aims to aid the development and optimization of TPP meat analogs with tailored textural qualities.

Key words Texturized plant protein, Meat analog, Extrusion, Operation procedure

1 Instruction

The abuse of veterinary drugs and zoonotic disease infection in the production or consumption of animal meats have led to public food safety issues, moreover, the dietary high meat consumption can induce dietary disorders or chronic diseases (e.g., obesity and angio-cardiopathy) [1]. Due to these and customers' awareness of environmental burdens, religious restrictions, animal welfare, and ethical/social aspects, meat analogs, especially plant protein-based, have evoked consumers' attention [1]. Meat analog includes in vitro cultured meat, microbial mycelium protein, and texturized plant protein (TPP) meat analogs [1]. Extrusion technology has been employed to produce TPP for over 50 years [2] and is increasingly gaining popularity due to the advantages of energy efficiency, high productivity, low cost, continuous operation, automatic control, versatility, adaptability, and environment-friendly [3].

Yonghui Li (ed.), *Plant-Based Proteins: Production, Physicochemical, Functional, and Sensory Properties*, Methods and Protocols in Food Science, https://doi.org/10.1007/978-1-0716-4272-6_32,

Extrusion is a process forcing a food material to flow, under conditions such as a combination of mechanical and thermal treatment like compression, mixing, shearing, kneading, and high-temperature cooking, through a narrow die at a predetermined rate to produce semi−/fully-cooked food with minimum nutrition losses [4]. As for TPP, it can be divided into low moisture TPP (LMTPP, containing about 20–40% moisture) and high moisture TPP (HMTPP, containing about 40–80% moisture) based on the moisture content [2]. While low moisture extrusion (LME) is the first-generation method for meat analogs manufacturing, high moisture extrusion (HME) is technologically more advanced and imparts to meat analogs a delicate fibrous structure that LMTPP cannot match [5]. HME technology, initiated in the 1990s, is developed on low moisture extrusion (started in the 1960s) and has still been in the research period [2]. The TPP meat analogs prepared using HME technology are similar in texture to animal meat with a dense fibrous structure, strong elasticity, and moisture content, and are considered to be an ideal substitute for animal meat with the advantages of low saturated fat, free cholesterol, high protein or essential amino acids content, and good digestibility [1]. The HME typically utilizes a specially designed machine, known as a twin-screw extruder, which is equipped with two long screws and a lengthy cooling die equal to the length of the barrel at the end. This setup helps in realigning proteins to form an anisotropic network in the flow direction [6, 7]. In contrast, the LME technology can employ either a single-screw or twin-screw extruder, both with a much shorter cooling die [1, 2]. At the die for LME, the release of pressure results in water evaporating and starch granules puffing, thus forming the air cells in LMTPP [2]. Twin screw extruder, which is more complicated than and prevails the single screw one, consists of two equal length parallel screws rotating inside the same barrel, of which the internal surface is smooth [3, 8]. Based on the screws position and rotation direction, the configuration of the twin screw extruder can be divided into co-rotating non-intermeshing screws and counter-rotating non-intermeshing screws [9]. As for the ingredient used in extrusion, soy protein by far is the most studied plant protein in extrusion and is used as a single ingredient by different authors or mixed with gluten by others [10]. Hence, these commonly used ingredients are to be adopted as experimental materials in this protocol.

The objective of this chapter is to offer a comprehensive guide for the production of both LM- and HM-TPP meat analogs using extrusion technology with a co-rotating intermeshing twin-screw extruder. This guide covers procedures from setting the parameters for the extruder to the methodology for its operation. The primary steps of the extrusion process encompass sample preparation, feed calibration, extrusion cooking, cleaning procedures, and extrudate analysis. Consequently, this chapter aims to standardize extrusion processing protocols and promote the use of TPP extrusion technology with the twin-screw extruder.

2 Materials

2.1 Main Instrument Configuration

This protocol is established using a Brabender TwinLab-F 20/40 (Brabender GmbH & Co. KG, Duisburg, Germany). The machine system of this equipment includes a co-rotating twin extruder, an integrated drive unit, an integrated temperature and speed control unit, and a water-cooling unit for cooling the control zones. The extruder screw has a diameter of 20 mm and a length-diameter ratio of 40:1 (Fig. 1). The extruder barrel is arranged into five separate heating zones. These zones are set to gradually increase until the final zone where a peak temperature is held. The cooling die at the end of the barrel is 315 mm long with a size of 7 mm × 25 mm (Details are listed in Fig. 2).

An independent CW Brabender feeder (DDSR20 2.0) is equipped which is composed of a stainless-steel screw trough, an agitator in the screw trough, a twin screw, a screw tube, a three phase AC motor, and an extension hopper with volumes of 10 dm^3 (0.35 cuft), shown as in Fig. 3. The screw feeder DDSR20 is a volumetric and gravimetric feeder equipped with a Brabender OP1-S operator interface.

2.2 Other Instruments or Devices

TCS system electronic platform scale (max 100 kg, graduation 20 g, GROMY®, OEM, Zhejiang, China), Explorer Precision Balance (Ohaus EX6202/E, USA), Ultra-Turrax T45 homogenizer (IKA WORKS INC., Germany), Commercial Stand Mixer (VEVOR, China).

2.3 Materials

Commercial soybean protein isolates (≥ 90% on dry matter basis, Solae LLC, St. Louis, MO, USA), gluten (≥75% protein content, Permolex LTD, Red deer, Alberta, Canada), salt (Iodized Sea Salt, MORTON®), vegetable oil (Cholesterol free, OurFamily®, distributed by Nash Finch Company), emulsifier (Polysorbate-80 Emulsifying Agent, Traverse Bay Bath and Body). All the materials used in this protocol can be substituted with alternatives; however, the protein content must meet the requirements specified above.

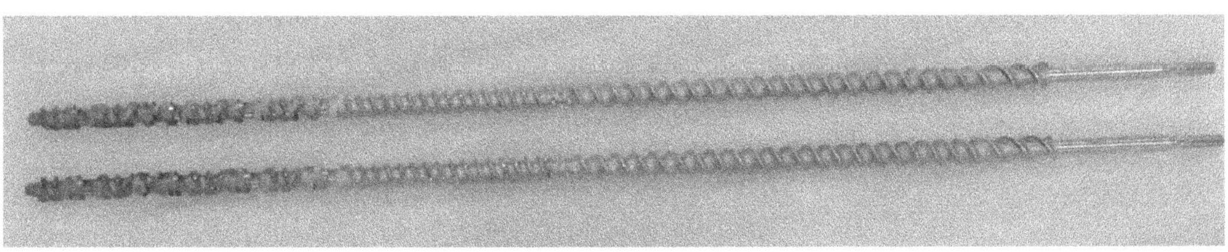

Fig. 1 Twin screw

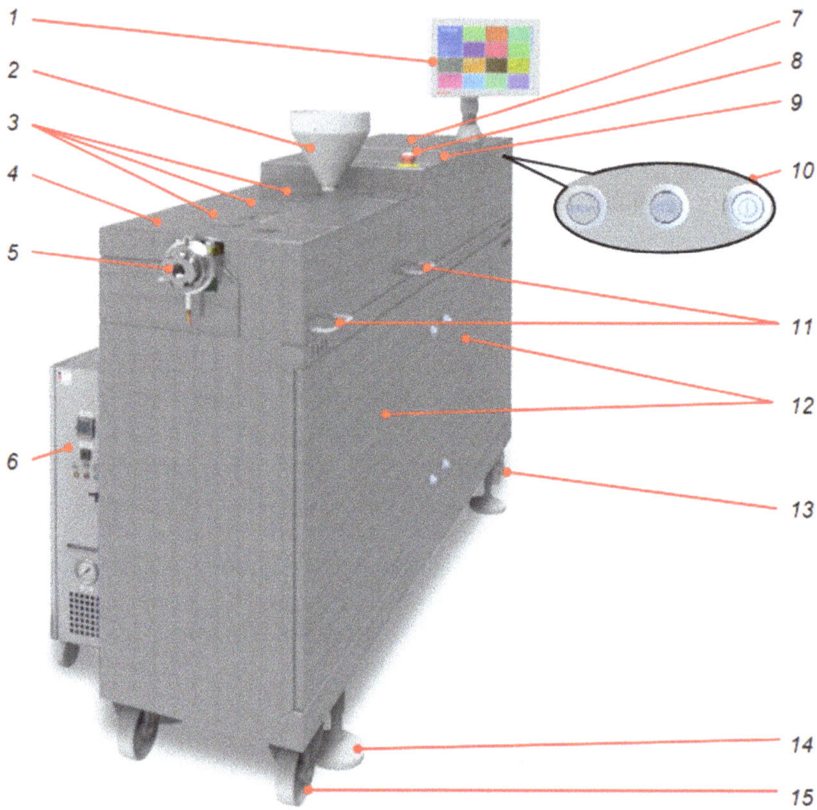

Fig. 2 Main parts of the Brabender TwinLab-F [11]. 1. Touchpanel with integrated MetaBridge software; 2. Feed hopper or position of main feeder; 3. Positions of additional feeders; 4. Processing unit, divided horizontally (under the cover hood); 5. Die adapter; 6. Water cooling unit (option); 7. Drive motor (under the cover hood); 8. Emergency motor stop button; 9. Clutch (under the cover hood); 10. Keys "Start"/"Stop"/"PC ON/OFF"; 11. Handles for opening the cover hood over the processing unit; 12. Doors to the switch cabinet; 13. Swivel castor; 14. Leveling foot; 15. Fixed castor. The picture is extracted from the manual

3 Methods

3.1 Instrument Mounting Before Experiment

Following the mounting process in Fig. 4. Care should be taken when mounting the agitator, *see* **Note 1**.

3.1.1 Feeder Mounting

3.1.2 Extruder Mounting

Follow the process in Fig. 5. Turn on the touch-panel by pressing the key "PC ON/OFF." Tapping the "New" tile opens the equipment setup window, shown as Fig. 6. Tap the temperature zone (Red square zone) and set the barrel temperature as follows, 40, 80, 100, and 120 °C from zone 1 to zone 4, respectively. These temperatures are pre-tested and are going to be applied to produce HME meat analog in the following extruding process. Tap "start

Fig. 3 Screw feeder DDSR20 gravimetric and volumetric

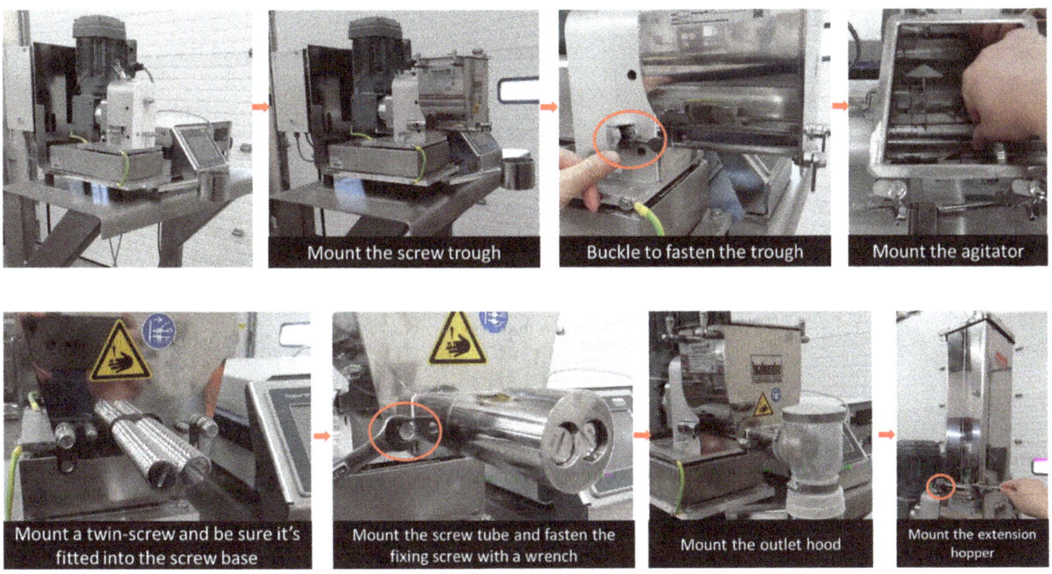

Fig. 4 Feeder mounting process

measurement" and "Data Recording Start/stop." Other parameters may be set in "Diagram settings." The temperatures could also be adjusted by tapping the temperature zone (Red square zone), shown as Fig. 7.

After the temperature settings, the temperature of the barrel begins to rise gradually, and the machine mounting could be started as following:

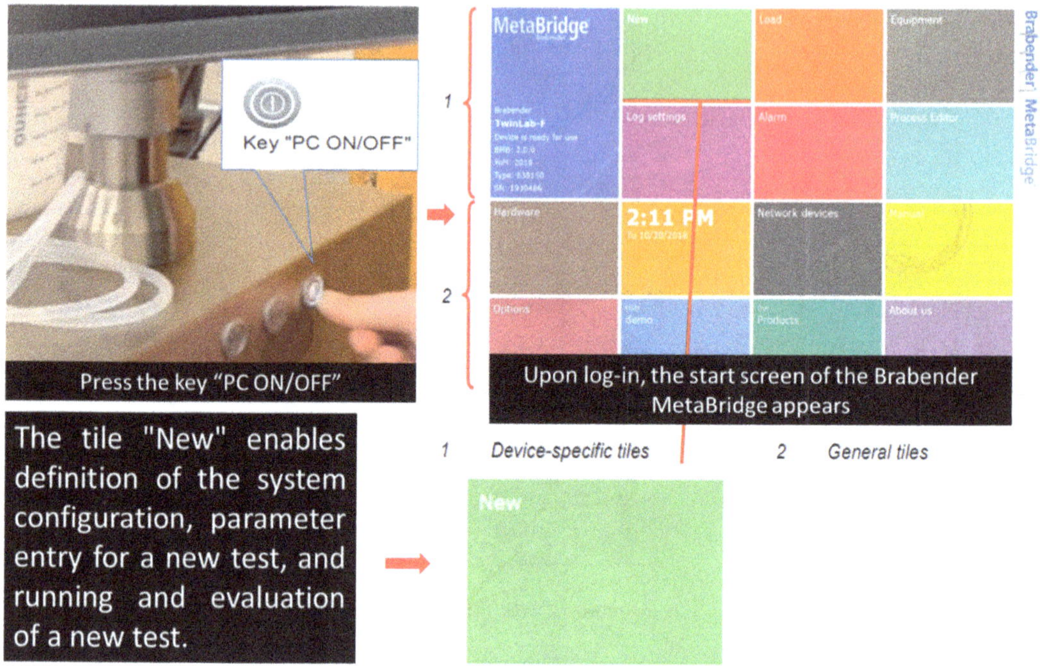

Fig. 5 Equipment setup window by tapping the tile "New" [11]. The picture is extracted from the manual and re-edited

1. *Mounting the twin screws as* Fig. 8. Roll the intermeshing screws over an even surface to check the screw configuration (Fig. 8). The screws must roll without any obstruction. When inserting the screws into the liner, take utmost care that the screws shafts and tips are exactly in line with each other (Fig. 8)! Insert the two screws into the screw shaft coupling with an offset of 90° towards each other and slightly turn the screws back and forth in order to push them completely into the location holes. *See* **Note 2**.
2. *Mounting the lateral/upper closing plugs, fixing screws M8 and straining screws as in* Fig. 9. Mount the lateral closing plugs into the side feeder openings, then mount the lower fixing screws of the lateral closing plugs and tighten them by hand (Fig. 9). Make sure that the insides of the two halves of the processing unit are clean, that no tools have been left on them, and that no cables or hoses can be squeezed when closing the two halves of the processing unit. Hold the upper half of the processing unit on the two handles and tilt it down so that the two halves of the processing unit are aligned with each other. Mount the upper fixing screws of the lateral closing plug and tighten them by hand. Insert the straining screws of the processing unit and tighten them by hand. Notes: Upper closing plugs with measuring bores should be mounted near the feed port (Fig. 11).

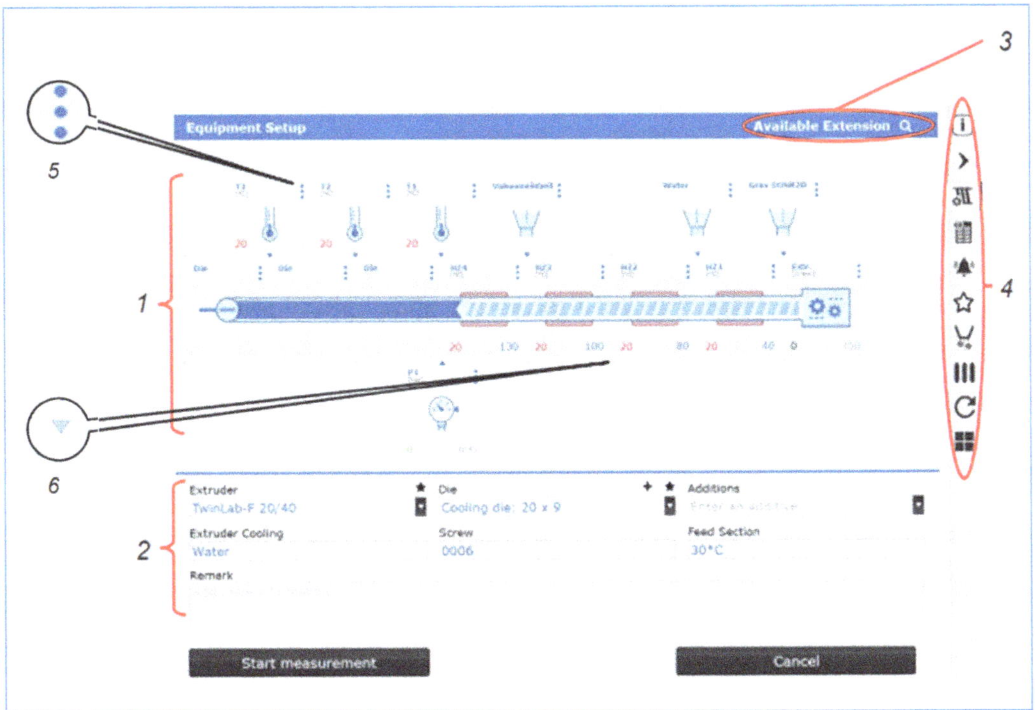

Fig. 6 Equipment setup window [11]. 1. Representation of the equipment setup; 2. Selection/entry fields for the system modules; 3. Opens a window with the modules recognized for positioning them as required; 4. Action bar; 5. Opens a window with options for the corresponding module; 6. Enables the selection of additional components. The picture is extracted from the manual

3. *Tightening fixing screws M8 and straining screws as* Figs. 10 and 11. Make sure that the screws are put in the right location. The closing plug with upper feed openings should be arranged next to the feed hopper port (Figs. 10 and 11). Use a torque wrench to tighten the screws of the lateral/upper closing plugs evenly in two passes first to 15 Nm, then to 30 Nm when the barrel is warm. Use a torque wrench to tighten the straining screws of the processing unit in two passes first to 30 Nm, then to 60 Nm when the barrel is warm, following the order marked in Figs. 10 and 11.
4. *Mounting the die adapter.* Hold the die adapter in front of the barrel plate of the TwinLab-F so that the two alignment pins of the die adapter engage into the corresponding bores in the barrel plate and the bursting pin shows downwards at an angle away from the control side (Fig. 12). Pressure transducer, melt temperature thermocouple, and bursting pin could be previously mounted in the die adapter (Caution, *see* **Note 3**). To tighten the measuring transducer firmly, wait until the system has reached its operating temperature! Always use a torque wrench for tightening the measuring transducer firmly!

Tapping the button "Diagram settings" allows individual adjustment of the measuring window and the displays therein.

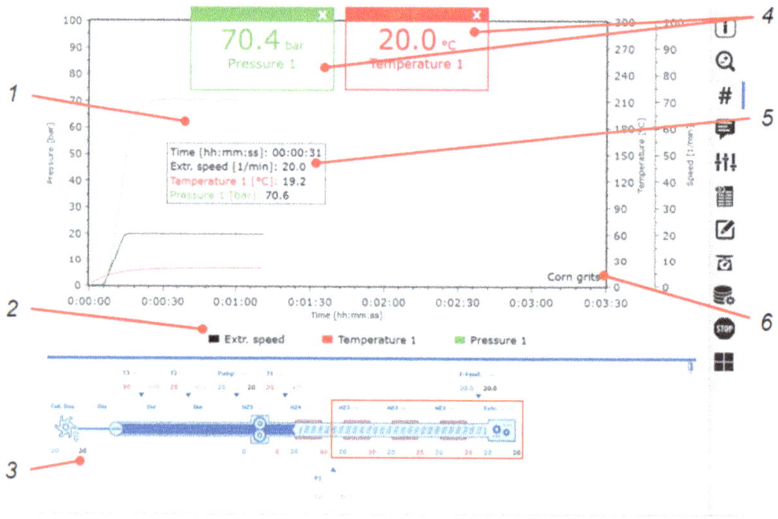

Fig. 7 Measuring window, here: with display of numerical details [11]. 1. Measuring diagram; 2. Legend; 3. Equipment setup with live ticker; 4. Measurement window (to show, tap the corresponding measuring value in the equipment setup); 5. Numerical details; 6. Sample name. The picture is extracted from the manual and re-edited

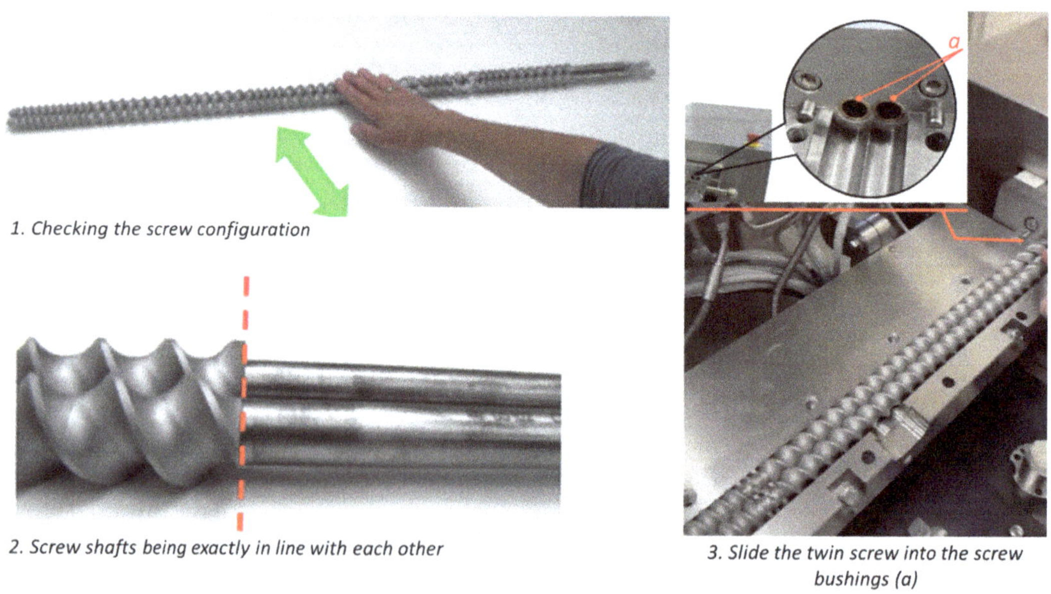

Fig. 8 Twin screw mounting [11]. The picture is extracted from the manual and re-edited

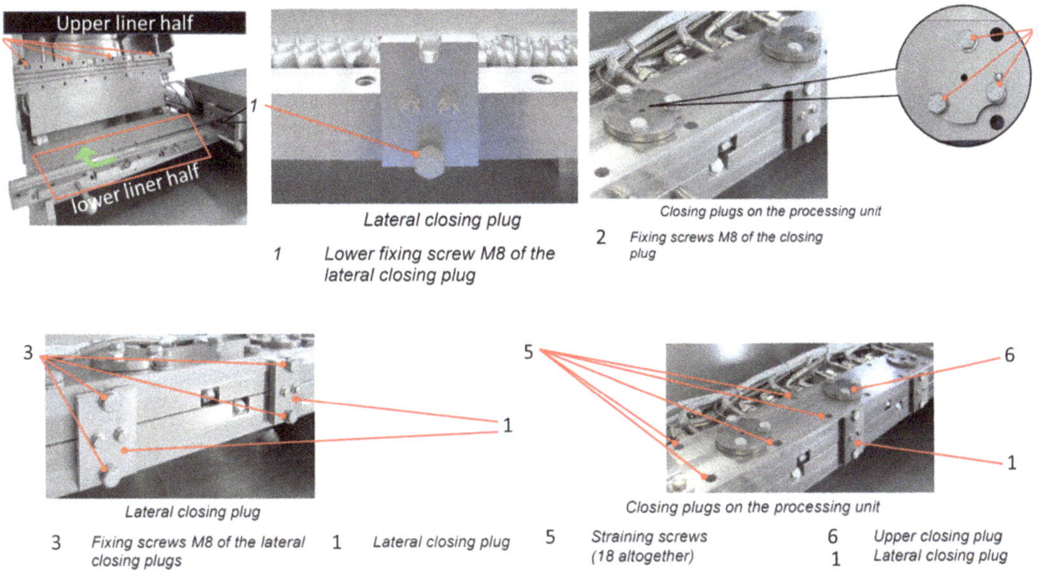

Fig. 9 Mounting lateral/upper closing plugs, fixing screws M8 and straining screws [11]. The picture is extracted from the manual and re-edited

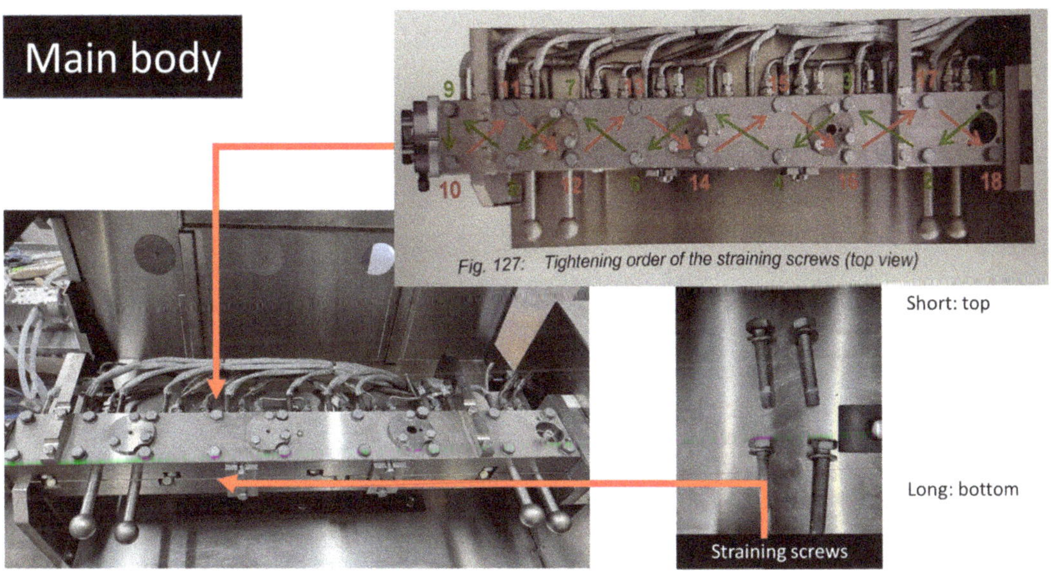

Fig. 10 Illustration of straining screws and fixing screws (M8) tightening [11]. The picture is extracted from the manual and re-edited

Do not exceed the max. Torque of 20 Nm in hot conditions! Tighten the four fixing screws of the die adapter crosswise by hand in order to align the two halves of the processing unit axially with each other so that they flush at the front. Use a torque wrench to tighten the four screws of the die adapter crosswise in two passes first to 15 Nm, then to 30 Nm.

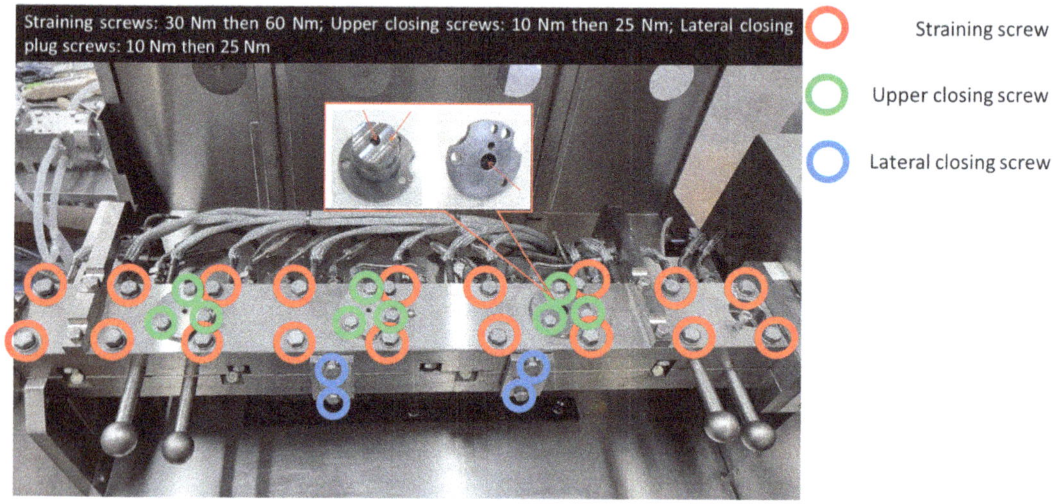

Fig. 11 Screw tightening layout [11]. The picture is partially extracted from the manual and re-edited

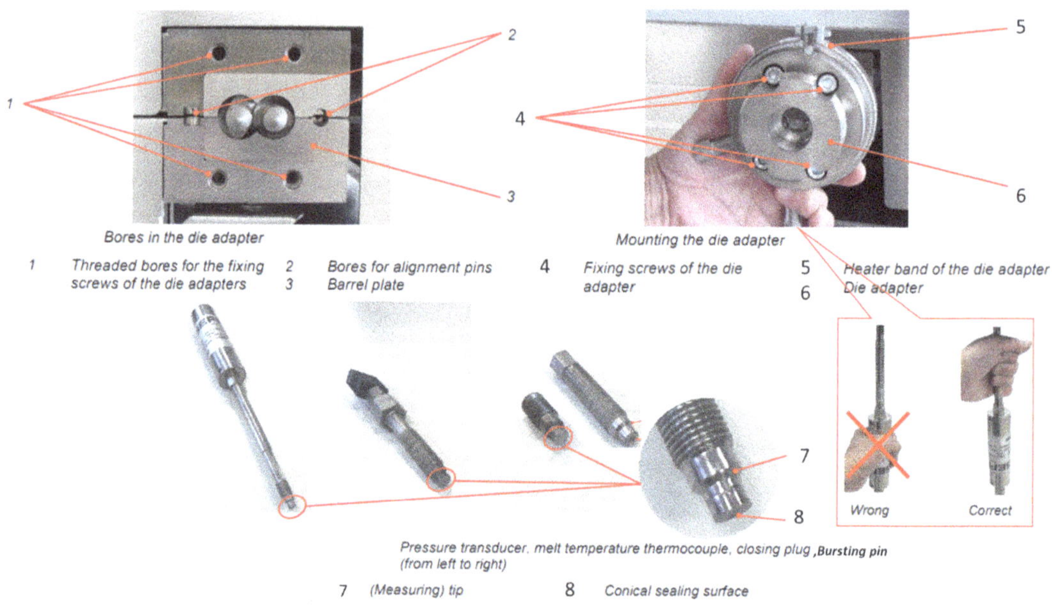

Fig. 12 Die adapter mounting [11]. The picture is extracted from the manual and re-edited

5. *Mounting the die head and the cooling die.* Step 1: Mount the die head to the long cooling die (Fig. 13) and torque evenly by hand. Step 2: Place a copper sealing ring (ID no. 2 14 106) on the sealing surface of the measuring extruder from the discharge side (Fig. 13). Step 3: Put on the die head and tighten the ring nut on the threaded ring of the die adapter by hand. The final cooling die should be mounted as Fig. 14.

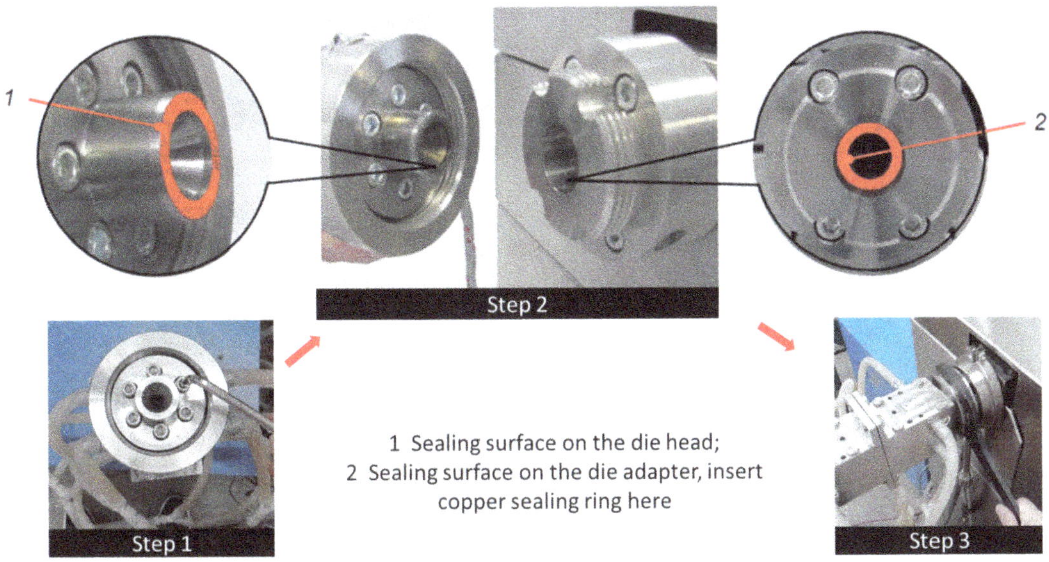

Fig. 13 Die head mounting [11]. The picture is extracted from the manual and re-edited

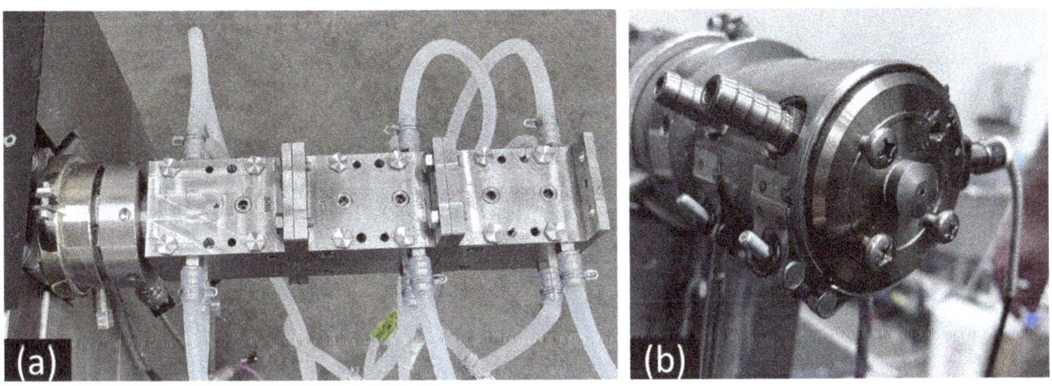

Fig. 14 Cooling die for high moisture meat analog (**a**) and the die for low moisture meat analog (**b**)

3.2 Sample Preparation

Solid ingredients preparation: Weigh 2.37 kg Soybean protein (79%), 0.60 kg gluten (20%) and 30 g salt (1%), stir well using a mixer (Fig. 15). After mixing, estimate the density of the powder. Weigh 250 mL of the solid mix using the electronic balance (Ohaus EX6202/E Explorer) and record the weight m (g). The estimated density of the solid ingredient (ρ, kg/L) can be calculated as following equation:

$$\rho = \frac{m}{250} \quad (1)$$

Determine the moisture content (α) of the mixed ingredients gravimetrically.

Fig. 15 Preparation of solid ingredients

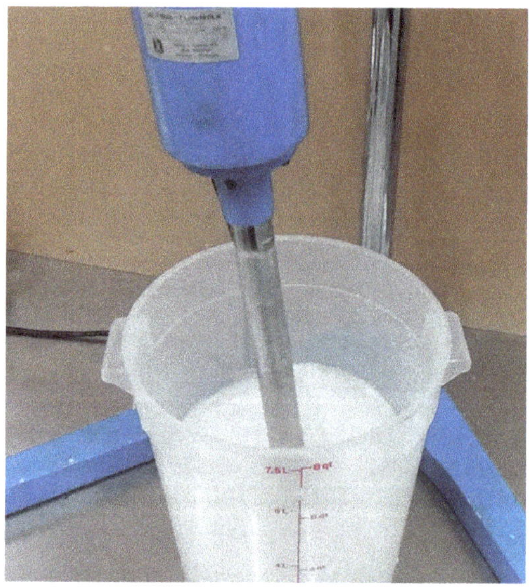

Fig. 16 Preparation of liquid ingredients

Liquid ingredients preparation: Liquid ingredient consists of water, oil, and emulsifier (with a ratio of 197: 1: 2 w/w/w). A total of 2 kg liquid ingredient is prepared using a homogenizer (Ultra-Turrax T45) at 20,000 rpm for 5 min (Fig. 16).

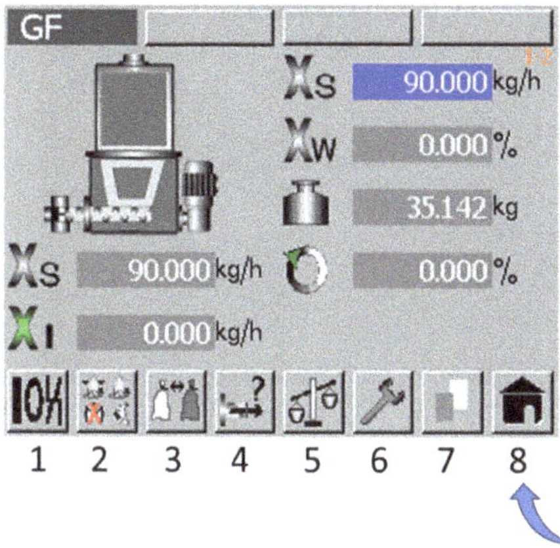

No	Signification
1	Calls the selection menu "Start/Stop".
2	Calls the selection menu "Mode".
3	Calls the screen page "Product parameters".
4	Calls the screen page "Checking max. output".
5	Calls the screen page "Calibration".
6	Calls the user level "Service" *This level is interlocked. The use is only allowed to trained service personnel.*
7	Scroll forward the screen page, if the page numbering shows that [number of pages] > 1.
8	Calls the start screen page (HOME function)

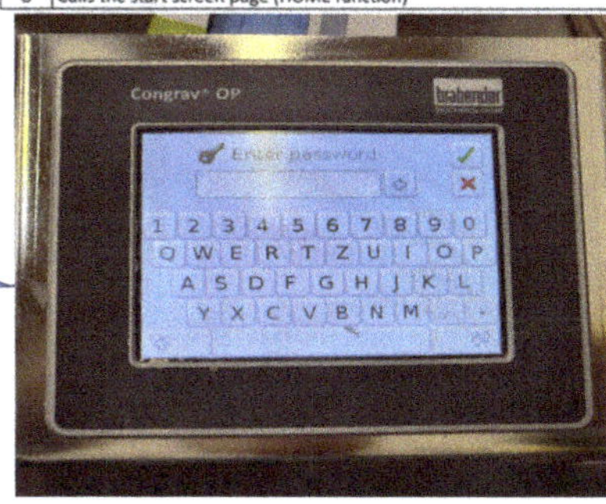

Fig. 17 Entering password to load homepage

3.3 Feeder Precondition and Solid Ingredients Feeding Calibration

3.3.1 Feeder Precondition

1. *Enter the password to log into the homepage* (Fig. 17).
2. *Check the "Service" settings before system operations.* Tap "Service (6)" icon, a window will show up as Fig. 9. And make sure all the indices in "Setup" "Parameter" "Error-Configuration" "I/O-Configuration" are correct (Fig. 18).
3. *Tare before adding in solid ingredients from the extension hopper.* Do not add any ingredients before taring. Tap the "Mode (2)" icon, and choose the gravimetric mode (Fig. 19). Then tap "Calibration (5)" icon, manually input the value (0 kg) of "Tare" marked in a red cell, or activate the "auto-tare" marked in b red cell (Fig. 20).
4. *Add solid ingredients to the extension hopper.* After taring to zero, solid ingredients can be added to the extension hooper of the feeder (Fig. 21) (Caution, *see* **Note 4**).

3.3.2 Solid Ingredients Feeding Calibration

1. Tap the "Mode (2)" icon, choose the gravimetric mode and then tap the "Discharge mode" (Fig. 22). Tap to select the menu "Start/Stop (1)" and press the "Start component" (Green), then the machine starts discharging to evacuate the air in screw tube. After the tube screw is filled with ingredients, hit "Stop component (Red)" to stop. During discharge, hold a cup under the discharge port.
2. Tap "Product parameter (3)" icon, set an alarm in the "Filling level minimum" and "Filling level maximum." Enter the estimated density of the mixed ingredient (ρ) (Fig. 23).

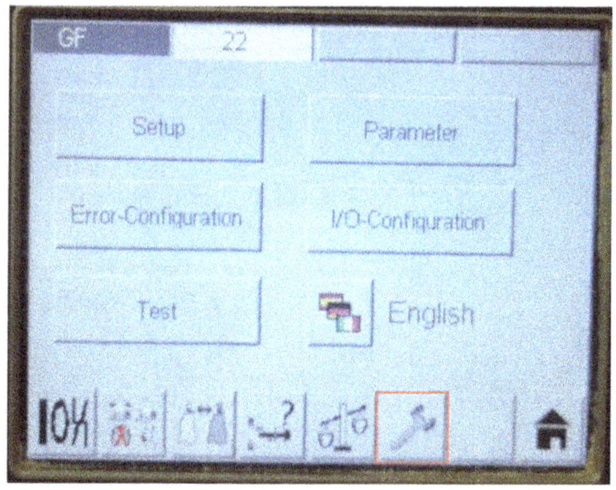

Fig. 18 Window of "Service"

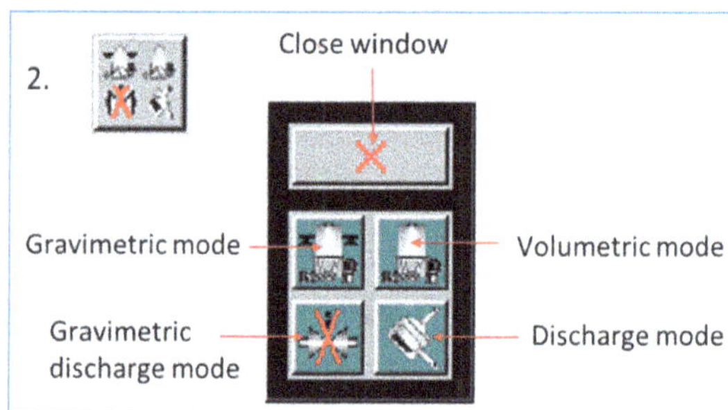

Fig. 19 Details in "Mode" icon

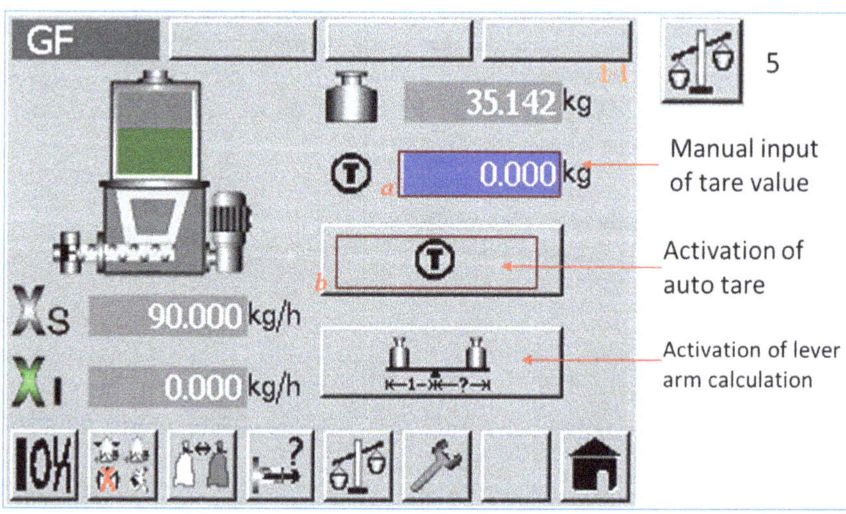

Fig. 20 Illustration of taring to zero

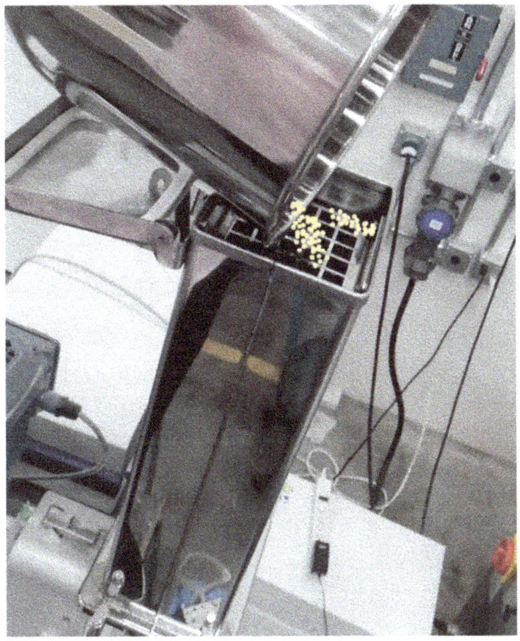

Fig. 21 Solid ingredients feeding into the feeder

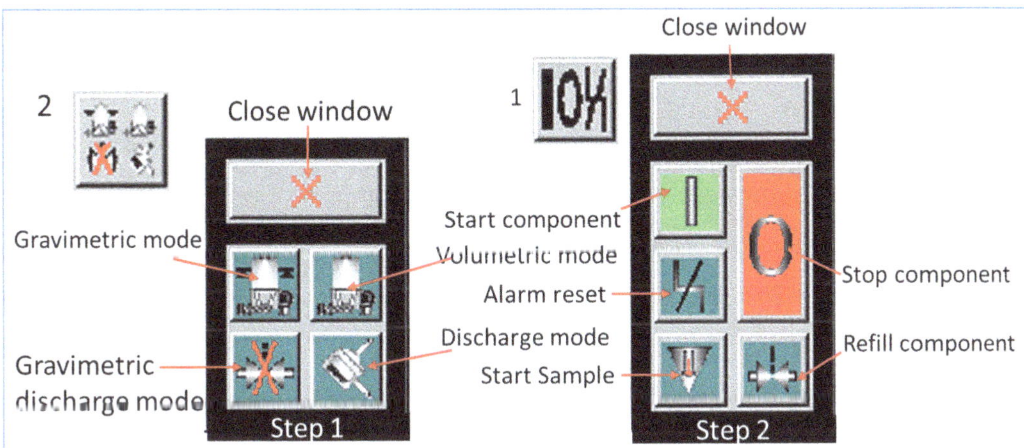

Fig. 22 Discharge procedure

3. Tap the "Check max. output," select upper speed for test, input duration of test, and then tap "Activate CM operating mode." Go to the "Start/Stop (1)" page and tap the green icon (Start component) to activate CM operating (Fig. 24). Hold a cup under the discharge port for 60 s (Fig. 25). This step is to make the machine calculate how much material it loses from the weight.

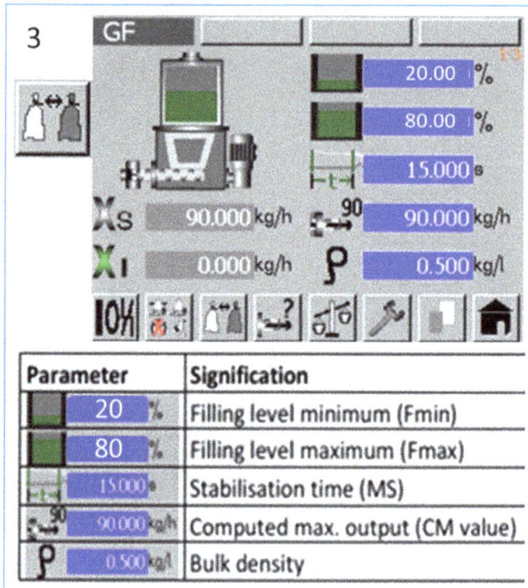

Fig. 23 Product parameter

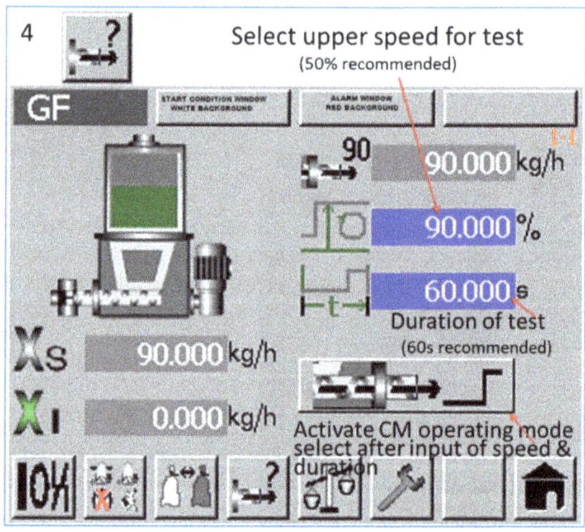

Fig. 24 Checking maximum output

4. Tap the "HOME" icon, set the feeding rate at 1.8 kg/h (F_S) in the "X_S" column (Fig. 17), which was previously tested by Dr. Xu's team [12]. Go to the "Start/Stop (1)" page. Pressing the green icon will start the ingredients feeding.
5. Place a cup in an electronic balance and tare. Hold the cup under the discharge port for 60 s, then manually weigh it on the electronic balance and record the weight (m_c, kg). The actual feeding rate F_V (kg/h) = 60× m_c. Adjust the feeding rate to 1.8 kg/h by entering a value higher or less than F_S in the "X_S" column if $F_V \neq F_S$.

Fig. 25 Holding a cup under the discharge port

3.4 Liquid Ingredients Feeding Calibration for HMTPP

1. Calculate the liquid feeding rate. The moisture content of the solid ingredient (α, %) is determined in Subheading 3.3.1. If the solid feeding rate is F_S, the liquid ingredient feeding rate is F_E and the moisture of the final products is β (%). The relationship between F_S and F_E is described as Eq. (1):

$$\frac{F_S \times \alpha + F_E}{F_S + F_E} = \beta \tag{2}$$

   Hence, the feeding rate of the liquid ingredient through the dosing pump should be $F_E = \frac{\beta - \alpha}{1 - \beta} \times F_S$. For example, if the solid ingredient moisture content is 5% (α) and to obtain the moisture content 55% (β) in final product. The feeding rate of the solid ingredient is 1.8 kg/h. Then the liquid ingredient feeding rate should be $F_E = \frac{55\% - 5\%}{1 - 55\%} \times 1.8 = 2.0$ (kg/h).

2. Turn on the dosing (Peristaltic) pump and set a running speed R (e.g., 80 rpm) and press to start. Collect the liquid ingredient for 60 s in a vessel at the end of feed bolt and weigh (m_w), (Fig. 26). For example, the actual liquid ingredient feeding rate is F_W (kg/h) = 60 m_w. If the actual liquid ingredient feeding rate is 1.6 kg/h, the final running speed could be adjusted to 100 rpm based on Eq. (3)

$$R_0 = \frac{R}{F_W} \times F_E \tag{3}$$

3. Mount the feed bolt into the closing plugs of the upper feed openings (Fig. 27).

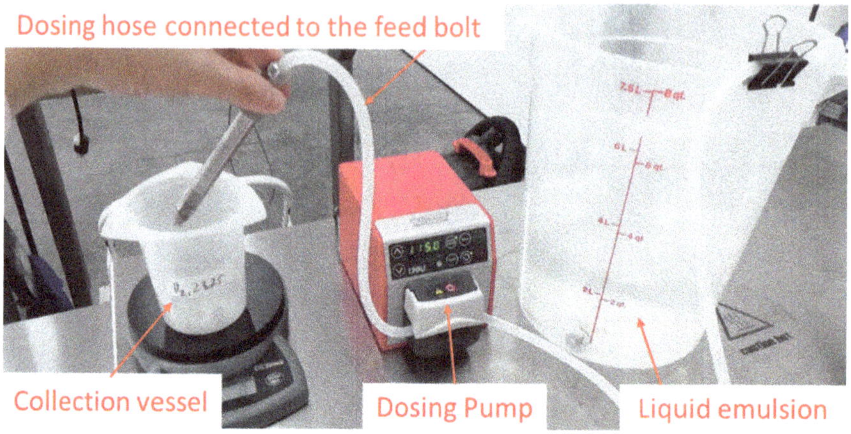

Fig. 26 Liquid ingredients feeding calibration

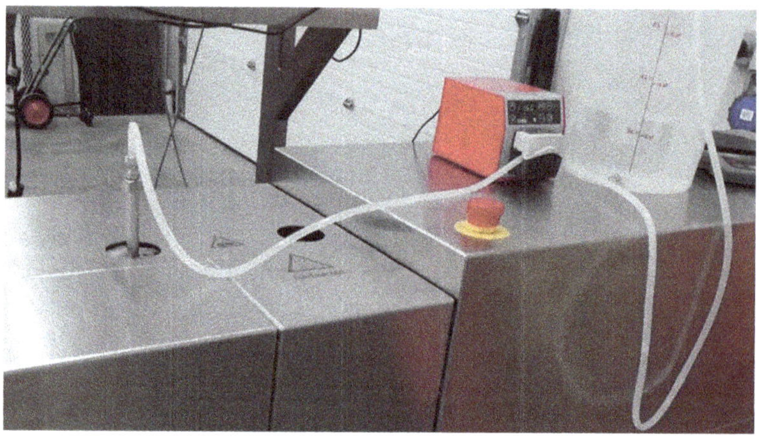

Fig. 27 Feed bolt mounting in the closing plugs of the upper feed openings

3.5 Extrusion Processing (HME Meat Analog Production)

1. Bring any feeders into ready mode, e.g., the Brabender screw feeder for solid ingredients feeding and the feed bolt for Liquid feeding (Fig. 28).
2. Switch on the tap water supply. Turn on the cooling circulator for the control zone (barrel). For producing high moisture meat analog, switch on the cooling water of die set the temperature at 40 °C.

   Hot products may have surging issue before stabilizing. Hence, it's better to arrange a baffle in front of the exit die (Fig. 29).
3. Press the "Start" key on the TwinLab-F in order to activate the drive unit. Set the system configuration in the software and set the process parameters in the red box (Fig. 30). Twin-screw extruder barrel temperature is set as follows, 40, 80, 100, and 120 °C (suggested) from zone 1 to zone 4, respectively. Zone

Fig. 28 Bringing feeders into ready mode

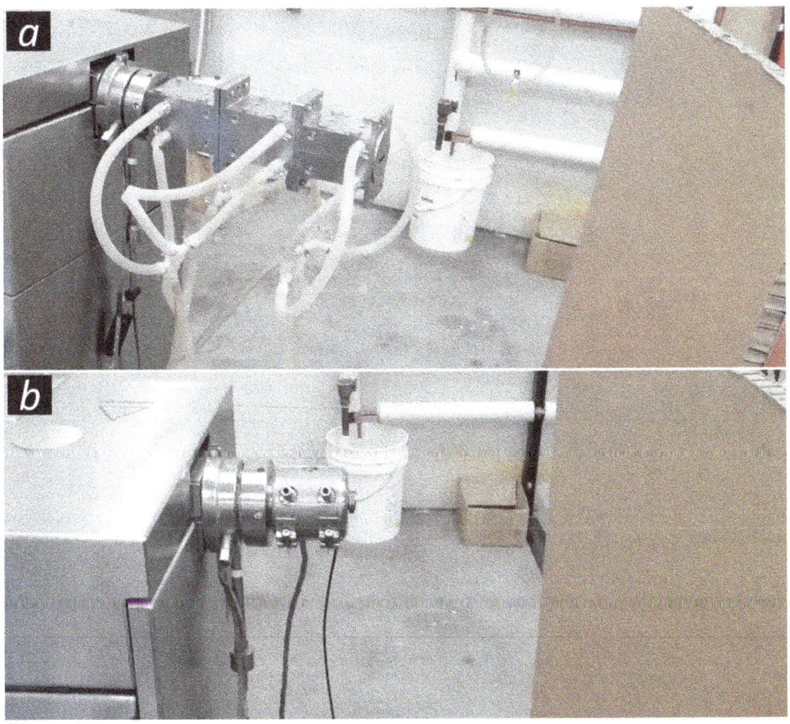

Fig. 29 Arrange a baffle in front of the exit die (**a** (HMTPP) or **b** (LMTPP))

5 is connected to the (7 mm × 25 mm) rectangular exit die. The texturization temperature of zone 5 is set at least 120 °C. The screw speed can be adjusted from 400 to 1200 rpm. Start the individual drive units one after the other through the software: Tap the individual drive units in the system configuration window below the online diagram (Fig. 31). In the subsequent window, enter the setpoint speed for the drive unit and activate

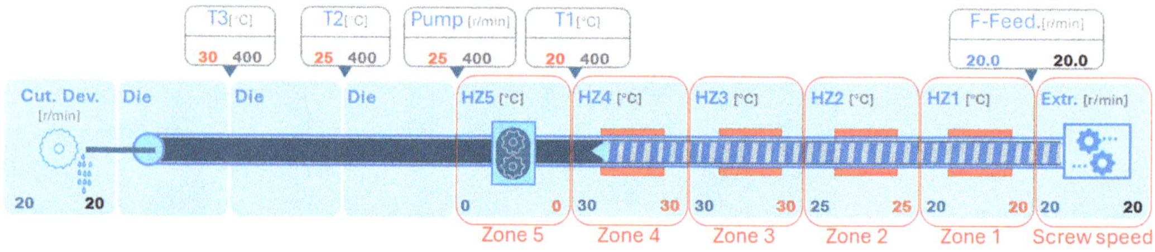

Fig. 30 Setting the process parameters [11]

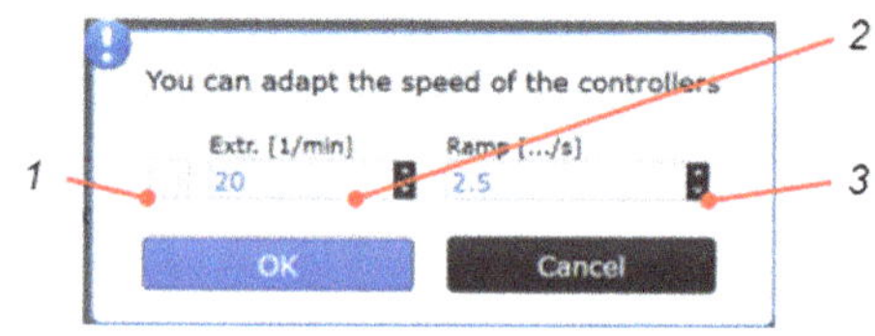

Fig. 31 Starting the drive units [11]. 1. Checkbox for starting the drive unit; 2. Entry field for setpoint speed; 3. Entry field for speed ramps. The picture is extracted from the manual and re-edited

the checkbox in order to start the drive unit. (Caution, *see* **Note 5**). The extruder drive motor will not start before the difference between the actual and setpoint temperature is <10 °C.

After warming up of the barrel, the dry ingredients feeding rates are set at 1–3 kg/h. For HMTPP, the liquid ingredient is calculated from Eq. (2). As for LMTPP, the liquid ingredient could be set at a ratio of 1:1 to solid liquid. Slowly feed the ingredients into the extruder: Feed the ingredients into the feeder and have the feeder run slowly. Continue feeding ingredients into the machine until product/melt comes out of the die. First carefully increase the speed of the TwinLab-F to the desired value, then carefully increase the speed of the feeder(s) as well. Always remember to start the liquid feeding first!

3.6 Emptying the Extruder After Experiment

Run the machine at operating temperature without feeding any more products until no more products exit the die head. Tap the button "Stop" in the action bar beside the measuring window in order to stop the measurement and the drive unit. Set the temperature of the barrel, die adapter, and die head to "0" in the software.

3.7 Feeder Cleaning

Close the extruder from the software, dismantle and clean the feeder as Fig. 32. After cleaning the feeder, air-dry it and disinfect the contacting surface using food grade disinfectant.

3.8 Cleaning Procedures After Extrusion

All the following steps should be done after the machine is cooled down. All the disinfectants should be food-grade. Caution should be taken! (*see* **Note 6–9**).

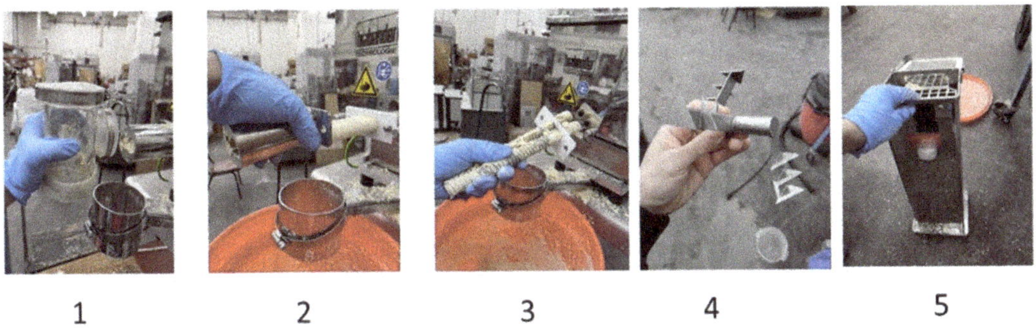

Fig. 32 Dismantling and cleaning the feeder

1. *Dismantling and cleaning the feed hopper.* Slightly turn the feed hopper to unlock it and lift it off. Clean the inner surfaces of the feed hopper with a clean damp, non-linting cloth. Apply a disinfectant to a clean, non-linting cloth and disinfect the products' wetted surfaces of the hopper with this cloth. Leave to air dry.
2. *Dismantling and cleaning the measuring transducers.* Unscrew the measuring transducers and/or the corresponding closing plugs from the die adapter and put them onto a suitable, heat-resistant base. Carefully remove any product residues on the measuring transducers and closing plugs with a soft, non-linting cloth. Have the measuring transducers cool down and soak the tip in a little bit of water. After soaking, clean the threads and the sealing surfaces of the measuring transducers carefully with a soft brass wire brush and a soft, non-linting cloth (Fig. 12). Clean the tip of the pressure transducers very carefully with a soft, non-linting cloth only. Clean the tip of the melt temperature thermocouples and of the closing plugs with a soft brass wire brush and a soft, non-linting cloth. Apply a food grade disinfectant to a clean, non-linting cloth and disinfect the product with wetted parts of the measuring transducers and closing plugs with this cloth. Leave to air dry.
3. *Dismantling and cleaning the bursting pin.* Unscrew the bursting pin from the die adapter. Remove any product residues on the tip of the bursting pin very carefully with a soft, non-linting cloth. Have the bursting pin soak for 15 min in a vessel with clear water. After soaking, remove any remaining product residues with a fine, soft brass wire brush and a soft, non-linting cloth. Apply a food grade disinfectant to a clean, non-linting cloth and disinfect the product's wetted surfaces of the bursting pin with this cloth. Leave to air dry.
4. *Dismantling and cleaning the die adapter.* Disconnect the heater cable and control the thermocouples of the die adapter from the connection panel. Unscrew the fixing screws of the die

adapter crosswise and remove them. Clean the die carefully—particularly the surfaces and threaded bores of the die—with a soft brass wire brush or round brush. Disinfect and leave to air dry.

5. *Dismantling and cleaning the die.* Unscrew the fixing screws of the die crosswise and remove them. Clean the die carefully—particularly the surfaces and threaded bores of the die—with a soft brass wire brush or round brush. Disinfect and leave to air dry.

6. *Dismantling and cleaning the closing plugs.* Remove any measuring transducers and/or feeders from within the tilting range of the cover hood of the processing unit. Lift off the vent dome from the vent port. Tilt the cover hood of the processing unit open. Dismantle the feed bolt. Loosen the fixing screws of the closing plugs of the upper feed openings (three screws M8 each, Fig. 9) so that the plugs no longer exert any force onto the sealing surfaces. If the closing plug is still tight after having loosened the fixing screws, carefully beat the closing plug out towards the top using a negative hammer (Fig. 33). Take out the closing plug and clean it immediately. Unscrew and remove the fixing screws of the lateral closing plugs and dismantle them. Carefully clean the closing plugs of the upper feed openings, the lateral closing plugs, and the feed bolt with a round

Fig. 33 Negative hammer screwed into a closing plug [11]

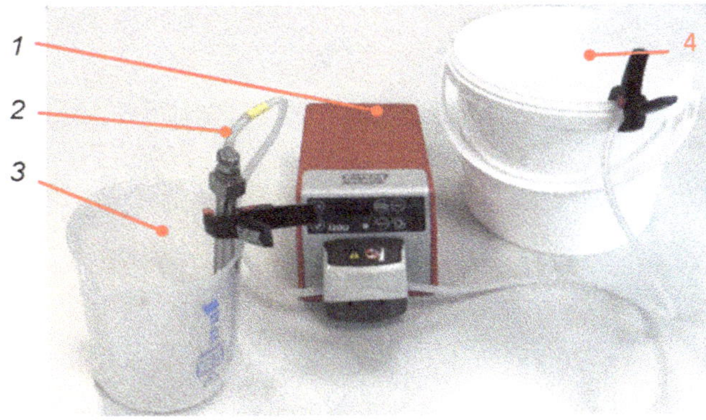

Fig. 34 Disinfection of feed bolt and dosing hose [11]. 1. Dosing Pump; 2. Dosing hose connected to the feed bolt; 3. Collective vessel; 4. Food compliant vessel with disinfectant. The picture is extracted from the manual

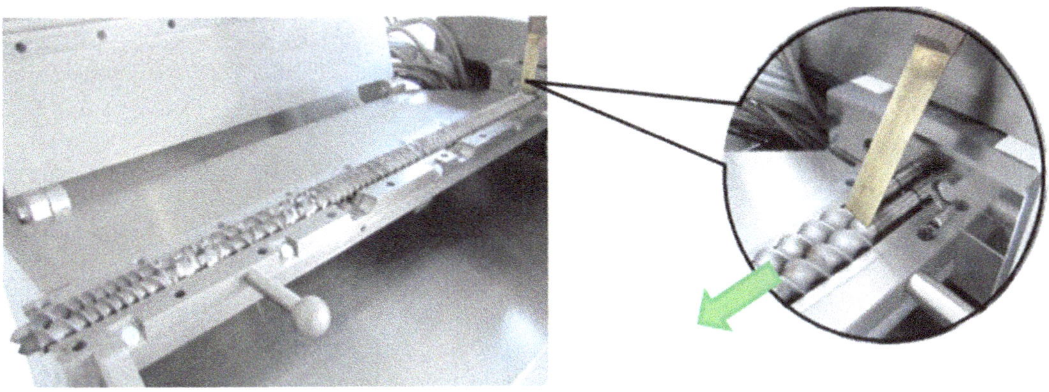

Fig. 35 Loosening and removing the screws [11]. The picture is extracted from the manual

brass brush. Put all parts into a food compatible cleaning solution for 15 min, then thoroughly rinse with clear water. Disinfect the product-wetted surfaces of all parts, then leave all parts to air dry.

7. *Clean and disinfect the feed bolt and the dosing hose* as Fig. 34. Rinse a food compatible vessel with clear water. Fill a food grade disinfectant into this vessel. Fix the dosing hose on the vessel with the disinfectant and connect it via a dosing pump to the feed bolt. Fix the feed bolt on a suitable collection vessel. Start the dosing pump so that the disinfectant flows through the dosing hose and the feed bolt into the collection vessel, and have it run for 5 min. Disassemble the assembly and leave the dosing hose and the feed bolt to air dry.

8. *Dismantling and cleaning the screws.* With the bronze spatula, carefully push the two screws towards the discharge side of the liner until the screw shafts are completely visible. Remove the pair of screws (Fig. 35). Clean the screws carefully with a soft

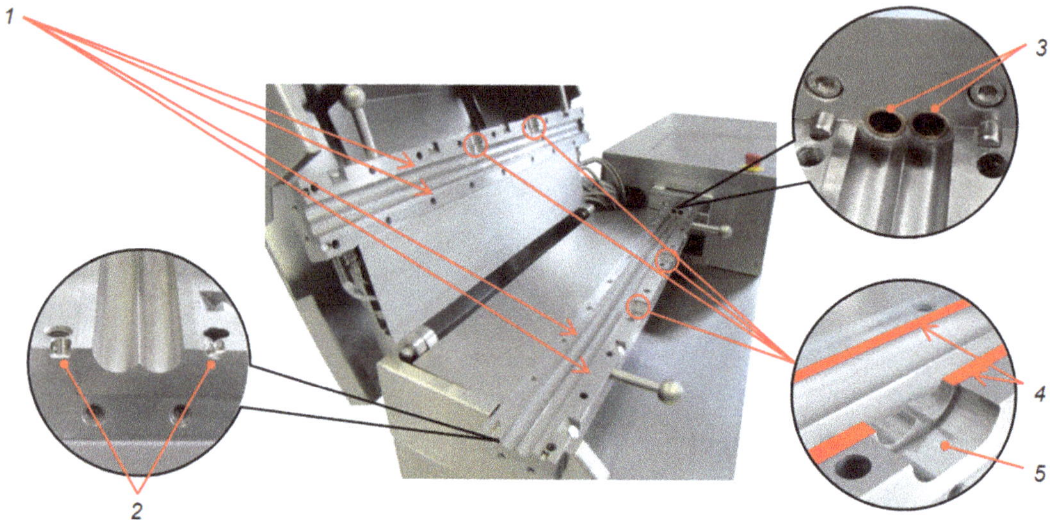

Fig. 36 Sealing surfaces [11]. 1. Sealing surfaces of the liner; 2. Positioning bores for the alignment pins of the die adapter; 3. Screw bushings; 4. Sealing surfaces (here: lower liner half); 5. Lateral feed opening. The picture is extracted from the manual

brass brush. Put the screws into a food compatible cleaning solution for 15 min. Rinse the screws thoroughly with clear water. Disinfect the screws, then leave to air dry.

9. *Cleaning the processing unit (sealing surfaces)* (Fig. 36). Clean the inside of the two halves of the processing unit with a soft brass brush. Clean the feed and vent openings with a soft round brass brush. Moisten the two liner halves with a food compatible cleaning solution and have them soak for 10 min. After soaking, wipe the two liner halves with clear water. Disinfect the product wetted parts of the two liner halves, then leave to air dry. Care must be taken, *see* **Note 10**.

   In the cleaning process, care must be taken in case of machine contamination, *see* **Notes 11** and **12**.

3.9 Calculation and Extrudates Texture Analysis

All the following tests are conducted on the extrudates of HMTPP, while Subheading 3.8, **step 1**, S_l of Subheadings 3.8, **steps 2** and **3**, can be tested on LMTPP. Figure 37 shows examples of HMTPP of different gluten ratios.

3.9.1 Specific Mechanical Energy (SME) Calculation

SME is the amount of work input into the mixture being extruded and is assumed to be generated as a result of friction between the screws and the mixture. SME is closely related to the screw configuration and the properties of the raw materials. A high SME value results in a large amount of mechanical and thermal energy applied to the mixture and leads to a high degree of denaturation of proteins and decomposition of molecules by mechanical

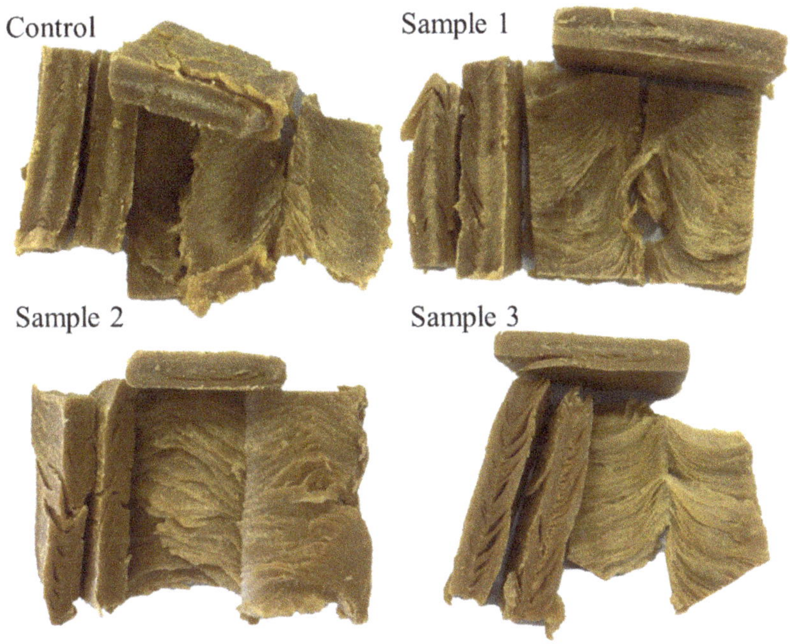

Fig. 37 Example of HMTPP of different gluten ratio

disruption, thus influencing the quality characteristics of the extrudates [5]. SME which qualifies the process intensity is defined in Eq. (4) [13]:

$$\mathrm{SME}\left(\mathrm{J}\cdot\mathrm{kg}^{-1}\right)=\frac{\text{Screw torque}(\mathrm{N}\cdot\mathrm{m})\cdot\text{Screw speed }(\mathrm{rad/s})}{\text{feed rate }(\mathrm{kg/s})} \quad (4)$$

3.9.2 Degree of Texturization

The cutting strengths of the TPP samples are to be analyzed using a texture analyzer (TA.XT plus, Stable Micro System Ltd. Godalming, UK) according to the modified methods of Jung-Soo Lee et al. [5] to evaluate the degree of texturization. A flat cuboid-shaped sample (25 × 25 × 7 mm) is cut using a knife blade (TA-42; 69.5 mm width × 74 long × 2.5 mm thickness with a 45°chiseled end) to 90% of its original thickness at a pre-test speed of 1 mm/s, test speed of 2 mm/s, and post-test speed of 10 mm/s. The cutting strength of a sample is evaluated transversely (transversal cutting strength: St) and longitudinally (longitudinal cutting strength: S_l) to the direction of the extrudate outflow from the extruder. Equal values of St and S_l indicated a uniform texture with low material anisotropy and no fibrous textures. In addition, the degree of texturization is expressed using the following equation. Thirty replicates of each analysis are to be performed to measure the St and S_l of each sample ($n = 30$).

$$\text{Degree of texturization} = St/S_l \quad (5)$$

where St = mean value of transversal cutting strength (N); S_l = mean value of longitudinal cutting strength (N).

3.9.3 Hardness Analysis

The hardness of extrudate samples is to be carried out [14] using a CT3-4500 Texture Analyzer (Brookfield Engineering Labs. Inc., USA). A square piece (25 mm × 25 mm) cut from fresh product strip is compressed using a cylinder TA 39 probe to 50% of its original thickness at a speed of 1 mm/s for 5 s. All the other processing parameters are the same as the texturization index measured. The hardness data are to be recorded three times.

3.9.4 Tensile Analysis

To better describe the tensile properties of meat analogs, we defined the tensile force along the extrusion direction as the longitudinal tensile force [15]. A sample shaped like the above is to be pulled using an A/TG probe at a speed of 0.5 mm/s until the strip was broken, and the tensile strength was recorded [16].

Note disclaimer in **Note 13**.

4 Notes

1. Care should be taken when mounting the agitator in case of sharp cut.
2. Take care not to offset the screws when inserting them into the liner!
3. Only tighten by hand when the transducer is in cold condition!
4. Care should be taken when lifting the steel bowl in case of falling.
5. Always start at a low speed before carefully increasing the speed in case of breaking the machine!
6. Only use brass or bronze wire brushes, soft and non-linting cloths, and wooden or brass spatulas!
7. Never use any sharp-edged or pointed tools or any other tools that might damage the surfaces of the instrument!
8. Never use any corrosive cleaning agents!
9. Clean all product wetted parts regularly. Dry all machine parts thoroughly after cleaning! Make sure that all edges, angles, and grooves are clean and dry.
10. Never let cross-linking products cool down or cure inside the machine!
11. Never leave residues of aggressive products like ceramics, distilled water, etc., inside the device over a long period of time!

12. Following a HACCP concept for cleaning and disinfection of the TwinLab-F.
13. Disclaimer: This chapter provides guidance based on the twin-screw extruder [size, company, country], readers should always consult the original manual from their instrument for comprehensive instructions and safety guidelines.

References

1. Zhang ZY, Zhang LJ, He SD et al (2022) High-moisture extrusion technology application in the processing of textured plant protein meat analogs: a review. Food Rev Int:1–36. https://doi.org/10.1080/87559129.2021.2024223
2. Zhang JC, Liu L, Liu HZ et al (2019) Changes in conformation and quality of vegetable protein during texturization process by extrusion. Crit Rev Food Sci Nutr 59(20):3267–3280. https://doi.org/10.1080/10408398.2018.1487383
3. Lazou AE (2022) Food extrusion: an advanced process for innovation and novel product development. Crit Rev Food Sci Nutr:1–29. https://doi.org/10.1080/10408398.2022.2143474
4. Prabha K, Ghosh P, Abdullah S et al (2021) Recent development, challenges, and prospects of extrusion technology. Future Foods 3: 100019. https://doi.org/10.1016/j.fufo.2021.100019
5. Lee JS, Kim S, Jeong YJ et al (2023) Impact of interactions between soy and pea proteins on quality characteristics of high-moisture meat analogs prepared via extrusion cooking process. Food Hydrocoll 139:108567. https://doi.org/10.1016/j.foodhyd.2023.108567
6. Pietsch VL, Emin MA, Schuchmann HP et al (2017) Process conditions influencing wheat gluten polymerization during high moisture extrusion of meat analog products. J Food Eng 198:28–35. https://doi.org/10.1016/j.jfoodeng.2016.10.027
7. Pietsch VL, Werner R, Karbstein HP et al (2019) High moisture extrusion of wheat gluten: relationship between process parameters, protein polymerization, and final product characteristics. J Food Eng 259:3–11. https://doi.org/10.1016/j.jfoodeng.2019.04.006
8. Choton S, Gupta N, Bandral JD et al (2020) Extrusion technology and its application in food processing: a review. Pharm Innov J 9(2):162–168. https://doi.org/10.22271/tpi.2020.v9.i2d.4367
9. Maurya AK, Said PP (2014) Extrusion processing on physical and chemical properties of protein rich products. J Bioresour Eng Technol 2:61–67
10. Wittek P, Zeiler N, Karbstein HP (2021) High moisture extrusion of soy protein: investigations on the formation of anisotropic product structure. Food Secur 10(1):102. https://doi.org/10.3390/foods10010102
11. Brabender GmbH & Co. KG (Edition 1219) TwinLab-F Instruction Manual 2020
12. Usman M, Swanson G, Chen B et al (2023) Sensory profile of pulse-based high moisture meat analogs: a study on the complex effect of germination and extrusion processing. Food Chem 426(15):136585. https://doi.org/10.1016/j.foodchem.2023.136585
13. Wolf B (2010) Polysaccharide functionality through extrusion processing. Curr Opin Colloid Interface Sci 15(1–2):50–54. https://doi.org/10.1016/j.cocis.2009.11.011
14. Wu M, Sun Y, Bi C et al (2018) Effects of extrusion conditions on the physicochemical properties of soy protein/gluten composite. Int J Agric Biol Eng 11(4):230–237. https://doi.org/10.25165/j.ijabe.20181104.4162
15. Xiao Z, Jiang R, Huo J (2022) Rice bran meat analogs: relationship between extrusion parameters, apparent properties and secondary structures. LWT 163:113535. https://doi.org/10.1016/j.lwt.2022.113535
16. Fang Y, Zhang B, Wei Y (2014) Effects of the specific mechanical energy on the physicochemical properties of texturized soy protein during high-moisture extrusion cooking. J Food Eng 121:32–38. https://doi.org/10.1016/j.jfoodeng.2013.08.002

Chapter 33

Sensory Analysis Implementation to Evaluate Plant Protein-Based Meat Products

Martin J. Talavera and Sajid Alavi

Abstract

Sensory analysis techniques can vary and include a wide array of tools such as descriptive panels (analytic) and consumer acceptance tests (affective) or be more qualitative in nature with focus groups and interviews. Similarly, plant-based meat products can be of various types including low moisture textured vegetable protein or its derivatives such as plant-based hamburgers and high moisture meat analogs. Each product has its own pre-processing step (for example, extrusion, shear cell, etc.) and preparation method before it can be consumed and/or evaluated. In this chapter, detailed steps for implementing a trained sensory panel, setting up a well-equipped sensory analysis facility, preparation procedures for evaluation of HMMA samples manufactured by extrusion, and execution of descriptive testing are described. This serves as an example for pairing the right sensory tool with specific plant-based meat product type based on the overall goals of the research team.

Key words Sensory, Descriptive, Attributes, Plant-based meat, Meat alternatives, Meat replacement

1 Introduction

An important role of sensory science is to help increase competitiveness of companies in a global market through product understanding and the development of new technologies [1]. The process that companies go through to discover and implement new technologies to develop food products of better quality is extensive. The term "better quality" can be related to a variety of aspects such as higher yield, resistance to weather and pests, longer shelf life, better health and nutritive aspects, and of course better sensory properties. Nevertheless, meeting all production goals still will not guarantee that the target consumer will want to purchase and consume the product repeatedly. The concept of food choice is complex and has several layers. Consumer satisfaction can be influenced by various factors, starting from sensory and perceptual features, information and environment, habits, experiences, liking

Yonghui Li (ed.), *Plant-Based Proteins: Production, Physicochemical, Functional, and Sensory Properties*, Methods and Protocols in Food Science, https://doi.org/10.1007/978-1-0716-4272-6_33,

and preference, moods and emotions, as well as including culture, economic, and political elements [2]. Therefore, understanding the characteristics of the product and understanding the perception profiles of consumers who will purchase the product is paramount to maximize the chances of product success in the market [3]. Sensory and Consumer science is a field of study that investigates the sensory properties of products (i.e., appearance, aroma, flavor, texture, and aftertaste) and how these modalities interact with human perception. Sensory science is officially defined as the "...scientific method used to evoke, measure, analyze, and interpret those responses to products as perceived through the senses of sight, smell, touch, taste, and hearing" [4, 5]. Because sensory science uses human subjects to measure perceptual responses, it is important to highlight that perception is defined as the sum of "sensation" and "interpretation" [6]. Interpretation can be a definite source of variability as each person is subjected to a completely different background of demographics, emotions, ethnic/culture and of course, their own personal likes and dislikes.

The industry of plant-based meat, meat alternatives, and meat replacement products is not a stranger to sensory science. There is a strong push in the industry to find ingredients and food items that provide options to consumers eager for alternatives to meat and animal protein sources, while also remaining palatable and acceptable to consumers. To accomplish this, it is important to match consumer expectations both from a concept and a sensory perspective (i.e., can I compare this product to anything I already know? Should I have any pre-conceived ideas about the flavor and texture of this product?). Matching those expectations will be crucial for success, and sensory research will be a critical step to understand if the sensory properties are a match to the concept and consumer expectations.

Sensory research is not one single method; it is an extensive toolbox of various approaches that individually, or as part of a broader research plan, can help answer specific objectives. More specifically, there are analytic and affective research approaches that can be qualitative and/or quantitative in nature. This chapter will focus on the procedure to set up, implement and execute a sensory study in the analytical branch, to obtain a detailed characterization of a product's sensory features.

Descriptive analysis uses expert assessors or panelists who are trained to describe and quantify the sensory properties of products in various modalities following pre-established procedures [7]. They serve as "human instruments" to measure multidimensional sensory relations that could, otherwise, be produced, in a more unidimensional and limited manner from other evaluation equipment. Below, we proceed to explain some of the important aspects to consider when conducting descriptive analysis, specifically focusing on meat alternative products.

2 Materials

2.1 Analytical (Descriptive) Testing—Trained Sensory Panel

If a one-time or limited time testing is needed, it is suggested to reach out to research agencies with descriptive capabilities on a fee-for-service system. This will ensure access to an experienced and well-trained panel, as well as the expertise of sensory scientists and researchers who can help in the design, set up, and execution of the study.

If a heavier project load is expected in the long term, and several studies are expected, then it is recommended to proceed to train a new panel. There are different approaches to training a panel depending on project needs: Flavor and Texture profile, Quantitative Descriptive Analysis (QDA), and Spectrum are considered the mainstream conventional sensory methodologies available [7]. Hybrid methods can depart from these initial approaches and be used depending on specific objectives.

For training purposes, it is recommended to follow established procedures such as the "Guidelines for Selection and Training of Sensory Panel Members" published by the American Society for Testing and Materials [8]. As established by these guidelines, "the use of panelists as measuring devices is analogous to the use of any scientific instrument" so careful consideration for selection and training must be followed.

Other standard documents such as "Basic Principles of Sensory Evaluation" (ASTM STP 433), and the manual on "Sensory Testing Methods" (ASTM STP 434) review different testing methods and can be used as a basis for screening and training descriptive panelists [9, 10].

2.2 Sensory Room

The Sensory room can take different shapes and sizes, but it must follow certain specifications to ensure that testing is not affected or biased by external factors unrelated to the product. Irrespective of specific designs, sensory areas must have certain features such as a well-ventilated testing space with negative air flow, sensory booths/table with neutral colors, temperature and humidity control, and a preparation area adjacent to the testing area. Figure 1 represents a generic layout for a sensory space as recommended by ISO 8589: 2007 [11]. Items expected to be in a sensory space are:

A. Food Preparation area (depending on the nature of the products to be evaluated)
   - Stove(s)
   - Oven(s)
   - Microwave(s)
   - Refrigerator

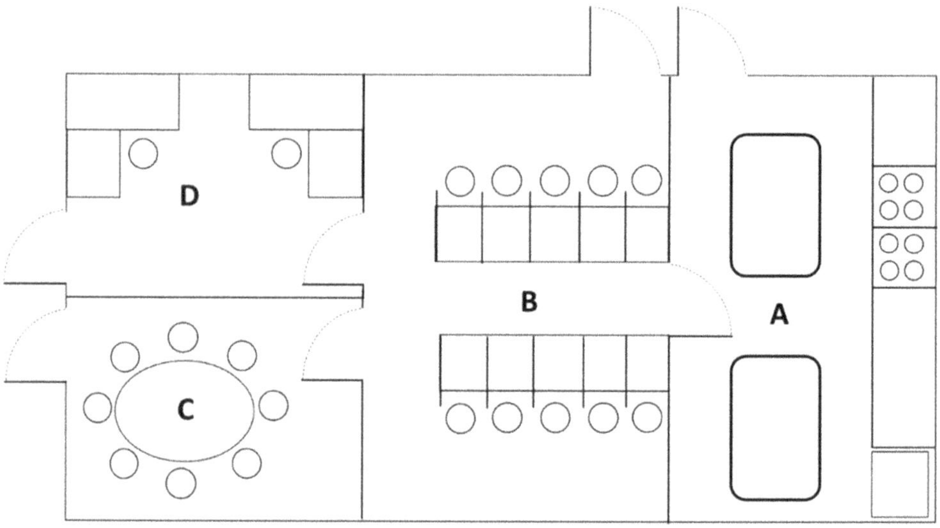

Fig. 1 Example floor plan for a sensory space. (**a**) Preparation area adjacent to testing area; (**b**) sensory booths; (**c**) Testing/meeting room with circular table; (**d**) Office space

- Food storage area
- Access to walk-in refrigerator and freezer

B. Well-ventilated room with booths for individual evaluation
- 8–10 booths (*see* **Note 1**)
- Booths equipped with red lights for appearance differences masking (if needed) and doors to pass samples through
- Adjacent to food prep area

C. Well-ventilated room with round table for group evaluation (orientation and training)
- To sit 8–10 people
- Adjacent to food prep area

D. Office space for panel manager/leader

E. Tablets or laptops with access to software for data collection should be available. Software and tools for data collection and analyses are further described in Subheading 2.3.

F. Additional space for storage of kitchen/serving materials/containers may be needed as well. These materials are listed in more detail in Subheading 2.4.

2.3 Data Collection and Analysis Software and Tools

Various software options are commonly used in sensory science for data collection and analysis. Some of these are Compusense (Compusense® Guelph, ON, Canada), SIMS (Sensory Computer Systems, Berkeley Heights, NJ), RedJade (RedJade Sensory Solutions, Pleasant Hill, CA), and Fizz (BIOSYSTEMES, Dijon, France) among others. Some of these softwares have direct statistical analysis capabilities while others create links to external softwares for evaluation. Softwares for statistical analysis are a must for data analysis and interpretation. The most commonly used are SAS software (SAS, Cary, NC, USA), XLSTAT (Lumivero, Denver, CO, USA), or IBM SPSS Statistics version 25.0 software (SPSS Inc., Chicago, IL, USA), among others. Sensory evaluation techniques use both parametric and non-parametric statistics to derive inferences starting from highly variable data. Variability is innate in sensory science and expected up to some degree because data is based on human perception. Panel training and testing experience aim to reduce variability by teaching terminology with definitions, measurement procedures and use of scales consistently, as well as calibrating the evaluation by means of references. Still, some variability is expected. Analysis techniques can be as simple as the use of descriptive statistics with arithmetic means and standard deviations to look at center points with spreads and presented using a tabular format or spider charts. Analysis of variance (ANOVA) can be used for mean separation and to establish the signal to noise ratio and the significant differences among treatments. The model can be built as a 1-way ANOVA (product effect only), 2-way ANOVA (with product and panelist effects), or 3-way ANOVA (including product, panelist, and replication effects). The later options will allow to look at the combination of interactions between product, panelists, and replications. In addition, multivariate techniques can be used for dimensionality reduction and visualization, and to uncover latent structures that can help explore patterns by producing a visual representation of the sensory space. The most popular technique is principal component analysis (or PCA). In general, various softwares are available to analyze sensory data.

2.4 Kitchen Utensils Needed for Product Preparation and Serving of Plant-Based Meat Alternatives

Various materials, utensils and product preparation, and serving tools should be available. Some of these materials and tools specific to plant-based meat products are listed below.

- Utensils appropriate for product manipulation (cutting, portioning, and serving)
  - Forks, knives, and spoons
- Vegetable oil
- 2% saltwater brine
- Aluminum foil
- Non-stick skillets *(for cooking products)*
- Stovetop *(for cooking products)*
- Microwavable bowls or containers *(for holding/warming products before serving)*
- Microwave *(for warming products before serving)*

- 96 ml odor-free plastic cups with lids and fork *(for product serving)*
- Styrofoam or plastic cup for water *(for mouth rinsing)*
- Unsalted crackers *(for mouth rinsing)*

3 Methods

The general methods for descriptive sensory evaluation of plant-based meat products are described in this section, including panel implementation and execution, example preparation methods used for products, and data output and presentation of results.

3.1 Descriptive Panel Implementation

The various steps for successful descriptive panel implementation are as follows.

- The researchers and sensory scientists should start with discussing objectives to finalize approach and research methodology for the descriptive panel.
- The sensory team should then select the most appropriate descriptive method based on the objective and resources available (i.e., Profile, QDA, Spectrum, or Hybrid) (*see* **Note 2**), scale (10-point, 15-point, line scale, etc.), modalities to be explored (i.e., appearance, aroma, flavor, texture, aftertaste, etc.), replications (consensus, duplicate, or triplicate), statistical analysis procedures (*see* **Note 3**), and use of attributes with definitions and references depending on the products and preparation selected for evaluation. Selection of attributes can start from the literature where previous work with plant-based protein alternatives has been conducted (Table 1). A well-trained and experienced sensory panel can also help develop new attributes with definitions and references to evaluate a specific set of products. In either scenario, a trained panel will need to review the attributes to make sure they fit the product set and are appropriate and comprehensive for evaluation. Sensory attributes should not be ambiguous (i.e., mean different things), and should be descriptive, differentiating, and singular (i.e., describe a unique sensory characteristic). Attributes should be accompanied by a definition, so all evaluators interpret the attribute in the same way, and referenced, so they can anchor their evaluation and aid in calibrated responses. Different methods and approaches differ in the use of references. Methods such as Sensory Spectrum rely on highly calibrated measurements, so references are required. Other methodologies such as QDA are more open to evaluation without references. This will affect testing reproducibility but will be much faster to implement and execute. Selecting which test to use will depend on resources available and study objectives.

Table 1
Texture and flavor lexicon for plant-based protein replacement products found in Plattner, 2022 [12]

Attribute	Definition	Reference
Springiness *(texture)*	Degree to which sample returns to its original height when compressed once partially with molar teeth and slowly released	Oscar Mayer Uncured Bun-Length Wieners
Denseness *(texture)*	The degree of compactness of the cross section	Oscar Mayer Uncured Bun-Length Wieners
Juiciness *(texture)*	The amount of liquid expressed from the sample at the maximum intensity from 5 chews	Hormel Cure 81 Ham
Residual particles *(texture)*	Particles remaining in mouth after mastication and swallowing. Maybe fibers, flakes, and/or granules	Hormel Cure 81 Ham
Firmness *(texture)*	The force required to bite completely through the meat pieces with the molar teeth	Hormel Cure 81 Ham
Chewiness *(texture)*	Difficulty with which the sample can be broken down with the molars for swallowing	Hormel Cure 81 Ham
Fiber awareness *(texture)*	The perception of filaments or strands of muscle tissue in product during mastication	Hillshire Farms Lit'l Beef Smokies
Tooth packing *(texture)*	The amount of sample packed in a between the molar teeth after swallowing	Wheaties
Astringent *(texture)*	A drying puckering or tingling sensation on the surface and/or edge of the tongue and mouth	Alum Solution
Starchy *(texture)*	Degree to which the sample mixes with saliva to form a starchy, pasty slurry that coats mouth surfaces after swallowing	American Beauty Elbo-Roni
Beany *(flavor)*	A slightly brown, musty, slightly nutty, and starchy flavor associated with cooked beans	Bush's Best Pinto Beans
Starchy *(flavor)*	The dry aromatics associated with starch and starch-based grain products such as wheat, rice, oats, and other grains	Bush's Best Pinto Beans
Grain *(flavor)*	A general term used to describe the aromatic which includes musty, dusty, slightly brown, slightly sweet and is associated with harvested grains and dry grain stems	Cereal Mixture (dry)
Green *(flavor)*	A green aromatic associated with fresh green peapods. May include beany, increased pungent, musty/earthy, bitter and astringent	Great Value Frozen Baby Lima Beans
Umami *(flavor)*	A general term for aromatics associated with juices from cooked seafood, meat, and/or vegetables	Button Mushroom Broth
Barnyard *(flavor)*	Combination of pungent, slightly sour, hay-like aromatics associated with farm animals and the inside of a barn	White Pepper in water
Cardboard *(flavor)*	The flat aromatics that may be associated with cardboard or paper packaging	Cardboard soaked in water

- Sensory scientists and panelists can also be involved in attribute development and selection, making sure that appropriate definitions and references are incorporated to calibrate the evaluations.
- Once the approach has been selected, proceed with panel implementation as specified above following standard procedures for panel selection and training as detailed in ASTM STP 758 [8].
- Before beginning evaluation, the panel must have agreed on the ballot with attributes, definitions, and evaluation procedures.
- Once the panel has been screened, appropriately trained, a ballot with terminology, definitions and references developed, and evaluation procedures established, then evaluation can proceed.
- Product preparation and presentation will have an effect on how the attributes are perceived so this will also have to be carefully implemented to make sure it represents the most appropriate product form (i.e., cooking/heating instruction, holding time, portion size, etc.). A more detailed explanation of the process to prepare plant-based meat prototypes for sensory evaluation is described in Subheading 3.2.

3.2 Preparation of Plant-Based Meat Prototypes for Testing

Plant-based meat products can be of various types (Fig. 2) including shredded textured vegetable protein formed into hamburgers and high moisture extrusion (HME) based products that are also called high moisture meat analogs (HMMA). The latter can be

Fig. 2 Different types of plant-based meat products—(top) shredded textured vegetable protein that can be formed into hamburgers and (bottom) high moisture meat analog chunks

prepared and consumed in the form of "chunks" or pieces that imitate meats such as chicken, beef, pork, or in other forms. These products require specific preparation methods post-primary processing before they can be consumed and/or evaluated for sensory characterization. The steps for preparation of HMMA chunks or pieces are described below.

- Procure/prepare samples using primary processing methods (e.g., extrusion) based on objectives established.
- HMMA prototypes usually have moisture greater than 60% (wet basis), thus they should be vacuum sealed and frozen (−23 °C) after preparation to minimize the changes in moisture distribution and prolong shelf life before evaluation.
- If the objective is to compare prototypes to real meat products, then procure beef/poultry/pork/fish products of interest to serve as comparison depending on textural expectations.
- Prototypes should be removed from packaging and thawed using a room-temperature 2% saltwater brine for 30 min.
- Thawed products should be quickly submerged in vegetable oil.
- Then transfer to heated non-stick skillet. Proceed to heat and cook samples and meat controls.
- Products should be heated approximately for 4 min at 177 °C, with samples being flipped once after 2 min.
- Animal meat controls should be thawed overnight in a refrigerator, and then transferred to a heated non-stick skillet and cooked to safe temperatures as recommended by USDA [13].
- After cooking, products should be transferred to microwaveable containers and sealed with aluminum foil to keep warm. They could be placed in warming ovens to maintain the temperature. Care should be taken to avoid losing too much moisture.
- Prior to serving, if no warming ovens are available, all samples should be cut into 1.25 cm cubes and warmed in a microwave on high power for 30 s.
- About 6–8 pieces for each product can be served to each panelist per replication, depending on the number of modalities being evaluated.
- Various containers could be used for serving. For example, 96 mL odor-free plastic cups with lids. Products could also be served over foil covered heated ceramic tiles to maintain a warm temperature.
- Each product should be served blind only identified with a 3-digit code.
- A Randomized Complete Block Design (RCBD) should be used if replications are being used and applicable to study objectives.

	Rep 1				Rep 2			
	Serving 1	Serving 2	Serving 3	Serving 4	Serving 5	Serving 6	Serving 7	Serving 8
Panelist 1	A	B	C	D	B	A	D	C
Panelist 2	B	A	C	D	A	B	D	C
Panelist 3	B	A	D	C	A	B	C	D
Panelist 4	A	B	D	C	B	A	C	D
Panelist 5	C	D	A	B	D	C	B	A
Panelist 6	D	C	B	A	C	D	A	B
Panelist 7	D	C	A	B	C	D	B	A
Panelist 8	C	D	B	A	D	C	A	B

Fig. 3 Example design for a sensory study following a Randomized Complete Block Design (RCBD) with four products, eight panelists, and replication as a blocking factor

- This design will help minimize bias due to presentation effect by ensuring that each product is presented in each position the same number of times.
- The example below (Fig. 3) shows 4 products (A, B, C, and D) served to eight panelists, and replicated twice. Because we are serving products in two replications, we are using replication as a blocking factor assuring that all four products of the first replication are evaluated before moving to the second replication. Another blocking factor could be the day of evaluation if evaluation will move across several days.
- Randomization is not applicable when using a consensus approach as there will be no replications and the same product will need to be served to all panelists simultaneously.

3.3 Descriptive Panel Execution

In the scenario where a descriptive panel is already trained as described in the sections earlier, the following steps should be followed for execution of descriptive testing before and during the sensory study after the plant-based meat samples are prepared and ready for evaluation.

- On evaluation day, the assigned placement for each panelist should be set up with a placemat, a napkin, unsalted crackers, a cup of water for mouth rinsing, and an additional cup for product elimination or "spit cup" (optional). Each placement will also receive a tray with references (Fig. 4).
- An orientation session should be conducted to ensure panel familiarity with samples evaluated and to select appropriate lexicon including definitions and references. Orientation sessions should last as long as needed by the panel to review products and finalize lexicon and ballots.
- After orientation, evaluation sessions will begin. Serve products monadically (i.e., one at a time), following a balanced design so

Fig. 4 Example of a room set up for sensory evaluation with six placements

all products are served in all positions the same number of times, to minimize carryover effects (see above for exceptions).

- Serve products warm and as soon as possible after preparation to minimize moisture loss.
- Each sample should be served blind with a 3-digit code and evaluated using appropriate utensils (spoons, forks, etc.).
- Products should be evaluated in duplicate or triplicate to improve test power. A separate descriptive session using the same panelists will be ideal for each replicate (*see* **Note 4**).
- Palate cleansers should be used between products to allow panelists to rinse and clear residual product and aftertaste. Palate cleansers used will depend on products being evaluated (e.g., water, unsalted crackers, carrots, apples, or mozzarella cheese for spicy products).

3.4 Data Output and Results

Data output should be presented in ways that represent product inter-relationships well, match the objectives and that are actionable (*see* **Note 5**). There are different approaches to show sensory data output, and all can be helpful depending on specific needs. This includes descriptive statistics (mean and standard deviations) with Analysis of Variance (ANOVA) and mean separation (Fig. 5). This can be useful to compare products numerically against benchmarks or internal action standards. Spider charts can be used to show relationships among products in a more graphical way (Fig. 6). However, this can be a limited option because it will not allow to include too many products or attributes. When more

Recipe	Springiness	Denseness	Juiciness	Residual Particles	Firmness	Chewiness	Fiber Awareness	Tooth Packing	Astringent	Starchy
A	5.19	6.94 ^{B}	5.94	4.25	8.94 ^{CD}	8.25 ^{B}	7.25	2.06	2.81	2.19 ^{AB}
B	4.31	7.94 ^{AB}	5.50	3.81	9.69 ^{BC}	9.00 ^{AB}	7.19	2.19	2.81	2.69 ^{A}
C	4.56	8.81 ^{AB}	5.00	3.63	10.56 ^{B}	9.31 ^{AB}	6.88	2.19	2.13	1.00 ^{BC}
D	3.19	8.31 ^{AB}	3.88	3.56	13.00 ^{A}	11.13 ^{A}	6.75	2.00	3.00	3.38 ^{A}
Meat 1	3.75	11.00 ^{A}	6.19	3.56	9.44 ^{BCD}	8.63 ^{B}	8.31	1.63	2.25	0.25 ^{C}
Meat 2	3.31	8.94 ^{AB}	5.63	3.19	8.25 ^{D}	7.75 ^{B}	7.88	2.19	1.69	0.25 ^{C}

Fig. 5 Tabular representation example of results including mean separation significance represented by superscript letters [12]. Samples A–D represent plant-based meat (HMMA) products and the samples labeled Meat 1 and Meat 2 are different animal meat benchmarks

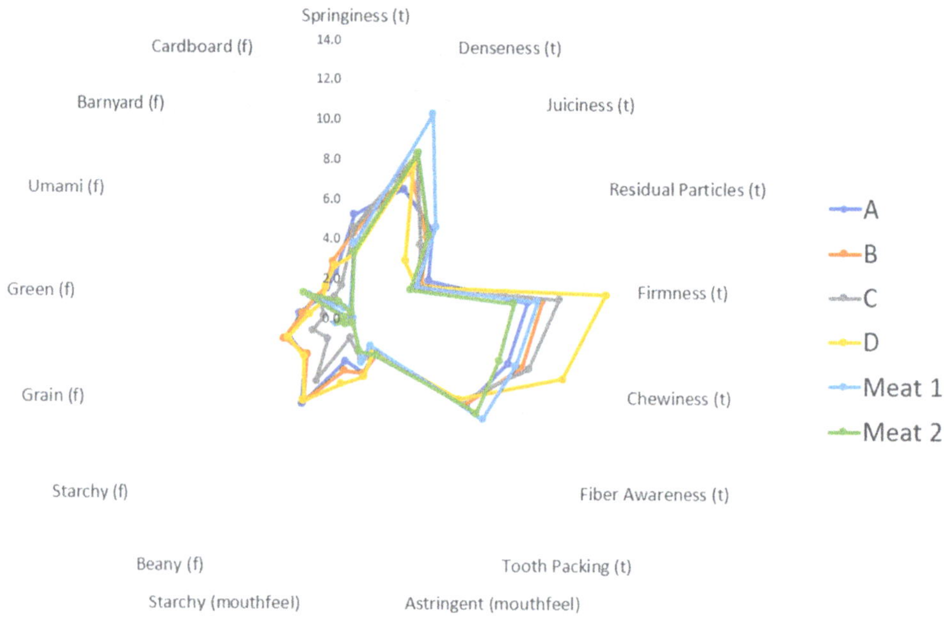

Fig. 6 Spider Chart example of various sensory profiles for six prototypes of plant-based meat alternatives, with animal meat controls [12]. Samples A–D represent plant-based meat (HMMA) products and the samples labeled Meat 1 and Meat 2 are different animal meat benchmarks

products or attributes are part of the analysis, a principal component analysis (PCA) or factor analysis are recommended (*see* **Note 6**). These are multivariate exploration techniques for data reduction that can help visualize trends and relationships among products and sensory attributes (Fig. 7). This technique is not recommended for use with less than four products.

4 Notes

1. Booths in sensory science are considered a “nice to have” but are not absolutely indispensable to execute sensory studies. They are mostly used to provide isolation for the evaluators to minimize distractions. However, sensory testing can still be

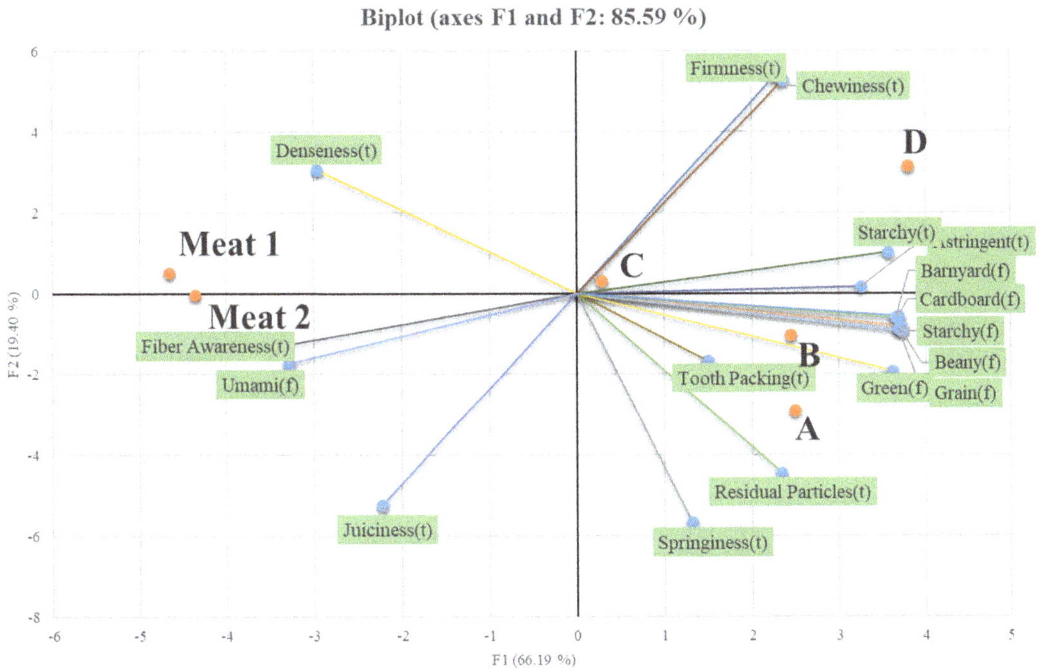

Fig. 7 Example of Principal Component Analysis (PCA) output for data visualization [12]. Samples A–D represent plant-based meat (HMMA) products and the samples labeled Meat 1 and Meat 2 are different animal meat benchmarks

conducted effectively without booths with appropriate spacing between evaluators and providing instructions and supervision to make sure judges will focus on their own evaluations. There are different opinions in the sensory community when it comes to the use of evaluation booths.

2. There are many sensory methodologies available out there that fit various objectives. Select the method or combination of methods that will best fit your own objectives. If very limited resources, there are various other "rapid" methods available that can be explored for product characterization [14]. Most of these methods do not require training and can be implemented fairly easily with consumer panels. These methods are not necessarily a good approach for the exploration of more technical attributes and have low sensitivity when looking for small differences among products. However, they can be very helpful for more broad characterization, categorization, and to generate consumer terminology to describe different modalities of consumer products in various categories.

3. Always a good idea to discuss test designs with a statistician if available, especially in the presence of more complex design scenarios (e.g., design of experiments, crossover designs, etc.). In the same way that ballots and evaluations procedures need to

be set up before evaluation, the design and the statistical procedures also need to be clear before execution, so the design can be implemented appropriately from the beginning.

4. Using replicated evaluations is a valid approach to increase statistical power and will allow the use of univariate statistical outputs based on variability such as means, standard deviations and analysis of variance (ANOVA). Other sensory methods also rely on the use of "consensus" evaluation where the output will be a single score for each attribute on each product resulting from panel discussion and agreement. Because this data does not involve statistical variability, only tabular presentation, graphs (i.e., bar charts, spider plots, etc.) and certain multivariate techniques such as PCA can be used to analyze and visualize the data. Discuss with a sensory professional to select the best option for evaluation and data output.
5. There is no single solution when it comes to sensory output. How we present sensory results should be in the way that makes most sense for the specific data being presented (i.e., number of products, number of attributes, group of variables or treatments, etc.). Be creative when it comes to sensory data, use colors and shapes that could help better visualize the information you want to present.
6. Multivariate analysis techniques are subjective and should be interpreted with caution. These techniques are very effective to aid in visualization of major trends but can easily over represent certain relationships and trends. Always a good idea to assess multivariate maps together with numeric results to make sure we only highlight meaningful relationships between attributes and products.

Acknowledgments

The authors would like to acknowledge Blake Plattner for work on plant-based meat research as part of his master's thesis at Kansas State University, from which certain descriptions of high moisture meat analog products have been derived.

References

1. Stone H, Sidel JL (1995) Strategic applications for sensory evaluation in a global market. Food Technol 2:80–89
2. Chen PJ, Antonelli M (2020) Conceptual models of food choice: influential factors related to foods, individual differences, and society. Food Secur 9:1898. https://doi.org/10.3390/foods9121898
3. Talavera M, Chambers EIV (2017) Using sensory sciences help products succeed. Br Food J 119(10):2130–2144
4. Institute of Food Technologists (IFT) Anonymous (1975) Minutes of division business meeting. Sensory Evaluation Division, IFT

5. Sidel JL, Stone H (1993) The role of sensory evaluation in the food industry. Food Qual Prefer 4(1–2):65–73
6. Bartoshuk (2019) Sensation and perception. In: Sternberg RJ, Pickren WE (eds) From the Cambridge handbook of the intellectual history of psychology. Cambridge University Press
7. Meilgaard MC, Civille GV, Carr BT (2016) Sensory evaluation techniques, 5th edn. CRC Press/Taylor and Francis Group, Boca Raton
8. American Society for Testing and Materials (1981) Guidelines for the selection and training of sensory panel members, ASTM STP 758. ASTM, Philadelphia
9. American Society for Testing and Materials (1968) Basic principles of sensory evaluation, ASTM STP 433. ASTM, Philadelphia
10. American Society for Testing and Materials (1968) Manual on sensory testing methods, ASTM STP 434. ASTM, Philadelphia
11. International Organization for Standardization (2007) Sensory analysis — general guidance for the design of test rooms, ISO 8589:2007(en)
12. Plattner BJ (2022) Impact of plant protein functionality and extrusion conditions on texture of high moisture meat analogs (HMMAs). Master's thesis, Kansas State University. https://krex.k-state.edu/handle/2097/42139
13. U.S. Department of Agriculture (2020) Cooking meat? Check the new recommended temperatures. USDA. https://www.usda.gov/media/blog/2011/05/25/cooking-meat-check-new-recommended-temperatures
14. Delarue J, Lawlor JB, Rogeaux M (2015) Rapid sensory profiling techniques and related methods. Woodhead Publishing Series in Food Science, Technology and Nutrition: Number 274

INDEX

Yonghui Li (ed.), *Plant-Based Proteins: Production, Physicochemical, Functional, and Sensory Properties*, Methods and Protocols in Food Science, https://doi.org/10.1007/978-1-0716-4272-6,

S

T

V

W

Z

www.ingramcontent.com/pod-product-compliance
Lightning Source LLC
Chambersburg PA
CBHW081545070225
PP16194600001B/1
9781071642719